AF516779

CLASS INSECTA

ENCYCLOPAEDIA OF LIFE SCIENCE SERIES

CLASS INSECTA

VOL. 2

BARON CUVIER

ANMOL PUBLICATIONS PVT. LTD.
NEW DELHI-110 002 (INDIA)

ANMOL PUBLICATIONS PVT. LTD.
4374/4B, Ansari Road, Daryaganj
New Delhi - 110 002

Class Insecta

First Edition, 1998

ISBN 81-261-0117-2 (Set)

PRINTED IN INDIA

Published by J.L. Kumar for Anmol Publications Pvt. Ltd., New Delhi and Printed at Tarun Offset Press, Delhi.

LIST OF PLATES

OF INSECTS.

LIST OF PLATES OF INSECTS.

The INSECTS (INSECTA)

have articulated feet, a dorsal vessel, which holds the situation of every vestige of heart, but without any branch for circulation;* respire through two principal tracheæ, which

* Anatomists are greatly divided with regard to this organ. Many consider it as a genuine heart; others (and such is the opinion of M. Cuvier, and one which appears to us to have been fully confirmed by the satisfactory researches of M. Marcel de Serres), entirely deny this position. According to the last mentioned author, the function of this vessel is the secretion of fat, which is afterwards elaborated in the adipose tissue which envelopes it. Lyonet says that it encloses a gummy substance of an orange colour. Some very recent observations appear to establish the existence of a few small vessels; but, besides that this circulation would be very partial, the insects must always differ greatly in this respect from the crustacea, in as much as the blood does not return to the heart. M. Straus, in reviewing a memoir of M. de Herold on the subject in the "Bulletin Universel," has given us his own opinion, founded on his own researches, on the *Melolontha*. "The dorsal vessel," he says, "is the true heart of the insect, being, as in the superior animals, the locomotive organ of the blood, which, instead of being contained in vessels, is spread through the general cavity of the body. This heart occupies the entire

extend parallel to each other through the whole length of the body. At intervals they have centres, from which pro-

length of the back part of the abdomen, and terminates anteriorly by a single artery not ramified, which transports the blood into the head, where it pours it out and from whence it returns into the abdomen, from the very effects of its accumulation in the head, to re-enter the heart. To this the entire sanguine circulation of the insects is reduced, which have thus a single artery without branches, and no veins. The *alæ* of the heart are not muscular, as Herold pretends: they are simple fibrous ligaments which keep the dorsal vessel in its place. The heart, that is to say, the abdominal portion of the vessel, is divided into eight chambers internally (*Melolontha vulgaris*), separated from each other by two convergent valvules, which permit the blood to proceed from behind, in front of one chamber, into the other as far as the artery which conducts it to the head, but which opposes its retrograde motion. Each chamber is provided laterally, in its anterior part, with two apertures, formed like transverse clefts, which communicate with the abdominal cavity, and through which the blood contained in this cavity can enter the heart. Each of these apertures is provided internally with a small valvule in the form of a semi-circle, which is attached on the aperture during the movement of the systole." From this short description it is easy to conceive, that when the posterior chamber is dilated, the blood contained in the abdominal cavity penetrates thither by the two apertures of which we have spoken, and which are named *auriculo-ventricular*. When the chamber contracts, the blood which it contains, not being able to return into the abdominal cavity, pushes the interventricular valvule, passes into the second chamber, which dilates to receive it, and which receives at the same time a certain quantity of blood by the proper auriculo-ventricular apertures. During the movement of the systole of this second chamber, the blood passes in the same manner into the third, which equally receives some by the lateral apertures; and it is thus that the blood is pushed from one chamber into the other as far as the artery. It is these successive contractions of the chambers of the heart, that one discovers through the skin of the caterpillar.

The heart of the decapode crustacea, of squillæ, limulæ, spiders, &c., also presents, according to the information which I have received from this profound observer, similar valvules. It is enclosed in a kind of sac or pericardium, which, according to him, stands in place of the auricle. These ceed a number of branches which correspond to the external apertures, or *stigmata*, for the entrance of air.*

divisions or chambers of the dorsal vessel, are what Lyonet denominates wings; and he has also seen the dorsal vessel prolonged as far as the head, and terminated there in the same manner; but he did not observe the apertures and valvules of which M. Straus has spoken. The definition of the dorsal vessel, given by this naturalist, whatever be the interior composition of this organ, proves evidently that it is not a genuine heart. These observations, moreover, do not teach us what the nature of the fluid is, nor how it is spread through the other parts of the body, so as to minister to their nutrition; at all events, it is certain, according to the observations of Lyonet, that all the parts of the body, communicate with the corpus adiposum by means of fibrillæ. The tracheæ throw out branches which extend even into the extremities of the different appendages of the body. It is possible that the action of the air may determine the ascension of the nutritive juices into the interstices, forming a sort of capillary tubes.

* The number of segments of the myriapodes being undetermined, that of their stigmata is so likewise, and it often exceeds twenty. In the hexapode insects it is often eighteen, nine on each side. This calculation, however, is rather founded on the larva than the perfect state of the animal. The caterpillars, the larvæ of the coleoptera, and those of a great number of other insects, have one pair of stigmata on the first segment, or that to which is attached the first pair of feet; the second and third have none, because, as I presume, the development of the wings, which takes place in these rings, renders useless in this place the presence of respiratory apertures. The fourth ring, and the seven following, are each provided with a pair; but in the coleoptera in a perfect state, besides the two anterior stigmata concealed in the cavity of the prothorax or corslet, and which are not discernible, we find two others situated between the origin of the elytra and that of the wings: these belong to the mesothorax. There are none at the metathorax, unless we consider the two of the first abdominal segment as supplementary ones of the thorax, like what takes place in the hymenoptera with pedicled abdomen, and the diptera, where these two stigmata, with the semi segment on which they depend, form a part of the thorax. Thus, in general, all the hexapode insects have eight pairs of stigmata to the abdomen; but the last two are frequently obliterated. In the crickets, the truxales, and the libellulæ, the sides of the metathorax are each provided with a stigma—those which

They all have two antennæ, and a head distinct. The nervous system of the majority of insects (the hexapodes) is generally composed of a brain formed of two opposite ganglia, united at their bases, giving but eight pairs of nerves, and two solitary nerves, and twelve ganglia,* all inferior. The first two are situated near the junction of the head with the thorax, and longitudinally contiguous; the anterior gives out nerves to the under lip, and to the parts adjacent. The second and the two following are proper to each of the first three segments, or those which in the hexapode insects compose the thorax. The other ganglia belong to the abdomen, so that the last, or twelfth, corresponds to its seventh ring, followed immediately by those which compose the several organs. Each of these ganglia give out nerves to the parts of their respective segments. The two last, very closely approaching each other, also give out nerves to the last rings of the body. The frontal region has three particular ganglia, designated by Lyonet under the name of *frontals*, and the first of which produces, posteriorly, a thick nerve with swellings: this is the longest of all the nerves, and he names it *recurrent*. The first ordinary ganglion, or the *sub-œsopha-*

M. Marcel de Serres names *trémaeres*. In these last mentioned insects as well as in the others with naked wings, or without elytra, the first two thoracic stigmata are placed above, between the prothorax and mesothorax. With the exception of the libellulæ, the thorax proper exhibits no distinct stigmata. I say the thorax proper, because, as we have remarked above, the two first of the abdomen are referred by many to the posterior extremity of the thorax. The metathorax of the pentatomata and scutellaria exhibits one pair of stigmata underneath. In the aptera, the second segment, or mesothorax, has none; but the following segment, or metathorax, has two pairs, one anterior, and which, being situated near the articulation of this segment with the preceding, may be deemed to belong to this last, and the other smaller, and placed very near that of the first abdominal segment.

* Divers lamellicorneous coleoptera in the perfect state form an exception.

gian, germinates, according to this author, four pairs of nerves, and the following, two pairs each ; so that, reckoning the eight pairs of the brain, the ten spinal bridles, which may be considered as so many pairs of nerves, we have in all forty-five pairs, independently of the two solitary nerves, or from a dozen to fourteen more than are in the human body. The two nervous cords, which, by their union, form the ganglia, are tubular, and composed of two tunics, the exterior of which presents tracheæ. A medullary substance fills the central canal. The excellent work of M. Herold on the anatomy of the caterpillar of the large cabbage butterfly, studied in its progressive growth, to its transformation into the chrysalis, shews us that the nervous system and that of the digestive organs undergo remarkable changes; that the nervous cords are in their origin longer and more separated, an observation which favours the opinion of one of the greatest zootomists of our age, Dr. Serres, on the origin and development of the nervous system. We have explained in the generalities common to the three classes of articulated animals with articulated feet, the various sentiments of physiologists on the seat of the senses of hearing and of smell. We shall confine ourselves to adding, that, with regard to the first, the small nervous ganglia situated on the forehead, of which we have spoken, seem to confirm the opinion of those, such as Scarpa, who place the seat of this sense near the origin of the antennæ. In some lepidoptera, I have observed two small holes situated near the eyes, and which, perhaps, are auditory conduits. If, in many insects, especially those which have filiform or setaceous and long antennæ, these organs answer the purpose of tact, it appears difficult to give a reason for the extraordinary development which they acquire in certain families, and more particularly in the males, if we do not admit that they are thén the seat of smell. Perhaps, also, in relation to taste, the palpi play in some cases, as when they are very much dilated at their extremity, the principal part. The tongue may also have something to do with this function.

A preparatory or buccal apparatus, the intestinal canal, the biliary vessels, also named *hepatic*, those which are called *salivary*, but which are less general, those free or floating vessels which have received tbe denomination of excremental, the epiploon, or corpus adiposum, and probably also the dorsal vessel, may be said to constitute the digestive system. It is singularly modified according to the diversity of aliments. or it forms a great number of peculiar types, which we shall describe in treating of the families of this class. We shall at present only say a word respecting the buccal apparatus, and the principal divisions of the intestinal canal, beginning with the last. In those insects, such as the carnassial coleoptera, in which it is more complicated, we may distinguish in it the pharynx, the æsophagus, the crop, the gizzard, the stomach, or chylific ventricle, and some intestines, which are divided into the slender intestine, the thick intestine, or cœcum, and the rectum. In the insects in which the tongue, properly so called, is attached to the anterior or internal face of the lip, or where it is not disengaged, the pharynx is situated on this same face, which is generally the case. We shall also add, that, with respect to the biliary vessels, a naturalist to whom we are indebted for the first valuable observations on the respiratory organs of the mygale, M. Gaëde, professor of natural history at Liege, does not consider these vessels as secretory, according to the common opinion; but his notion does not seem to be sufficiently well founded, and the observations of M. Leon Dufour even appear to destroy it.*

* This last mentioned naturalist, whom I shall often have occasion to quote, has explained in the most minute manner every thing which has

A few insects, and always without wings, such as the *myriapodes* or *millepedes*, approximate to many crustacea either in the quantity of rings of the body and of their feet, or in some traits of analogy in the conformation of the parts of the mouth; but all the others have constantly but six feet; and their body, the number of segments in which never exceeds a dozen, is always divided into three principal portions, the *head*, the *trunk*, and the *abdomen*. Among these last, some have no wings, and preserve during life the form which they had when born, merely growing and changing their skin.* In this respect they have some relation to the animals of the preceding classes. The other insects with six feet are almost all winged; but the wings, and frequently even the feet, do not appear at first, and are not developed until after a series of changes, more or less remarkable, called *metamorphoses*, and which we shall shortly describe.

The head† is the seat of the antennæ, the eyes, and the mouth. The composition and form of the antennæ vary much more than in the crustacea, and are often more developed in the males than in the females.

The eyes are either complicated or smooth: the former, according to the researches of Baron Cuvier, Marcel de Serres, and others, are formed, first, of a cornea divided into a mul-

reference to the digestive system of insects, in a series of excellent memoirs, which have contributed to enrich the annals of natural sciences. M. Victor Audouin has given a well-executed recapitulation of them in his article Insects, in the "Dictionnaire Classique d'Histoire Naturelle."

* These are what I name *homotenes*, (alike to the end,) or the *ametobolia* of Dr. Leach.

† Its surface is divided into many regions, which are named the *nose*, the *forehead*, the *vertex* or *top*, and the cheeks. The French term used for the nose *(chaperon)* being equivocal, M. Latreille, has changed it to *epistoma* or sub-mouth; this part serves for the insertion of the *labrum* or upper lip.

N.B. *Nose* is the term given by Mr. Kirby —Ed.

titude of small facets, by so much the more convex as the insect is more carnassial, clothed at its external face with a substance not very fluid, opake, variously coloured, but in general black, or of a sombre violet; secondly, of a choroid, fixed both in its extent and by its edges to the cornea, covered with a black varnish, exhibiting a multitude of air-vessels proceeding from tolerably thick trunks of tracheæ situated in the head, and whose branches form around the eye a circular trachea. This is wanting, as well as the choroid, in divers lucifugous insects. Thirdly, of nerves, which originate from a thick trunk, proceeding immediately from the brain, expanding afterwards into the form of an inverted cone, the base of which is on the side of the cornea, while its radii or threads, traversing the choroid and the covering of the cornea, end each of them at one of its facets. There is neither crystalline nor vitreous humour.

Many insects besides the compound, or complicated eyes, have simple ones, or such as have the cornea altogether of a piece. These are usually three in number, and disposed in a triangular form on the top of the head. In most of the apterous insects, and of the larva of those which are winged, they supply the place of other eyes, and are often united in a group. If we may judge by those of the arachnida, they must serve the purposes of vision.

The mouth of insects with six feet is in general composed of six principal pieces, four of which are lateral, disposed in pairs, and move transversely. The other two, opposite to each other, in a contrary direction to that of the preceding, fill the vacancies comprised between them. One is situated above the upper pair, and the other below the lower. In the *grinding* insects, or such as feed on solid substances, the four lateral pieces perform the office of jaws, and the other two are considered as lips: but, as we have already observed, the two upper jaws have been distinguished by the peculiar denomination of *mandibles;* the two others alone have preserved that of jaws. They have, moreover, one or two articulated threads, which are called *palpi* or *antennulæ*, a character which, in this class, the mandibles never exhibit. Their extremity is often terminated by two divisions or lobes, the exterior of which is named, in the orthoptera (in French), *galète* (*galea*, Fabr.). We have already mentioned that the upper lip is termed *labrum;* the other, or the lip properly so called (*labium*) is composed of two parts. The more solid and under part is the chin (*mentum*); the upper, which very frequently has two palpi, is the tongue (*lingua*).*

In the sucking insects, or those which take nothing but fluid aliment, these divers organs of manducation present themselves under two sorts of general modifications. In the first the mandibles and jaws are replaced by small laminæ, formed like blades or lancets, composing, by their union, a sort of sucker, which is received into a sheath, the substitute for a lip, either cylindrical or conical, and articulated in the

* See what we have said on this subject, in the generalities which precede the particular account of each class. The lower lip appears to us to be but a particular modification of the second jaws of the decapode crustacea combined with their tongue. The changes which these parts gradually undergo in the crustacea, the arachnida and the myriapoda, would lead us to believe it; on this hypothesis, the six thoracic feet would be the analogues of the jaw-feet, and this has been already recognized in relation to the crustacea of the genus *apus.* Then the first five segments of the abdomen in the hexapod insects, would represent those to which, in the decapode crustacea, are attached the feet, properly so called, or the third and fourth following ones of the amphipod and isopod crustacea. All the researches which have been published on the thorax of insects, though otherwise very useful and laudable, will necessarily undergo essential alterations, when naturalists come to compare this part of the body in the three classes of articulated animals with articulated feet. Nomenclature is very far from being fixed in this particular.

form of a beak (*rostrum*), or membranous or fleshy, unarticulated, and terminated by two lips (the *proboscis*). The labrum is triangular, vaulted, and covers the base of the sucker. In the second kind of organization, the labrum and mandibles are almost obliterated and extremely small. The lip is no longer a free body, and is distinguished only by the presence of two palpi, of which it is the support. The jaws have acquired an extraordinary length, and transformed into two tubular threads, which, uniting at their edges, form a sort of proboscis, which rolls into a spiral form, and is usually named *tongue;* but which, to avoid all equivoque, it would be preferable to call *spiral proboscis* or *tongue* (*spirignatha*).* Its interior exhibits three canals, the middle one of which is the conduit of the nutritive juices. At the basis of each of these threads is a palpus, usually very small, and but little apparent.

The myriapodes, or millipedes, are the only insects whose mouth shews a different type of organization, which I shall describe when I come to treat of them.

The trunk† of insects, or that intermediate part of their body to which the feet are attached, is generally designated by the Latin name of thorax, which has been rendered into

* Kirby calls it *antlia.*—Ed.

† This denomination is here synonymous with that of thorax. I am of opinion, however, that to avoid all embarrassment, we should apply the first term only to the apterous insects of Linnæus, which have more than six feet, and in which these organs are attached to segments proper to them; that is, in which the head is distinct from the trunk. With regard to the crustacea in which these parts of the body are confounded, the thorax should take the name of *thoracida,* and that of *cephalothorax* in the arachnida; animals which exhibit the same character, but in which the trunk or thorax is more simple, and provided with less numerous appendages. The entomostracea even approximate, in this respect, to these latter animals; but as they belong to another class, the term *thoracida* should be preserved for them—that of thorax should be exclusively reserved for the hexapod insects.

our language (French) by that of *corselet* (*corslet*). It is composed of three segments, which at first were not well distinguished, and whose relative proportions vary. Sometimes, as in the coleoptera, the anterior one, by much the largest, separated from the following by an articulation, mobile, and alone discovered, appears at the first glance to compose of itself the whole trunk, and bears the name of thorax or corslet. Sometimes, as in the hymenoptera, the lepidoptera, &c. much shorter than the succeeding one, it has the form of a collar, and constitutes, with the other two, a common body attaching to the abdomen by a pedicle, or intimately united with it in all its hinder breadth, and this is again called *thorax*. These distinctions, established on this particular, were insufficient and often ambiguous, seeing that they did not rest on a ternary division which I have strictly announced in the first edition of this work as a character proper to the hexapod insects, Mr. Kirby having already employed the denomination of *metathorax*, to distinguish the hinder thorax,* those of *prothorax* and *mesothorax*, the ter-

* This segment ought not to be restrained in the hymenoptera, to this superior division of the thorax, very short and transverse, on the sides of which the second wings are inserted; it is formed, besides, of that thoracic portion which extends hindwards, as far as the origin of the abdomen, and this is what evidently proves the position of the last two stigmata of the trunk, since they are placed on the sides of this extremity, behind the wings and above the last two feet. I am even of opinion, that this observation should apply to all winged insects. Their metathorax will be divided, at least in the upper part, into two portions or demi-segments, the one bearing in the tetroptera, the second wings, and without stigmata, and the other being provided with them. This last sometimes appears to depend on the abdomen, as in almost all the insects, with the exception of the hymenoptera, with pedicled abdomen, the rhipiptera, and the diptera; sometimes it is incorporated with the trunk or thorax, and closes it behind, as in these last insects: it is on this account, that I have named the second division of the metathorax, the mediary segment;

nary division being once established, naturally presented themselves to the mind, and it was the celebrated Professor Nitzch who first adopted them. Some naturalists have since given the name of collar (*collare*) to the prothorax or anterior segment, which bears the first two feet. Desirous of preserving the denomination of corslet, but at the same time of restraining its application within just limits, we shall use it in all cases where this segment considerably surpasses the others in size, or where the latter are united with the abdomen, and appear to constitute an integrant part of it. This is peculiar to the coleoptera, the orthoptera, and to many hemiptera. When the prothorax, being short, forms with the following segments a common and visible mass, the trunk thus composed of three united segments, shall preserve the denomination of thorax. We shall continue to call the lower surface of the trunk, the breast dividing it according to the segments into three areas, the fore-breast, the middle-breast, and the hind-breast. The medial line shall be the sternum, which again we shall divide into three parts, the fore-sternum, the middle-sternum, and the hind-sternum.

The teguments of the thoracic segments, as well as those of the abdominal segments, are generally divided into two rings or semi-rings, one dorsal or upper, the other lower, and united laterally by means of a soft and flexible membrane, which is in fact nothing but a portion of the same teguments,

thus all the segments of the thorax will have each one pair of stigmata, but of which those of the metathorax are not very visible, or are obliterated in the hymenoptera or diptera, and the two posterior or metathoracic ones, are situated on the segment which comes immediately after that which bears the second wings. In the orthoptera, the hymenoptera, the lepidoptera, and the diptera, the two anterior or prothoracic, are placed between the prothorax and the mesothorax. The abdomen will be composed of nine complete segments, the last three of which compose the organs of generation.

but less solid in many insects, especially in the coleoptera. We discover at the junction of these rings a small space more firm than they, or of the same consistence, and each having a stigma, so that the sides of the abdomen present a longitudinal series of small pieces, where each segment is, as it were, divided into four. Other pieces, equally corneous, occupy the lower sides of the mesothorax and the metathorax, and immediately under the origin of the elytra and the wings, which are themselves supported on another piece disposed longitudinally. The relations of these parts, the size and form of the first articulation of the haunches, the manner in which they articulate with the semi-ring on which they depend, the varying extent and direction of these semi-rings, and the thorax, considered under this point of view, present a combination of characters peculiarly advantageous for classification. Some naturalists, Knoch in particular, had already employed them, but without any fixed principle, and with arbitrary denominations. A necessary preliminary would have been carefully to study the composition of the thorax, and to pursue it comparatively into all the orders of this class. The late Lachat, at my suggestion, had commenced this labour. His friend, M. Victor Audouin, has pursued these researches, and has presented to the Academy of Sciences a memoir on the subject, which has obtained its suffrages. But it is also known by the general sketch given of it by Baron Cuvier in his Report,* and by the extract

* "An exposition of the parts of the thorax, and a fixed nomenclature expressly created for them," says the Baron in his Report, "should naturally be placed at the head of the work. The trunk of the insect may always be divided into three rings, each of which bears a pair of feet, and which M. Audouin names, according to their position, *prothorax*, *mesothorax*, and *metathorax*. Beside these feet, the mesothorax bears the first pair of wings, and the metathorax the second; each of these three segments is composed of four parts: one lower, two lateral,

which the author gives in the article INSECTS of the "Dictionnaire Classique d'Histoire Naturelle." Before adopting

(the three forming the breast,) and one upper which forms the back; the lower one takes the name of *sternum*. The lateral part or *side*, is divided into three principal pieces: one which attaches to the sternum, and is called *episternum*, another placed behind this last, and to which the haunch is articulated, is named *epimeros*. A small moveable piece, which serves for the union of the latter with the haunch, hitherto unknown, is named *trochantin*, from its opposition to the trochanter; the third piece of the side, which is the mesothorax and metathorax, placed in front of the *episternum* and under the wing, is called *hypopteron*; sometimes there is, moreover, around the stigma a small corneous piece, named *peritreme*. The upper part of each segment, which the author names *tergum*, is divided into four pieces, named according to their position in each ring—*præscutum, scutum, scutellum*, and *post-scutellum*. The first is often, and the fourth almost always, concealed in the interior. Naturalists have generally distinguished only the *scutellum* of the metothorax, which is often remarkable for its size and configuration, but its analogue may be found in the three segments; thus the trunk of insects may be divided into twenty three principal pieces, and if the hypoptera be reckoned, the number of these pieces may extend to forty-three, more or less visible in the interior. One portion of these pieces, moreover, gives out, internally, divers productions which also merit names, in consequence of their importance and uses. Thus from the lower part of the sternum of each segment arises within a vertical apophysis, sometimes figured like a Y, and which M. Audouin names *entothorax*. It furnishes attachments to the muscles, and protects the medullary cord. Its analogue is found in the head, and sometimes in the first rings of the abdomen. Other internal prominences result from the elongation of the neighbouring external pieces, united by a sort of synoptosis. M. Audouin names them *apodemata*. Some give an attachment to the muscle, others to the wings. Finally, there are some more small moveable pieces, either internally between the muscles, or at the base of the wings, which the author names *epidemata*. We have said that the principal pieces, or their vestiges, are always found, but it by no means follows, that they are always separate; many among them are even always united in certain genera and certain orders, and are only distinguished by the traces of sutures"——M. Audouin has since changed, in his article of INSECTS, in the "Dictionnaire Classique this nomenclature, and making a general application of it, we must wait until his researches, and the figures which are to accompany them, shall be published. In practice, the denominations already introduced may suffice. In relation to this subject, justice as well as friendship compels us to call the attention of naturalists to the labours of M. Chabrier, ancient superior officer of artillery, respecting the flight of insects. They form a part of the Memoirs of the Museum of Natural History, and also constitute a separate work. The figures are on a very large scale, like those of a memoir of Jurine the Father, on the wings of the hymenoptera, a work of the most patient research, as well as the preceding.

As insects inhabit every where, they are provided with all sorts of organs of motion, with wings and feet, which in many species answer the purposes of fins and oars. The wings are membranous, dry, and elastic pieces, usually transparent, and attached to the sides of the back of the thorax. The first, when there are four, or when there is but one pair, are placed on the sides of the second segment, and the second on those of the following, or of the metathorax. They

des Sciences Naturelles," the denomiuation *hypopteron,* to that of *paroptere.* That of the enthorax will change also, under some circumstances, and should be called *entocephalon,* (relative to the head) and *entogaster* (in relation to the abdomen. He remarks that the head of insects is composed of many segments. We have also observed that the bill of the cricket representing the lower lip, is not attached to the head, but to the membrane which unites it with the thorax. Thus, the two medullary chords form underneath the mouth two contiguous ganglions. On this principle let us consider the first segment of the body of the scolopendra, that which supports the two hooks, as a division of the analogous head. It appears that Knoch has distinguished the epimeros under the denominations of *scapulæ* and *parapleura,* the hind chest by that of *acetabulum,* whilst the middle chest is the *peristæthium.* The first articulation of the four posterior haunches forms, in most of the coleoptera, a transverse lamina, articulated in the flank, and that is, as it appears to me, the piece he calls *nicerium.*

are composed of two membranes applied one over the other, and traversed in various directions by nervures more or less numerous, which are so many tracheal tubes, and form sometimes a net-work, sometimes simple veins. A celebrated naturalist, the elder Jurine, has made an advantageous use, for the purposes of classification, of the disposition and crossing of these nervures. The dragon-flies, bees, wasps, butterflies, &c., have four wings; but those of the butterflies are covered with small scales, which, at first view, resemble dust, and give them the colours with which they are adorned. They may be easily removed with the finger, and the portion of the wing which has lost them is transparent. Through the microscope we can see that these scales, of figures very various, are implanted by means of a pedicle, and disposed in a graduated series, like tiles upon a roof. In front of the upper wings of these insects, are two species of epaulettes (*pterygodes*), which are prolonged behind along a portion of the back, on which they are attached. In certain insects the wings remain straight, or fold back upon themselves. In others they are doubled or folded longitudinally, like a fan. Sometimes they are horizontal, sometimes inclined. In many they cross on the back, being separate elsewhere.* The insects with two wings, of the order of diptera, have under the wings two small moveable threads, terminated like a club or bat, and which according to the most common†

* The insect is supposed to be in a state of rest. The rapidity of the vibrations of these organs appears to me to be one of the principal causes of the humming noise made by divers animals of this class. The explications which have been given of it are far from satisfactory.

† Appendages, in my opinion, of the tracheæ of the first abdominal segment, and corresponding to that space, pierced by a small hole adjacent to the anterior side of an aperture with a membranous and internal diaphragm, which is seen on each side at the same segment, in many crickets, or *crydia*.—See my Memoir on the Articulated Appendages of Insects, in the Collection of Memoirs of the Museum of Nat. Hist.

opinion seem to replace the two wings which are wanting. They are named *balancers;* other insects with two wings, and some of the most extraordinary, have also two balancers, but situated at the anterior extremity of the thorax, and which we shall name, to distinguish them from the other, *prebalancers.* Above the balancers, is a small membranous scale, formed of two pieces united by one of the edges, and resembling the two lids of a bivalve shell; this is the *winglet.* Some aquatic coleoptera have these below their elytra, and inserted at the base of the latter.

Many insects, such as the May-bugs, cantharides, &c., have, instead of the two upper or anterior wings, two kinds of scales, more or less thick, and more or less solid. They are opake, open and shut, and the wings are folded transversely under them in a state of repose. These sort of cases have received the denomination of elytra.* The insects which are provided with these, are called *coleoptera*, or insects with cases; these pieces are never wanting in this order, but it is not always so with the wings. In other insects, the extremity of these scales is altogether membranous like the wings; they are then named *half cases*, or *hemelytra.*

The shield is a piece usually triangular, situated on the back of the mesothorax, between the attachments of the elytra, or the wings; it is sometimes very large, and then covers the greater part of the upper portion of the abdomen. Divers hymenoptera have behind it, on the metathorax, a small space which is named hinder shield, or false shield.

* See, for their chemical composition, a memoir, already cited, of M. Odier, inserted in the Collection of Memoirs of the Society of Natural History in Paris, and the article INSECTS in the fore-mentioned "Dict. Classique d'Hist. Nat."

The limbs are composed of a haunch with two articulations, of a thigh, of a leg, with a single articulation, and of a toe which is usually named *tarsus*, and which is divided into several phalanges. The number of its articulations varies, from three to five, which depends much on the changes which the first and the last but one undergo in their proportion. Although their calculation may sometimes embarrass, and this numerical series is not always in relation with the natural order, it nevertheless furnishes a good character for the distinction of genera. The last articulation is commonly terminated by two crotchets or hooks. The form of the tarsi is subject to some modifications, according to the habits of insects. Those of the aquatic species are usually flatted, much ciliated or provided with small hairs, and in the form of oars.*

The abdomen, which forms the third and last part of the body, is confounded with the corslet, in the myriapods—but it is distinct from it in all other insects, or those which have but six feet. It encloses the viscera, the sexual organs, and presents from nine to ten segments, some of which, however, are often concealed, or very much abridged. The parts of generation are situated at its posterior extremity, and issue through the anus. The iuli and libellulæ alone form an exception to this. The last rings of the abdomen form, in many females, an oviduct, which is retractile, or always projecting, more or less complicated, and serving the purpose of an auger. It is replaced by a sting in most of the females of the hymenoptera. The fecundating

* Mr. Kirby, in his Monograph of English Bees, denominates the two anterior tarsi, *hands*. The first articulation is the *palm*. In conjunction with Mr. Spence he has published the Elements of Entomology in a very detailed and complete style.

organ of the male is almost always accompanied by hooks or pincers.* The two sexes generally couple but once, and this copulation even suffices, in some genera, for the fecundation of many successive generations. The female soon lays,† and deposits her eggs in the manner most favourable to their preservation, so that the young on coming forth, find suitable aliment within their reach; sometimes she provisions them herself. These maternal cares frequently excite our surprise, and more particularly unveil to us the "instinct of insects." In very numerous societies of several of these insects, such as the ants, the termites, the wasps, the bees, &c., the individuals composing the major part of the population, and which by their labours and vigilance maintain these societies, have been considered as neuters, or devoid of sex. They have also been designated as *workers* and *mules*. It is known at present that they are females whose sexual organs or ovaria, have not received a perfect elaboration, and which may become fruitful, if an amelioration in their nourishment shall develope these same organs at a certain period of their youth.

The eggs sometimes open in the womb of the mother: she

* Genitalia *mascula* senet; penis, canalis excretorius, vesiculæ seminales, vasa deferentia, testes; prehensores et semen—*Prehensores* sunt organa figurâ variâ quibuscum mas in coitu feminæ, anum corripit. Feminæ, genitalia sunt; ovarium, receptaculum suâ basi formatum, et oviductus. *Vide* (Siquid amplioris desideratur) M. Dufour, *Annales des Sciences Nat.*—et dissertatio Latiné Scripta, Dom. Hegetschweiler. Zurich, 1820.

† M. Audouin supposes that, in the case of a great number of insects the eggs are fecundated, on their passage, in a pouch situated near the anus; but this opinion requires confirmation from experiment. M. Leon Dufour, one of those naturalists who has most particularly studied the anatomy of these animals, does not adopt it.

is then *viviparous.* The number of annual generations of a species, depends on the duration of each of them; most frequently there is but one or two in the year. A species, *cæteris paribus*, is so much the more common, as the generations succeed each other with more rapidity, and the female is more fruitful.

A female butterfly, after copulation, lays her eggs, from which spring, not butterflies, but animals with a very elongated body, divided into rings, with a head provided with jaws, and several small eyes, having very short feet, six of which are scaly and pointed, placed in front, the others variable in number, membranous and attached to the final rings. These animals, known under the name of *caterpillars*, live for a certain time in this state, and change their skin many times. Finally, there arrives a period when from this *skin* of *the caterpillar* issues a being totally different, of an oblong form, without distinct limbs, and which soon ceases to move, remaining a long time apparently dead and dried up, under the denomination of *chrysalis.* When we examine it pretty closely, we discover, in relief, on the external surface of this *chrysalis*, certain lineaments which represent all the parts of the butterfly, but in proportions different from those which these parts will one day possess. After a period more or less long, the skin of the *chrysalis* opens, and the butterfly comes forth, humid, soft, with short and flaccid wings. But in a few moments it dries, its wings grow, become strong, and it is in a state to fly. It has six long feet, antennæ, a spiral proboscis, and complicated eyes; in a word, it resembles in nothing the caterpillar from which it has sprung, for it has been verified that the changes of state are nothing else but successive developments of the parts contained one in the other.

All this is what is termed the *metamorphoses* of insects.

Their first state is named *larva;* the second, *nympha;* the third, *perfect state* (*imago*). It is only in this last that they are capable of reproduction.

All insects do not pass through these three states. Those which have no wings leave the egg with the form which they are always to preserve.* These are called insects *without metamorphosis.* Among those which have wings, a great number undergo no other change than this acquisition. Their *larva* resembles the perfect state, with the exception only of the wings, which are wholly wanting. The *nymph* does not differ from the larva but in the rudiments of wings, which in the last change are developed to place the insect in its perfect state. Such is the case with bugs, locusts, &c. Finally, with the rest of the winged insects in which the metamorphosis is complete, the first state is a *larva*, formed like a caterpillar or worm, the second, a motionless *nympha*, in which, however, all the parts of the perfect insect exist, but in a contracted state.

These parts are free, although closely approximating to each other, and fixed against the body, in the nymphæ of the coleoptera, the neuroptera, the hymenoptera, &c.; but they are not so in those of the lepidoptera and of many insects with two wings. An elastic skin, or one of a consistence tolerably firm, is moulded over the body and its external parts, or forms for it, as it were, a sort of case.

That of the nymphæ, or chrysalids of the lepidoptera, consisting only of a simple pellicle, applied on the external organs, following all their contours, and forming for each of them so many special moulds, like the envelope of a mummy, enables us directly to recognize and distinguish them. (*pupa obtecta*, Lin.) But those flies, &c. formed of the dried skin

* The *flea*, the females of the *mutilla*, and the *working ants*, with a very few other insects, excepted.

of the larva, has only the appearance of a small egg-shell. It is a sort of capsula or case in which the animal is enclosed. (*Pupa coarctata*, Lin.)

Many larvæ, before they pass into the nymph state, prepare for themselves with silk, which they draw from within, or with other materials which they combine together, a shell in which they enclose themselves. The perfect insect issues from the nymph, through a cleft or division made on the back of the corslet. In the nymphæ of flies one of the extremities is detached, coming off like a cap for the passage of the insect.

The larvæ and the nymphæ of insects, which undergo but a semi-metamorphosis, differ from these insects in a perfect state only with relation to the wings. The other external organs are identical. But in the complete metamorphosis, the form of the body of the larva has no invariable relation with that which these insects are destined to possess in this perfect state. It is usually more elongated; the head is often very different both in its consistence and figure; it has only the rudiments of antennæ, or wants them altogether, and never exhibits the complicated eyes.

In the organs of manducation there is still greater disparity, as may be seen by comparing the mouth of a caterpillar with that of a butterfly, the mouth of the larva of a fly with that of the completely developed insect.

Many of these larvæ have no feet; others, such as the caterpillar, have many, but which, with the exception of the first six, are all membranous, and have no claws at the end. Some insects, as the ephemera, exhibit in their metamorphosis a singular exception to the rest. When arrived at the perfect state, they are again stripped of their wings.

The insects which compose our first three orders preserve during their existence the form which they had at birth. In the myriapods, however, we perceive a faint sketch of a metamorphosis. They have at first but six feet, or even none at all, according to M. Savi; the others, like the segments on which they depend, are developed with age.

There are few vegetable substances which are sheltered from the voracity of insects; and as those which are necessary and useful to our wants are no more spared than the others, they do us infinite mischief, particularly in such years as are favourable to their multiplication. Their destruction depends much upon our knowledge of their habits, and much upon our vigilance. There are some of them omnivorous, and of this number are the termites, ants, &c. whose ravages are but too well known. Many of those which are carnassial, and the species which live either on cadaverous, or excrementitious substances, may be considered as a benefit from the hand of the Author of Nature, and as compensating in some measure for the lossess and inconveniences we experience from the others. Some insects are employed in medicine, in the arts, and in domestic economy.

They have also many enemies. The fish destroy a great quantity of the aquatic species. Birds, bats, lizards, &c. deliver us from a portion of those which inhabit the earth and the air. The majority of insects endeavour to escape by running or flying from the dangers which menace their existence; but there are some which employ for the same purpose peculiar stratagems, or the arms with which nature has supplied them.

When these animals are arrival at their final transformation, and in full possession and enjoyment of all their faculties, they hasten to propagate their race, and when this object is accomplished they soon cease to exist. Accordingly in our climates, each of the three fine seasons of the year present us with several species peculiar to itself. It appears, however, that the females and the neuter individuals of those which live in societies enjoy a longer career. Many individuals, born in Autumn, escape the rigours of winter and re-appear in the spring of the following year.

The species of insects, like those of vegetables, are circumscribed within geographical limits. Those, for example, of the New World, with the exception of a small number, which are all northern, are essentially peculiar to that quarter of the globe. It also produces many particular genera. The ancient continent possesses some in its turn which are unknown in the other. The insects of the south of Europe, of North Africa, and of the western and southern countries of Asia, have many mutual relations. The same is true of those of the Moluccas, and the most eastern islands, those of the South Sea included. Many northern species are to be found in the mountains of the south. Those of Africa differ much from those of the opposite countries of America. The insects of South Asia, proceeding eastward from the Indus or the Sind, as far as the confines of China, have strong traits of inter-resemblance. The intertropical regions, covered with immense forests, whose trees are watered with a superabundant dew, are by far the most rich in insects. In this point of view, those of Brazil and Guiana are the most favoured.

All the general methods relative to insects are essentially reducible to three. Swammerdam has taken the metamorphoses as the basis of his system. Linnæus established his on the presence and absence of wings, their number, their consistence, their superposition, the nature of their surface, and the existence or absence of a sting. Fabricius, in his plan, only employed the parts of the mouth. The crustacea and arachnida, in all these distributions, form a part of the insects, and they even occupy the last place in that of Linné, which has generally been adopted. Brisson, however, separated them; and his class of crustacea, which he places before that of the insects, contains all of these animals which have more than six feet, that is the crustacea and arachnida of M. de Lamark, or the insects *apiropodes* of M. Savigny. Although this order was more natural than that of Linné, it was not followed, and it is only in recent times, that anatomical observations, and the rigorous exactitude of the applications which have been made of them, have brought us back to the natural method.*

I divide this class into twelve orders, the first three of which, composed of insects without wings, not changing essentially their forms and habits, subject only either to simple moultings, or a mere outline of metamorphosis which increases the number of feet and rings of the body, correspond to the order of *Arachnides antennistes* of M. de Lamark. The organ of vision in these animals is usually only an assemblage, more or less extensive, of smooth eyes, formed like small grains. The following orders compose the class of insects of the same naturalist. From its natural relations, that of the suckers, which contains but a single genus, the flea, ought to terminate the class. But as I place at the head the insects which have no wings, this order, for regularity of method, should succeed immediately to that of the parasites.

Some English naturalists have established, on the consideration of the wings, certain new orders. But I do not see the necessity of admitting them, with the exception, however, of the *stresiptera* (twisted or crossed wings), the denomination of which appears to me to be vicious, and which I shall call *rhipiptera* (fan-wings).

* Cuvier, Tab. Elem. de l'Hist. Nat. des Anim., and Lecons d'Anat. Comparée; Lamark, Système des Anim. sans Vertèbres; Latreille, Precis des Caract. Génér. des Insects, et Gener. Crust. et Insectorum. For more details consult the excellent Introduction to Entomology, by Messrs. Kirby and Spence, p. 312.

The first order, the MYRIAPODS, has more than six feet, (twenty-four and upwards,) disposed through the whole length of the body on a series of rings, each of which have one or two pairs, and the first of which, and in many the second too, seem to constitute part of the mouth. They are apterous.

The second order, the THYSANURA, has six feet, and the abdomen furnished on the sides with mobile pieces in the form of false feet, or terminated by appendages proper for leaping.

The third order, the PARASITES, (PARASITA,) has six feet, wants wings, has for visual organs smooth eyes; their mouth is, in a great degree, internal, and consists only in a muzzle enclosing a retractile sucker, or in a cleft situated between two lips, with two crochet-formed mandibles.

The fourth order, THE SUCKERS, (SUCTORIA,) have six feet, want wings;* their mouth is composed of a sucker enclosed in a cylindrical sheath of two articulated pieces.

The fifth order, the COLEOPTERA, has six feet, four wings, of which the two upper are case-formed, mandibles and jaws for mastication; the lower wings folded, simply across, and the crustaceous cases always horizontal. They undergo a complete metamorphosis.

The sixth order, the ORTHOPTERA,† has six feet; four wings, of which the upper two are case formed; mandibles and jaws for mastication (covered at the extremity by a

* They undergo metamorphoses, and acquire locomotive organs which they did not possess at their birth. This character is common to the following orders, but in the insects under consideration the metamorphosis developes another sort of locomotive organs, the wings.

† De Geer had established this order, and given it the name of *dermoptera,* which Olivier has changed, not much to the purpose, into that of *orthoptera.* We preserve, however, this last, because French naturalists have generally adopted it.

galea); the lower wings folded in two directions, or simply in their length, and the cases usually coriaceous, and most frequently crossed at their internal edge: they undergo merely a semi-metamorphosis.

The seventh order, the HEMIPTERA, has six feet; four wings, of which the upper two are in the form of crustaceous cases, with the extremity membranous, or similar to the lower, but larger and stronger: the mandibles and jaws are replaced by hairs, forming a sucker, enclosed in a sheath of a single piece, articulated, cylindrical or conical, and formed like a beak.

The eighth order, the NEUROPTERA, has six feet; four membranous and naked wings; mandibles and jaws for mastication; their wings are finely reticulated, and the lower are usually of the size of the upper, or more extended in one of their diameters.

The ninth order, the HYMENOPTERA, has six feet; four membranous and naked wings; mandibles and jaws for mastication; the lower wings smaller than the upper; the abdomen of the females almost always terminated by an auger, or a sting.

The tenth order, the LEPIDOPTERA, has six feet; four membranous wings, covered with small coloured scales, resembling dust or powder. A corneous piece like an epaulette thrown backwards, inserted in front of each upper wing; the jaws replaced by two tubular threads, united and composing a kind of spiral tongue.*

The eleventh order, the RHIPIPTERA, has six feet; two

* *Spiritrompe*.—See the generalities of the class. The thorax of the lepidoptera has more analogy with that of the heuroptera than with that of the hymenoptera, the segment which I have named medial appearing to constitute part of the abdomen, while in those last, and the diptera, it is incorporated with the thorax.

membranous wings, and folded like a fan; two mobile, crustaceous bodies, formed like small elytra, situated at the extremity of the thorax,* and for organs of mastication, simple jaws, in the form of bristles, with two palpi.

The twelfth order, the Diptera, has six feet; two membranous and extended wings, accompanied in almost all of them, by two mobile bodies, formed like balancers, situated behind them; and for organs of manducation, a sucker, composed of a variable number of bristles or hairs, enclosed in an inarticulated sheath, most frequently formed like a proboscis, and terminated by two lips.

* Formed, as we would presume, by pieces analogous to the epaulettes or *pterygodes* of the lepidoptera.

SUPPLEMENT

ON

INSECTS IN GENERAL.

Of all the classes of the Animal Kingdom, that of which we are now to attempt a general and brief sketch, is, by universal acknowledgment, allowed to be the most various and extensive. In diversity of conformation it surpasses every other, and does not yield even to botany in the number of species. Those which are known amount to upwards of twenty thousand, and there can be little doubt that numbers have escaped observation, and have never been described. It is possible, notwithstanding the zeal with which entomology has been recently cultivated, that this may be the case to some extent, even with species indigenous in Europe; and with respect to exotics, we may well believe that our catalogues are far from being complete. Travellers have generally contented themselves with collecting, on some isolated portions of the globe, such species as were remarkable either for the singularity of their forms, or the beauty of their colours.

These animals are many of them so minute, that their forms are not to be recognized but by the aid of the microscope. Magnitude and extension are, however, but of little consequence in the eye of the philosopher. If we may be allowed to make any distinction between His works, which are all wonderful, we would say, that the wisdom of the Supreme Being never is more manifest than in the conformation of organic beings which are not to be scanned by our feeble senses. In such, there appears to be a concentration of divine wisdom and power. Infinity exists in small as well as in great things, and the insignificance of an atom, the imparting to it, under such minute dimensions, so many organs, capable of a variety of sensations, is a greater marvel than the production of those colossal animals on which we look with terror and astonishment. But vulgar prejudice, which always despises minuteness, and idolizes magnitude, for a long period retarded the study of these interesting animals; they were considered as of little import in the scale of creation, and totally unworthy of any fixed attention.

It is impossible to repeat too frequently, that mankind are the fools of their senses as to *great* and *small*. Even philosophers themselves, who are, or ought to be, best aware that these are but comparative terms, that they are merely simple relations, often yield unconsciously to the impressions that magnitude is calculated to produce on our external senses:—" Wherefore," says one of the most philosophical observers of nature, " should we fear to bestow too much praise on the works of the Supreme Being? A machine is the more admirable, and does the greater honour to its inventor, in proportion as it is simple in relation to its destined object, though complex as to the number and variety of its parts. The union and concurrence of so many different and necessary pieces to the production of one given end, impress us with a high idea of the genius of the mechanist. He who has formed those animated machines which we term *insects*, has assuredly admitted no unnecessary parts into their composition. Notwithstanding their minuteness, they cannot fail to excite our admiration in a much greater degree than larger animals, when we consider that there are many more component parts in their body, than in the enormous living masses of the elephant or the whale. In the production of the butterfly, and of every insect which undergoes a true metamorphosis, the equivalent at least of two animals is produced."

The prodigious variety of forms in the insect world, through all its numerous tribes, presents a grand and most interesting spectable to the eye of reflection—what diversity in the general figure of the body, in the structure of the organs of locomotion, manducation and generation, and of all the external parts! There are doubtless, however, many wonders concealed from our inspection in the insect frame, by the difficulty of examining its internal conformation. Much, notwithstanding, has been discovered, which exhibits the most surprising mechanism, and is calculated to excite our curiosity, and stimulate our zeal in the study of those wonderful little beings. M. Latreille conjectures, and perhaps with probability, that a completer knowledge of the anatomy of insects might lead us to the solution of many problems relative to the anatomy of the human body.

But even though the study of an organization, so singular and varied as that of insects, should be deemed to possess but little interest for the general reader, still there are many other phenomena connected with these animals, that cannot fail to engage observation and rivet attention. Among these are the surprising transformations of their early age; their stratagems, address, skill and industry, in pursuing their prey, or procuring their food of whatever kind; the ingenious resources which they employ for defence against their enemies; all the marvels connected with their reproduction, and especially those operations to which our ignorance has given the name of ***instinctive***, but which, whatever be their cause, are so superior in precision and completeness of result, to those originated by human reason; their attentive cares for the preservation of the species; and, finally, the destructive action which numbers of them exercise on our properties or ourselves, and the important advantages which we derive from so many others.

If time and space permitted us, we might enlarge on the advantages which man has derived, and may yet derive, from the study of the insect tribes. We might notice that the idea of many of the most useful and necessary arts, have been owing even to a superficial observation of this class of animals, and their curious operations. We might remark, that many superstitious and vulgar prejudices have been dissipated by an extended knowledge of their species and peculiarities, and that our notions on natural science have been rectified by the same cause. The last mentioned particular is well deserving of attention. Before Redi, Malpighi, Swammerdam, and Reaumur had pursued their invaluable researches on the animals in question, many operations of nature in general, were most grossly misunderstood. The absurd theory, for instance, of *equivocal generation*, from which other absurdities were questionless derived, was overturned by the attention given to this part of insect physiology by the illustrious observers just cited. There can be little doubt that the opinion of the ancients, that insects sprung from the corruption of material substances, retarded, for a long time, the progress of entomological science. When it was believed that they were thus produced, it naturally followed that the most curious part of their history, all that appertaining to a perpetuation of their species, must be neglected and unknown. The case was similar respecting the transformations of insects, as long as it was supposed that they were any thing but simple developments. It is, therefore, most absurd to think, that to have correct notions on a subject like the present, can be matter of indifference to a philosopher. It is the nature of error to propagate itself, as it is of truth to conduct to the discovery of truth ; for one error may be said, without any exaggeration, to be connected with a multitude of others, and the same relation to exist among all important truths.

As soon as it was recognized that sexual intercourse was necessary to the re-production of insects, the general law of nature, which is applicable to other animals, was, of course, applied to them. However, without a careful and consecutive study of the subject, who would have imagined the possibility of their multiplying without the repetition of the fecundating act for a series of generations ? Is it not a new and most interesting phenomenon in animal physiology, to find that the insects to which we have just alluded (*the aphides*) are viviparous in the fine season, and oviparous at the approach of winter. Nor are these the only insects that should teach us the presumption of limiting nature within the narrow circle of our knowledge or conceptions. From the *hippoboscus* we learn, that it is possible for the young of some animals, at the instant of their birth, to be nearly as large as the mother.

As the appearance of different species of insects, on the return of fine weather, has a direct relation with the temperature of the air, they may be considered, in some sort, as natural thermometers. Some require but a moderate degree of heat for their development, others one much more considerable. Some insects may even answer the purpose of barometers. Thus, we may expect storm or rain when we behold the bees returning to their hives with unusual haste ; when the ants conceal their larvæ, or their nymps ; or the butterflies do not soar to their usual elevation. Such phenomena have not escaped the observation of the great poet of agriculture—

> " Sæpius et tectis penetralibus extulit ova,
> Angustum *formica* terens iter.

And again, of the bees, he says—

> " Nec vero à stabulis pluvia impendente, recedunt
> Longiùs aut credant cœlo adventantibus Euris;
> Sed circum tutæ sub mænibus urbis aquantur,
> Excursusque breves tentant.

But it becomes the less necessary to dwell on the pleasure and profit derivable from this delightful study, as they have been already amply demonstrated by far abler pens than ours; and, as the prejudices which have existed against it, have been considerably dissipated by the zeal and ability of its votaries. For those who cultivate it, it proves an inexhaustible source of instruction, a pleasure of perpetual recurrence, which may be indulged in to an extreme, without the chance of satiety, or the possibility of remorse. Its happy influence can withdraw us from the storms of passion, and console us under the accidents of life, and the reverses of fortune. It elevates the soul above sordid cares and sensual gratifications, to the contemplation of divine power and wisdom, and the lofty calm of philosophical religion. But it may not be unnecessary to remark, that it is not by a superficial study of this, or any other department of the works of nature, that such happy effects can be produced. Those whose exclusive object is amusement, and who take up a work on entomology as they would a romance, will be certain to meet both disappointment and disgust. The pleasure experienced from this, as well as every other science, is the result of the zeal and assiduity with which it is cultivated; and it is, or ought to be, unnecessary to add, that the improvement of the mind entirely depends upon them. Yet these remarks are not, perhaps, wholly superfluous in an age which professes to convert the acquisition of knowledge into sport, to make men wise without the trouble of thinking, and learned without the toil of study.

In this general sketch of insects, we propose to confine our observations as much as possible to what is common to the class, without a repetition of the text, while we reserve what is peculiar to each order for the succcessive supplementary remarks on each. We shall commence with some physiological reflections, which seem necessary to be premised for the purpose of determining the rank to which insects are entitled in the animal creation.

It is perfectly known, that the general laws by which inert bodies seem solely regulated, are counteracted or modified in those which possess life. The phenomena resulting from this sort of struggle, evidently depend on so many apparatus of organs, or instruments, with which such bodies are especially provided, for this very purpose. This character of the constitution of all living beings, which enables them to resist the constant action of natural laws, continually tending to their destruction, and the re-union of their component parts with the common mass of the elements, has been termed *vital* power, or *life.* This is a conventional term, by which is expressed a series of actions, very different from each other, though generally concurring to one and the same end —the preservation of the individual, or of the race.

Physiologists have given the term of functions to each of the principal actions of life, performed by systems of organs, or series of instruments, often altogether different from each other in their structure and mechanism, but productive of the same effect.

The vital functions are all referable to two grand series of phenomena. The first is produced by those organs which endow the beings that possess them with two faculties; namely, that of growing, or being developed by the incorporation of foreign substances, which, for a time, participate in the vital action; and that of reproducing or engendering individuals similar to themselves. The second series consists in the means accorded to living beings of changing place at will, either wholly or in part, and those by which they can perceive or feel the action which other bodies, by their qualities, can exercise upon them. These are the four grand functions of nutrition, generation, locomotion, and sensation.

Nutrition and generation take place in certain organized beings, independently of the other two functions. This is a more simple mode of existence, but the beings who so exist, possessing fewer faculties, must of necessity remain, and be developed in the same place where their germs were deposited.

The faculties of motion and sensation never exist in the same isolated way in any living beings, because they must be nourished and reproduced. Bodies thus organized, are more complicated, for they unite the two kinds of life. It is by the organs which give rise to the faculties of motion and sensation, and perform those functions in such a variety of modes, that animals are distinguished from vegetables, which cannot change place, and which, as far as we can discover, have no vestige of sensation.

It may be further observed, that these organs of motion and sensation, have modified the other two functions. An animal with the power of changing place altogether, or in part, and by its own proper motion, must, if the medium which it inhabits be not liquid, or if other beings do not provide it with nourishment, proceed in search of its food, introduce it into an interior cavity, so organized as to distribute the nutritious fluids which it contains, into all parts of the body. It must, also, by the aid of certain instruments, for the purpose of denoting the qualities of bodies, attract or repel, approach or avoid, certain substances. How the generative functions have been modified by the organs of motion and sensation, it is only necessary to allude to.

The consequences of these truths, properly followed up, will assist us in developing the elements of the natural classification of animals. Having once established that the existence of the organs of motion and sensation is the characteristic distinction of animals, it is certain, that the more that these faculties are developed in animals, the more are the latter removed from vegetables, and *vice versâ.*

The degree of this development is easily observed. If we find animals endowed with the faculty of reproducing themselves by germs or slips, of being dried up, and preserving the characters of life through the influence of moisture, light, or other natural agents; being able to exist only in a liquid medium; often fixed upon a point of space in the very midst of their aliment, which many of them absorb through external pores; shewing but few vestiges of any thing like motion—such animals assuredly exhibit the nearest possible relation to plants. They have neither distinct nerves, nor organs of sense, with the exception of passive feeling; no alimentary or digestive tube, no articulated appendages for motion, no distinct organs of respiration. These are the *zoophytes*, the last class of animated nature.

We next find animals condemned for the most part to live in water, whose motions are slow, and often hardly perceptible. They are, therefore, destitute of several of the organs of sense, though possessing nerves. They have no articulated limbs. Their mode of generation sometimes resembles that of plants. Sometimes there is a triple complication of distinct individual sex; sometimes the sexes are united in one and the same being—such are the *mollusca*.

The living creatures next in the scale, cannot be considered much more perfect than the last. They are either inhabitants of the water, or some constantly obscure and humid medium, and are destitute of almost all the organs of the senses. Their body, it is true, is divided into rings, which facilitate locomotion, but it is unfurnished with those articulated appendages which constitute limbs. Their nerves are well distinguished and knotty, and from each of the knots or ganglia, radiations of threads proceed towards the organs. The sexes are united. These are the *worms*.

The beings which belong to the two following classes, have the trunck formed of distinct and articulated levers, and are furnished with limbs of lateral appendages destined for various motions, according to their mode of existence. Those which live in the water have organs appropriate to that medium, being provided with gills: these are the *Crustacea*. In the others, the air penetrates into the various parts of the body, through apertures conducting into aerïferous tubes, named tracheæ: these are the *Insects*. They are far more animalized, if we may be allowed the phrase, that is, farther removed from the vegetable existence, than any of the preceding classes. They are endowed with sight, hearing, smell, taste, and touch. They enjoy all the various modes of motion on the water, on the earth, and through the air. In the organs destined for nutrition and generation, they are fully on a par with animals of a more elevated order.

It would be beside our purpose to trace these physiological relations any higher. We have said sufficient to shew the elevated rank which the insects should hold in the classification of the animal kingdom. They should certainly be placed immediately after the lowest of the vertebralia, over even which, they may be said to possess more advantages than one.

From the observations of the Baron it follows, that the *crustacea*, and the *araneïda*, differ from the insects proper, and especially from those which are winged, in having a complete system of circulation, and a different mode of respiration. From these motives, in his lessons on comparative anatomy, he formed a particular class of the crustacea, and subsequently admitted that of the arachnida into the "Regne Animal," comprising however, in this last, those species only which have no antennæ. The insects are, accordingly, distinguished from the crustacea, by M. Latreille, by the absence of peculiar circulating vessels, and by their tracheæ; from the arachnida, by having two antennæ, by the tracheæ forming two parallel cords through the whole length of the body, with centre of branches, like verticillæ at certain intervals, and by the stigmata not being limited to a particular space.

We shall now add to the text a few brief remarks on the structure of insects in general.

There are three principal regions of the body to be considered in insects: the head, the corslet, or thorax, and the abdomen, or belly.

The head is constantly articulated with the thorax, but the mode of articulation varies according to the orders, families, and genera. All the true insects have the head articulated, and mobile on the trunk, and furnished with antennæ.

The mouth is a very important portion of the conformation. Its parts differ not only in the different orders, but by slight modifications in all the genera, and more than probaby, in all the species. On these modifications, some entomologists who have studied them deeply, have established both orders and genera.

It cannot be expected that we should enlarge much on the anatomy of this part. It will be sufficient to say that with regard to it, all insects may be referred to two grand divisions: the species with free mandibles and jaws, disposed isolatedly in mobile parts; these are the masticating or grinding insects, and belong to the orders, coleoptera, orthoptera, neuroptera hymenoptera, and most of the apterous families. But we find that in the hymenoptera, especially in some particula· families, the jaws became elongated and flatted, and form with the assistance of the under lip a sort of tube and tongue, which imparts to these insects the double faculty of masticating their food, and imbibing it by a kind of suction.

By the suctorial insects properly so termed, no aliments can be swallowed but such as are in a liquid state. But the organs which produce this suction are greatly diversified in the different orders. In the hemiptera, this organ is an articulated beak, a kind of tube composed of many pieces which diminish in thickness from base to point, and in the interior of which are contained fine and sharp bristles, usually three in number.

In others, as in the lepidoptera. the mouth consists in a peculiar instrument, usually rolled back spirally, to which the name of *spiral tongue* or *proboscis* is given. This tongue forms a canal composed of two semi-sheaths, which correspond to the jaws of other insects, but exceedingly elongated. At the base of these the palpi are found, often well supplied with hairs, and also all the rudimen:s of the other parts of the mouth. For a further description of this tongue or *antlia*, as Mr. Kirby more properly terms it, must be reserved for the proper place; here we can afford but a hurried glance over insect conformation.

In the diptera, the tongue sometimes forms a fleshy proboscis terminated by two lips, which perform the office of a upping-glass, at the centre of which is the orifice of the alimentary canal. The genera in which this kind of instrument is observed, are forced to take their aliment, such as they find it on the surface of bodies, or to liquefy it so as to be able to swallow it. In others, this instrument is what entomologists generally have called a *sucker*. This is a sort of proboscis not widened at its free extremity, and containing bristles, which are vulnerary instruments used by the insect, to pierce the skin of organized beings, on whose fluids it subsists.

The appendages to the head of almost all insects called the antennæ, are extremely curious and very various in their structure in the different orders and genera, and even in the different sexes of the same species. Their use or office may be said to be still unknown to us.

The word antenna, which in Latin signifies the yard of a ship's mast, was adopted probably from an idea that these organs were useful to insects in flying or directing their flight, a notion long since exploded. Aristotle, with a better analogy calls them κεϱας, or horns, which word has been compounded with others, by some of the modern naturalists to form their generic names.

Insects with reference to their antennæ are divided primarily into dicerous, or such as have two* antennæ, and acerous, or such as have none.

With regard to their situation, the antennæ are generally inserted above the eyes in the upper surface of the head, but sometimes they are between the eyes, in the *canthus* of the eyes, or under the eyes, and sometimes they are on the rostrum. They are also more or less remote from each other at the base, and are sometimes united there. They are also various as to length, some being larger than the whole insect, and others shorter even than the head. Some are perfectly straight, and others pliant, but without an angle, while others have an angle or elbow more or less acute; some curve out-

* Some of the species of the genus *otiocerus*, seem to have four antennæ, and some six; but Kirby and Spence consider the supernumaries as mere appendages, and not as true antennæ.

wards, others inwards, downwards, upwards—some are bent back on the body, while others bend forward—some are convoluted spirally, and others are stiff and inflexible.

In some the antennæ when the insect reposes are hidden under the head or trunk, and in others they are never concealed.

The termination of these antennæ is also very various: this termination is sometimes a minute joint, a bristle, a capillary joint, a short point or sinew, a kind of hook, a claw; sometimes it is thicker at the point forming a knot, and sometimes it is divided into laminæ, which the insect can open and shut.

Some antennæ are feathered or hairy in various forms, as on both or one side in whorls, plumose, or flowing and ciliate, or straight and stiff.

Kirby and Spence name no less than thirty-two varieties of general figure and size in these organs. The figures of the several genera will display these varieties, and we insert here some greatly magnified figures of the antennæ found in three species of lepidopterous insects, to display their more minute details. Fig. 1, represents the antennæ of *Pap. Io.* 2. The same with a portion of its hairy exterior coat removed. 3. The knot or capitulum laid open, displaying the appearance of vessels like the pith of elder. 4. A transverse section of a portion of the antennæ. 5. The tubular insertion, and 6, the same more magnified.

Fig. 7, shews a front view of the head of *Bombyx Caja*, magnified by the lucernal microscope, with one of the antennæ removed. 8. Is part of the antennæ seen vertically covered with an imbrication of white scales or feathers, under which the pectens issue on each side. 9. Is one of the antennæ entire, with its pectens, and a part of the imbrication of white scales or feathers removed at *a* to shew the articulations. 10. Is the extremity of the antennæ more magnified. 11. Is the

P 10

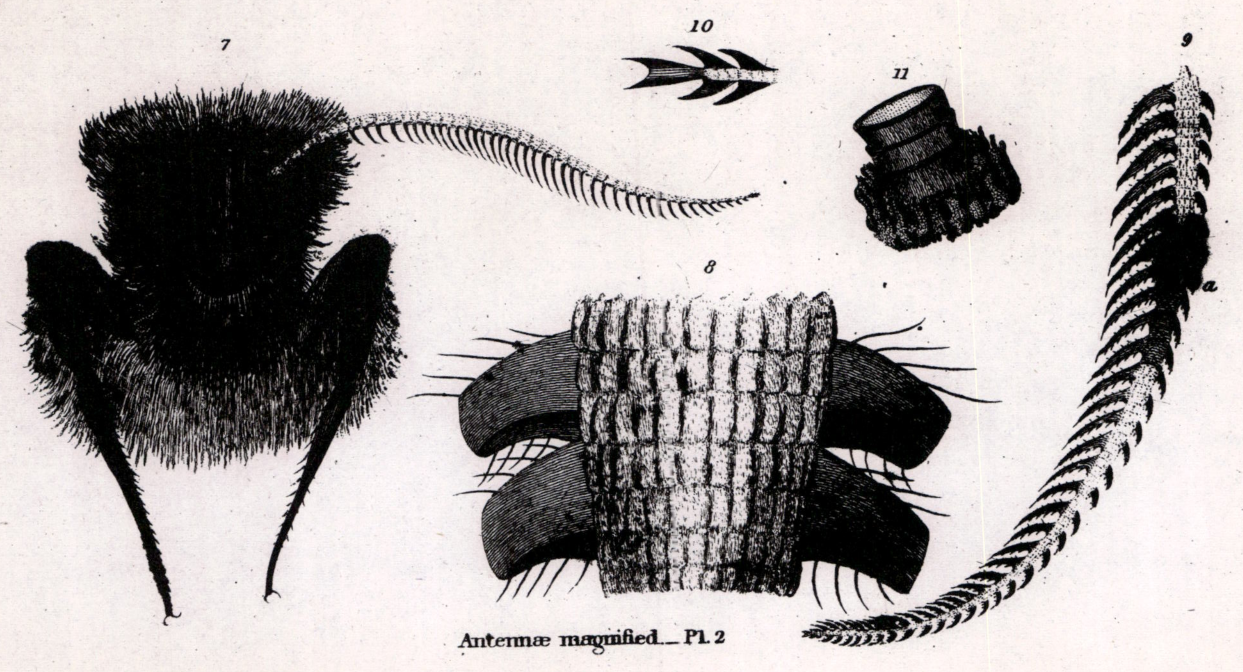

Antennæ magnified.—Pl. 2

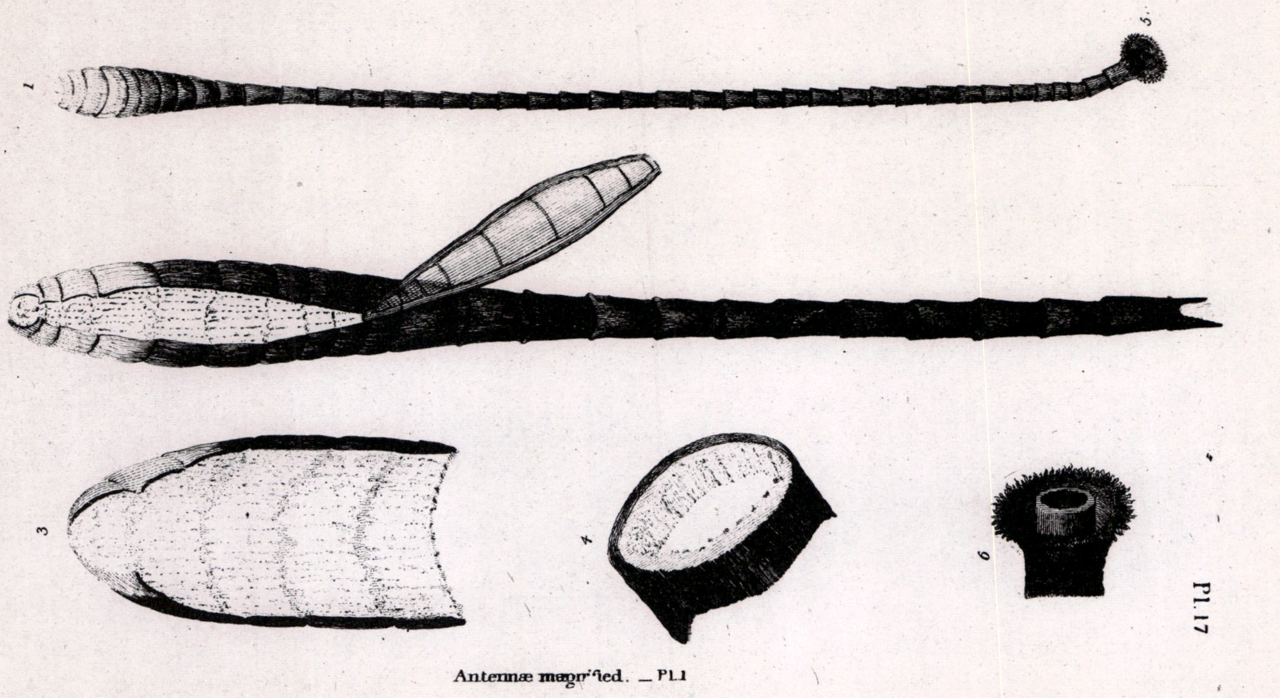

Antennæ magnified. — Pl. 1

Pl. 17

Pl.18

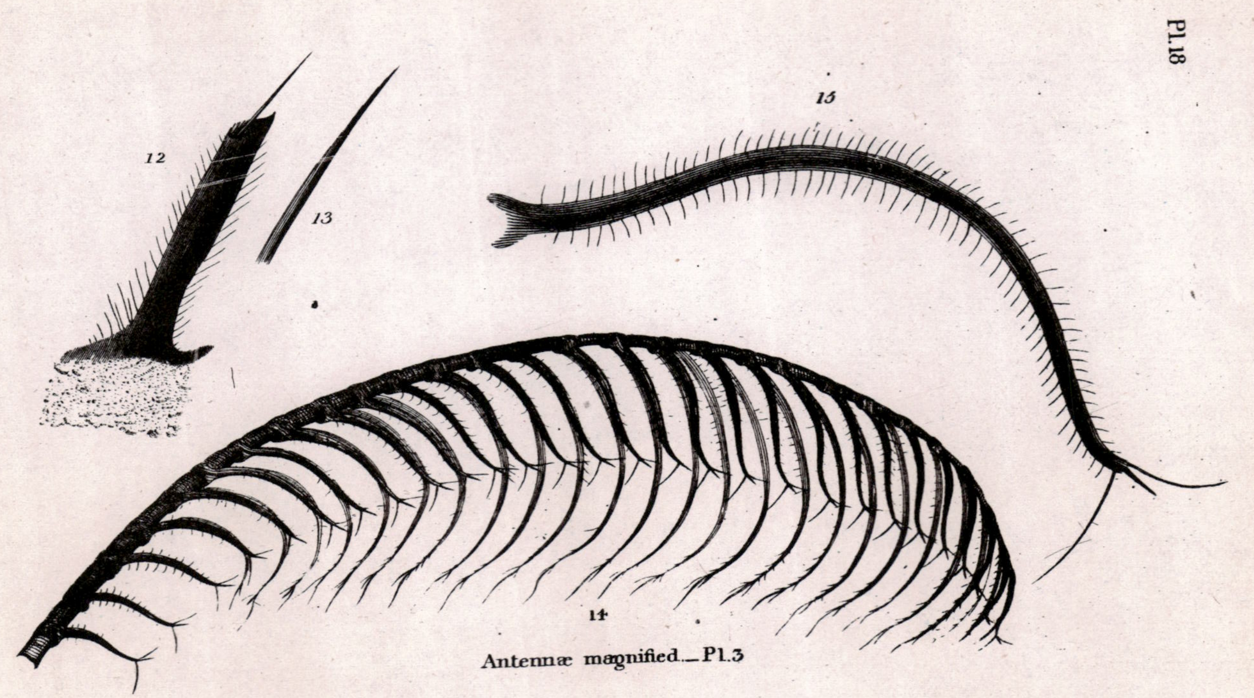

Antennæ magnified.—Pl.3

seat of the antennæ above the eye. 12. Shews the extremity of one pecten greatly magnified; and 13, the bristle thereof, at the termination more magnified. 14. Is the antennæ of *Bombyx antiqua*, separated and greatly magnified; and 15, is a separate pecten.

The substance of these organs seems regulated in some degree by the nature of the integument of each insect being softer or harder; accordingly they are not spoken of as vascular, nervous, or porous; there is, however, every reason to conclude that they are so: though the vessels are too minute for detection, we know that they consist in general of moveable tubular joints, varying greatly in number in the genera.

With respect to the uses to which insects apply their antennæ, naturalists are absolutely in the dark; some have considered them as organs of scent, others of hearing, but the majority have referred them to the sense of touch.

It is obvious that insects enjoy the sense of smell, and that in all probability in a very high degree, for we cannot otherwise account for many of the species so easily finding their proper food. One great office of the existence of a very large portion of insects, is evidently the more speedily to promote the dissolution of dead organic matter; accordingly, we find them rapidly attracted by putrescent matter, either as a nidus for their eggs, on which their almost countless larvæ feed; or as food for themselves when in the perfect state; and that this attraction is effected by the medium of this sense, may appear the more unquestionably, from the curious circumstance of the common flesh-fly frequently depositing its eggs in flowers, which possess a strong smell like carrion, *strapelia hirsuta*, by which the fly is deceived. There seems, however, no ground whatever for supposing that the antennæ are connected with this faculty, beyond the mere analogy of these organs being frequently placed in the local situation of the nose of other animals.

Experiments have indeed satisfactorily proved that insects possess the faculty of hearing; but others have as clearly shewn that it is not through the instrumentality of the antennæ that they enjoy it, and with regard both to this and the last mentioned faculty, it may be observed, that insects destitute of antennæ naturally, or deprived of them by accident, or design, seem to possess both their smell and hearing unimpaired.

That the antennæ are organs of tact seems more probable than the preceding hypothesis; but even this conjecture, in its ordinary mode of acceptation, is subject to material objections. The majority of insects in walking certainly carry the antennæ in a forward direction, and seem to feel every new object they approach by their means; but there are other insects which seem with great care to lay their antennæ backward, as if to avoid their coming in contact with any thing they may meet in their advance, and this, notwithstanding the antennæ in those same species are much elongated, and very flexible. The shortness of the antennæ in some species, and their total absence in others, militate also against the presumption in favour of touch.

It seems probable, on the whole, that tact is one of the functions of these antennæ; but it is equally so from the manner especially in which bees, and the social insects, employ these organs in the very inexplicable communications, which take place among the individuals of those social creatures to which we shall have occasion to refer more at large hereafter, that the antennæ are the medium of some faculty beyond that of touch, undiscovered by us, and perhaps without affinity, or even analogy in mankind. We know that the bat when blinded, for instance, can avoid every object artfully put in the way of its flight; and it seems probable that the extended membrane with which the head of this animal is furnished, may be the means of conveying to the senses some communication from surrounding objects by the elasticity of the air. But this is mere conjecture, however probable, and we are unable to apply this modus operandi to our own faculties; there may however, be something similar to this in insects; they appear to us to see with opaque eyes, and they may smell, hear, and feel, without the aid of particular organs appropriated to each of these faculties, or their antennæ may serve them in a way unknown to us in the enjoyment of most or all of their senses. When we consider the extent and almost universality of antennæ among insects, we may at least fairly conclude that in some way or other, or probably in many, they are essential to the enjoyment of their existence.

The eyes of insects are widely different from those of the other classes of the Animal Kingdom. They are divisible into simple and compound eyes; the former are very large, and the whole surface is reticulated, with a minute tubercle or lens between each reticulation; the latter, like the eyes of the vertebrata, form but one tubercle or lens of the entire eye.

All the coleoptera, and the diurnal lepidoptera, have two compound eyes only, but these are sometimes divided across, and appear as if double, as in the *gyrini*. The *orthoptera*, *hemiptera*, *hymenoptera*, *neuroptera*, and *diptera*, have, with a few exceptions, two compound with three simple eyes placed between them. The exceptions comprehend the *ephemera*, some species of the *phrygania*, which have only two very large simple eyes. The *hemerobia*, and the antlions, moreover, have not simple eyes: no winged insect therefore is deprived of compound eyes.

The larvæ of those insects whose metamorphosis is incomplete, have their eyes the same as in the imago state; but those whose metamorphosis is complete have only simple eyes, which vary in number in the several species. Caterpillars, for example, have six on each side, while the larvæ of bees, the saw-flies, &c., have only two: many of the last mentioned kind of larvæ, are entirely destitute of eyes.

The simple eyes of insects are in general too small for dissection, the wonders therefore of their organization are beyond our reach; these wonders can hardly however be supposed to surpass, even should they equal, those which the microscope displays to us in the compound eyes of these curious animals: the structure of these has been fully examined, and is tolerably well understood, though the mode in which the image is formed on the organ is by no means so intelligible to us, as in the case of the larger and vertebrated animals.

The structure indeed of the compound eyes of insects is so different from that of the eyes of other animals, even those not far removed in many respects from the mollusca, that we might have had reason to doubt whether these eyes were indeed the organs of sight; but experiment as well as analogy will determine that they are: thus, if the large hemispherical eyes of the dragon-fly be covered with any opaque matter, the insect will strike itself forcibly against the first object in its way; and if the compound eyes of the wasp be in like manner rendered impervious to the light, the insect will mount immediately in a vertical direction into the air, regardless of any object in its way; and if the simple eyes of the same wasp be afterwards covered the animal will remain motionless, or at least will not attempt flight. It seems, therefore, by this experiment, in addition to the general observation, that all winged insects have compound eyes either alone or with others, as if the former were essential to the use of wings, but unimportant to the animal while using its legs only for motion.

It is hardly possible to convey a sufficiently intelligible description of these compound eyes, without the aid of figures drawn with the assistance of a powerful microscope, with which we shall endeavour to illustrate the subject. Fig. 1, of these eyes, pl. 28, represents the head of *Papilio Urticæ*, with the spiral tongue in situ (to which organ we shall refer more at large when treating of the lepidoptera in particular), with the left palpus removed from the side of the tongue; this is drawn from the reflected image in the lucernal microscope, and is intended generally to display the outline, and particularly the straight filiform hairs or ciliæ, with which the whole hemisphere of the eye is beset at right angles with its surface, and which act probably as eye-lashes to defend the lenses.

The eye itself presents under the microscope a vast number of apparently hexagonal facets or lenses slightly convex, and separated from each other by slight ridges, forming as many sockets, from which proceed the eye-lashes; the lenses are, in fact, circular, and their sockets appear in some lights hexagonal, but this hexangular appearance seems to be occasioned by the intersections of the edges of the lenses. Fig. 2, shews these lenses magnified two hundred thousand times in surface. Fig. 3, represents the sockets and the circles, and Fig. 4, the mode in which the circular lenses display an hexagonal surface.

These facets or lenses, and their sockets, form altogether a hard elastic membrane, which is of itself very transparent, and constitutes the exterior coat or cornea of the eye; each facet may be considered a distinct cornea, or the whole of them may be said to form one aggregate cornea. They are convex on the outer surface and concave within, but thicker in the middle than at the edge, and what is very extraordinary, these lenses form the only apparently transparent part of this complicated organ. The number of lenses varies considerably in the different genera: seventeen thousand, three hundred and twenty-five have been counted in the eye of a butterfly, but they are said to be much more numerous in the beetle of Mr. Mac Leay's genus *Dynastes*. Leeuweenhoeck counted almost twelve thousand in the eye of the dragon-fly. In the genus *Xenops*, however, these lenses do not exceed fifty; they vary in magnitude in the different genera, and sometimes in the same eyes.

Immediately behind the transparent lens and its socket, is an opaque plaster, varnish or tunick, which varies considerably in colour in the different species, and which frequently forms in the eye itself dark spots and configurations. This plaster covers the entire concave surface of the whole aggregate cornea, and consequently the inner surface of each particular lens, without leaving the slightest aperture for the passage of light; how therefore the image of the object, after being refracted through the lenses, passes this apparently opaque substance, remains to be explained: at present, it seems most probable that we are deceived as to the real opacity of this tunick.

The object of the configurations formed by the different colours, and probably by the different degrees of opacity and transparency in the tunick of the eye, is not ascertained, though it seems not improbable that it may be intended to absorb a portion of the rays of light, and thereby diminish the too great sensibility of the optic nerves: if this be so, it is not perhaps the only instance in which Nature seems to have had recourse to especial means to remedy her own luxuriance. Fig. 5, shews the aggregate cornea of the eye of *Pha. bombyx lubriciperda*, with the coloured tunick altogether removed, and Fig. 6, the general appearance of the

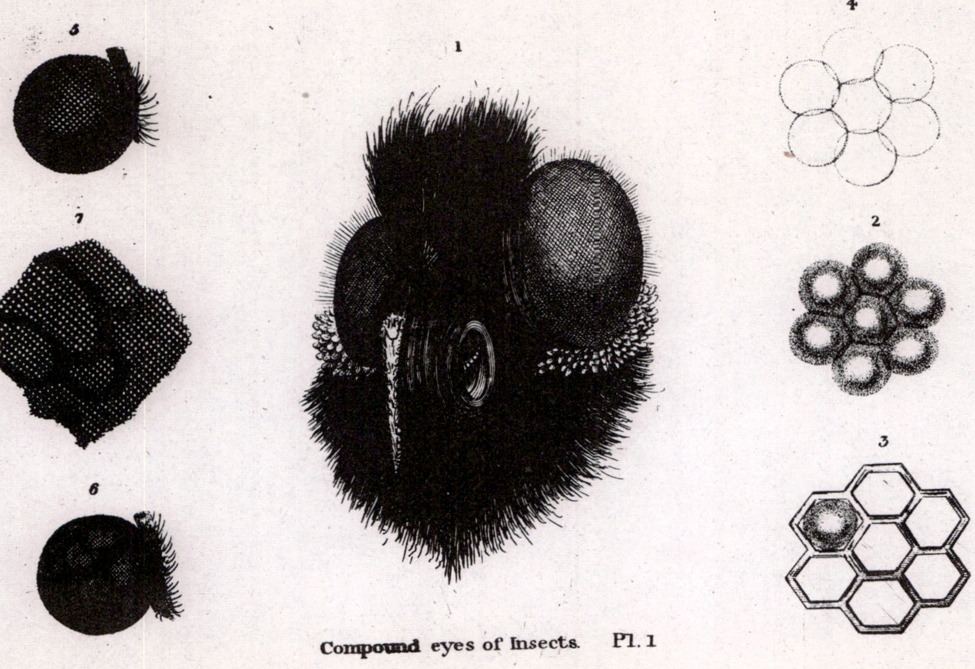

Compound eyes of Insects. Pl. 1

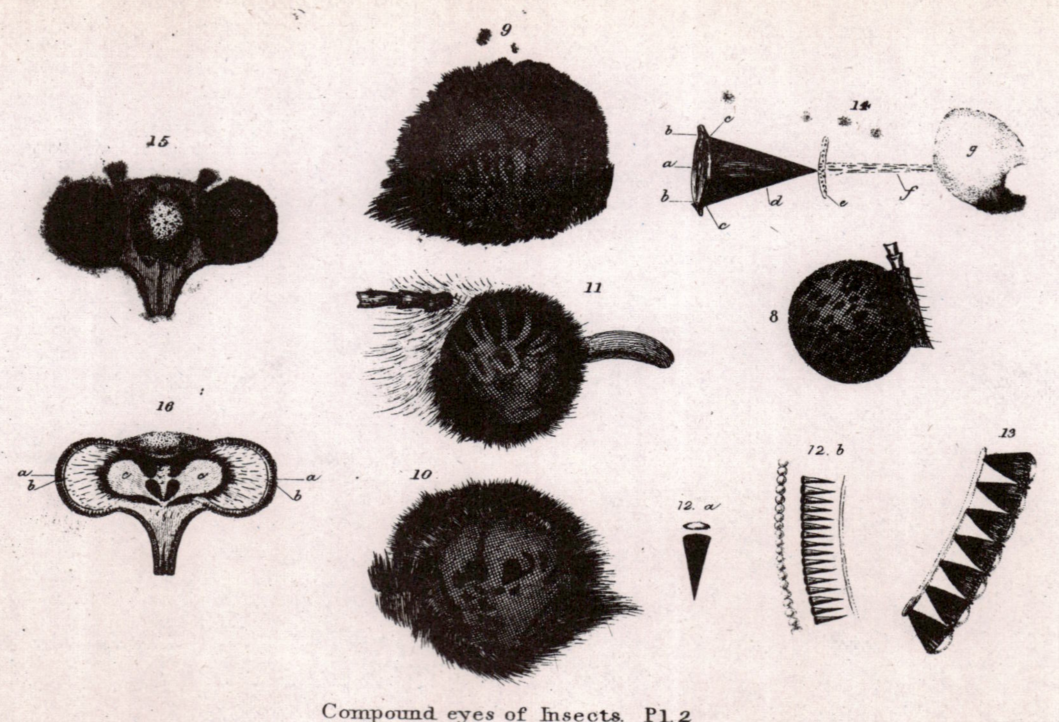

Compound eyes of Insects. Pl. 2

eye, of the same with the tunick and the configurations formed there upon. Fig. 7, represents a portion of the same more magnified, the second plate of the compound eyes, pl. 41. Fig. 8, the eye of *Pha. Geom. Grossulariata*, with its peculiar configurations. Figs. 9 and 10, shew the different forms of these configurations, found in the two eyes of the same individual insect in the *Pha. bombyx Caja*, and Fig. 11, the peculiarity in this respect of *Pha. bombyx Crategata.*

From behind this coloured plaster or tunick, corresponding with the socket of each lens, proceed a compact bundle of fibres: these form a case, with its base applied to the socket of the lens, and seem to be of a nervous character; "if so," says the Baron, in his Comparative Anatomy, "we may consider each of them as a retina to its particular facet or lens; but we have still to explain how the light can act on such a retina behind an opaque tunick." Figs. 12 a. and 12 b. shew these fasciculi, or conical bundle of nerves, a little removed from the lenses, and Fig. 13, a series of several of them more magnified, with the lenses in situ: the fasciculi may be observed at their points to come in contact with a membrane running parallel with the lenses.

This membrane is extremely fine, and penetrated with very minute air-vessels, which produce branches still more minute, passing between the pyramidal fasciculi to the cornea; it is called by Cuvier, from analogy, the *choroid.*

Behind this membrane, is a thin expansion of the optic nerve, which is a true nervous chord, perfectly similar to the retina of animals with red blood. It appears that the filaments which form the pyramidal fasciculus or retina to each lens, are produced from this nervous chord, after having pierced the choroid by a multitude of almost imperceptible pores. Fig. 14, displays, entire, the component parts of these compound eyes; *a*, the lens in its socket; *b*, *b*, the socket itself; *c*, *c*, the tunick; *d*, the pyramidal fasciculus; *e*, the membrane, called by Cuvier the choroid; *f*, the nervous chord communicating directly with *g*, a lobe of the brain.

To these figures we have added, Fig. 15. The head of a moth with the plumage removed, shewing the cranium horny and granulated, and the configurations on the eyes; and Fig. 16, a vertical section of the same to display the general appearance: *a*, *a*, the pyramidal fasciculi as attached to the cornea, forming a point where they come in contact with *b*, *b*, the choroid membrane which they pass through, and proceed in the shape of nervous chords or fibres to *c*, *c*, the lobe of the brain.

What are here delineated may be called the grosser parts of this wonderful organ, for besides these, we have those vessels incident to its existence, passing from the first ganglion called the brain, in various directions, through which the air is conveyed, a fluid which seems indispensable to the energy and application of the whole organ. The utmost powers of the microscope with difficulty disclose to us the largest of these airducts, but their almost infinitely minute ramifications elude all human powers of investigation. Much may be demonstrated to our senses, but much is at the same time left to imagination.

How little is it that we discover of the grand economy of organized beings! a very small portion only of its grosser forms. The more minute and curious details defy our most laborious research, and act silently and unseen, save only to their omnipotent mechanic. Limited faculties soon arrest our progress, even in physical science, but the very entrance to metaphysical knowledge is hermetically closed against us; our utmost researches cannot ascertain, nay, even the most vivid imagination cannot conceive the means by which perception is produced. We imperfectly puzzle out, how the grosser parts of organs are disposed inter-relatively, but this is all we can do; for matter, however arranged or modified, cannot produce intellect, cannot of itself create that perception, which the organ excites; or even should we adopt the absurd creed of the materialists, and admit that matter may be so organised as of itself to perceive and think, do we in the slightest degree diminish the difficulty before us, or gain one atom of knowledge on the subject?

There are, moreover, on the head of insects divers regions, whose development, colours, depressions, and other peculiarities have given characteristic distinctions to the naturalist. It will be sufficient here to name them. The *occiput*, which is articulated with the corslet, sometimes by a single condyle, sometimes by two, is sometimes truncated, rounded, flatted, depressed, prolonged into a sort of neck, &c. The *vertex* or summit of the head; the front (*frons*), the cheeks (*genæ*), between the eyes and mouth; the chin (*mentum*), on which the under lip is articulated, &c.

The corslet, or thorax, is that part of the trunk situated between the head and the abdomen. It has been quite sufficiently described in the text for all general purposes. From the description there, it is easy to judge that the form and extent of its various parts, must differ greatly in the different orders. Thus, the mesothorax is but little developed in the coleoptera and orthoptera, which have certain elytra of but little usage in flight. In the cigalæ, the epimeros is prolonged under the first ring of the abdomen, to form the great concave plate which covers the instrument of song in these insects. The four regions of the back are more sensible in the lepidoptera, hymenoptera, and diptera. The episternum is most developed in the libellulæ, in the coleoptera, or the metathorax.

The abdomen, the third region of the trunk in insects, has been sufficiently described. It is sometimes articulated with the metathorax, in the hinder region, by a broad surface. Sometimes, on the contrary, the articulation is reduced to a stalk, or pedicle, as in wasps, &c. The free extremity of the abdomen is most generally pierced by the anus. The last ring varies considerably in form: it is often conformed, so as to favour the approximation of the sexes, to facilitate the exclusion of the eggs, or their introduction into such substances as are proper to receive them. In this last instance, it forms what Mr. Kirby has denominated the *ovipositor*. It is frequently organized, so as to become a weapon offensive or defensive. A variety of instruments, destined to peculiar uses, often constitute a part of this region of the trunk. We may add, that in each of the rings of the belly, the under, upper, and lateral regions are distinguished, as to their form, structure, spots, and movements, which furnish very good characters, not only for genera, but even for species, and difference of sex.

The wings consist of pieces, articulated on the mesothorax and metathorax, in the interior of which are placed very powerful muscles, by which all the variations in their movements are produced. They are, in truth, light, but solid oars, formed by membranes, and sustained by radii or strings, variously disposed, to give them the necessary degree of suppleness, resistance, and mobility.

No insect is born truly winged, and some, which we term *apterous*, never have wings. Some have but two, and then they are named *dipterous*. Those with four wings are termed *tetrapterous*. But these distinctions have already been sufficiently pointed out.

Respecting the articulation, the composition, the shape, the clothing, the colours, and divers other peculiarities of the wings of insects, much might be written. But the necessity of constant reference for the purposes of illustration, to the orders, genera, and species, render the subject improper to be dwelt on here, to say nothing of the imperious obligation of confining this general essay within narrow bounds. We shall, therefore, dismiss all further consideration of the external forms of insects, and proceed to a review of their principal functions.

We have already insisted on the elevated rank to which insects are entitled in the classification of the living world. Under the relation of *animality*, they certainly, as we have said, stand next to the vertebrated animals, being provided with an articulated body, supported by articulated limbs, and being capable of all the various species of motion. In proportion to their mass, they exhibit both strength and activity in a most astonishing degree. They are endowed equally with the faculty of perceiving with strength, and at a distance, by means of the organs of the senses, the qualities of bodies, and better, perhaps, than we can appreciate them ourselves. The organs of respiration, in insects, are placed in immediate contact with the fluids, to render them proper for the excitation of life: a circumstance which compensates for the defect of circulation, and probably supplies its place with greater energy, so that the organs of nutrition and generation are not less energetic, nor less perfect, than those of the crustacea, the annelides, the mollusca, and the zoophytes.

Motions, in insects, though greatly varied, are performed without a correspondent complication of structure. As the parts of their body are, in general, very symmetrical, we find at one side, what is observed at the other, so that the study of one half of their body gives the idea of the correspondent part. In the trunk, the head and its annexes alone are capable of much motion. The three regions of the thorax are all moved by the limbs, without themselves contributing much to the movement. The rings of the abdomen are, in general, articulated on each other in the same manner, so that the muscles on one segment, are pretty nearly the same on the preceding, and on the following.

As most of the articulations are performed in ginglymus, two muscles are sufficient to produce them—an extensor, in general the smaller, and a flector or abductor, much more voluminous. These muscles are always placed in the cavity of the articulations, so that all the corneous pieces of the limbs, for example, are cases for the muscles.

There is this difficulty in the study of the muscles of insects, that they are truly circumscribed, and distinct only at their insertion, or the termination of their fibres on a solid tendon, or articulated prolongation of the piece which they are intended to move. As the insects are destitute of vessels and cellular tissue, these fibres are not united together, and when separated from their insertion, or fixed point of attachment, they remain floating like tufts, which renders their study very difficult. In the soft insects, as the grasshoppers, in the orthoptera, in the diptera, but especially in larvæ, and caterpillars, this study is much more easy. Lyonnet, in his excellent treatise on the anatomy of the caterpillar of the cossus, has given capital figures of these organs of motion. Swammerdam has also furnished us with very exact descriptions and drawings of the muscles. More ample details on this subject than we can afford to give, will be found in the admirable work of Messrs. Kirby and Spence. The Baron's "Comparative Anatomy" may also be consulted with advantage, though the magnifiers used at the time of its publication (1805) were insufficient to detect the ultimate fibre, an achievement since performed by M. Bauer, with improved glasses.

Insects are evidently endowed with a nervous system, and this system is absolutely the same as that of the crustacea and annelides. It consists in a nervous marrow, tolerably homogeneous, most usually composed of twelve ganglia, placed in lineal succession through the length of the body, from the head to the opposite extremity of the trunk. From these ganglia constantly proceed two nerves, which go to join the succeeding ganglion, and moreover other nerves, variable in number, which proceed, radiating, to join the surrounding organs, and whose thickness, or elongation, is proportionate to the development of these organs, or their distance from the ganglion. The principal ganglia are generally thus disposed; the first, which has been considered as a brain, is situated in the head above the mouth and the origin of the alimentary conduits. Besides the threads which it furnishes to the different parts of the mouth, and which it is not necessary to examine here, it sends some of the thickest to the eyes and the antennæ, and two hindwards, which connect the first ganglion to the succeeding. These two threads constantly embrace the œsophagus, and thus form a sort of collar, through which the aliments must pass. The series of the other ganglia remains then under the intestines and in the lower parts of the body. There are three in the breast; one for the prothorax, which gives out the nerves of the fore feet; one for the mesothorax, which furnishes the nerves of the upper wings, or of the elytræ, and those of the middle feet: finally, in the metathorax, the correspondent ganglion, which is the fourth of the series, furnishes the nerves of the lower wings and of the hinder feet. Each of them give out two threads, which establish the series of the ganglia. In the abdomen, this series has as many ganglia as there are rings, and these ganglia furnish the nerves of the muscles, and those which accompany the tracheæ, the genital, digestive, and secretory viscera.

The larvæ have the nerves differently arranged; nevertheless, they are absolutely the same as those which shall be manifested in the perfect insect, with this difference, that the ganglia are remote from, or approach each other according as the larva, being short, produces an elongated insect, or *vice versa*. The antlion is an example of the first, and the scarabeus of the second.

There cannot be the least doubt that the parts of which we now treat are the instruments by which the insect perceives its sensations, and that these nervous threads transmit into the organs the sensibility with which they are endowed, thus connecting together all the parts of the body. Positive experiment would have demonstrated this, even if it were not evident from analogy. We shall now proceed to examine the modes in which insects perceive their sensations, omitting all mention of the eye, which has been already sufficiently treated of.

All naturalists are persuaded that insects are possessed of the faculty of perceiving sounds, or vibrations of the air, since many insects produce them under such circumstances of their lives, as render it important that they should reciprocally manifest their existence to each other. The various sounds produced by insects, and which it is quite unnecessary to enumerate here, were undoubtedly destined to be perceived by some specific organ. That insects hear, even in the larva state, is proved by some observations of Bonnet, though that naturalist denied them this faculty. The sound of his voice, and the ringing of a small bell, evidently affected some caterpillars, which Bonnet referred to their sense of touch. Flies will move their legs at distinct sounds, and the social insects evidently intercommunicate by peculiar noises. Some orthoptera and hemiptera, whose males are vocal, afford the best proof of the existence of this sense. The females have been observed to attend to their call, and a tolerably loud sound will stop their chirping. But the seat of this sense yet remains undiscovered ; "perhaps," says M. Dumeril, "because, from analogy its existence has been always sought for somewhere about the head." It is but justice, however, to remark that Mr. Spence (we believe) has advanced strong probabilities in favour of the *antennæ* being the auditory organ, notwithstanding what we have already cited in treating of those appendages.

"As to the organ of smell, when we reflect on the nature of this perception," says M. Dumeril, "we are surprised that physiologists have desired, by no very judicious analogy, to find towards the head of insects the instrument intended to arrest odours, and appreciate their qualities. Mammifera, birds, and reptiles, are organized like man in relation to the olfactory sense, as they all respire by lungs, and the air which penetrates into their body for this purpose, can travel but by one road, which is the double entrance of the nostrils. It is on this forced passage, and on the very orifice, that the trial of the nature of this air must be made, to advertise the animal of the danger of admitting it, or the necessity of expelling it.

"Odours, in fact, have the greatest analogy with tastes. They consist materially of the particles of bodies held in suspension, some in gases, some in other fluids. The elastic fluids continually dissolve bodies at their surface, by which means, they are charged with some atoms of their constituent parts, and they retain them thus suspended in a sort of solution, disposed however, to abandon them when they shall have a greater tendency to unite with other substances. Under some circumstances, very volatile substances, and often in consequence very odorous, assume momentarily the form of vapours, or impermanent gases, which possess most of the properties of the air, or of the elastic fluids with which they mingle. It is in this point of view, and as gaseïfied corpuscles, or æriform fluids, that the operation of odours ought to be studied."

"Necessarily transmitted by the air, which is their only vehicle, odours have a tendency to penetrate along with it into the body of the animal; arrested in their passage in a sort of *custom-house*, *(bureau de douane,)* where they are to be promptly visited and analyzed, they are put in contact with a humid surface, with which they have some affinity. They immediately combine with it; but at the same time they touch, and give notice of their presence to the nerves distributed over these same parts which report to the brain, of which they are the elongation, the chemical or physical action, in a word, the sort of sensation which they denote, or which perhaps they have experienced."

"Smells, therefore, like all other physical sensations, are a sort of touch in which the body, whatever be its nature, comes to the organ, and transports itself on the only part of the animal where its action can manifest all its properties. In the final analysis, all our sensations are thus reducible, either to a passive *taction*, i. e. to the action of being touched, or to an active *tact*, which gives us the faculty of carrying our bodies or some parts thereof to the surface of objects, for the purpose of ascertaining their qualities."

"By this admirable arrangement we experience the action of the majority of bodies. It is thus that light, an imponderable fluid, which is so variously modified on the surface of objects, transmits their image into the eye by being applied exactly on the nerve of the retina; that the matter of heat or caloric, is placed in equilibrium with our bodies, is applied to, or withdrawn from them, thus manifesting its presence or absence; that the vibrations communicated to bodies are transmitted either directly by contact, or by the intervention of air or gases, to a small quantity of air contained in one of our organs, with which they unite in perfect harmony to enable us to appreciate sounds and produce hearing; that, in fine, the substances which are susceptible of being dissolved in liquids, first manifest their qualities on that part of the animal where it was most required that they should be appreciated, before coming into the interior of his economy, for taste is one of the qualities of food."

"In the final analysis, all the organs of sense are constituted by physical or chemical apparatus, true criterions to which the nerves lead to produce, on the instant, a complete idea of the sensation."

M. Dumeril, after these physiological observations, which he owns to be necessary for explaining the mode in which the sensation of odours is produced in insects, goes on to cite the facts which prove that they do possess this sensation. His observations are interesting.

"It appears that the object of Nature, in endowing with existence this innumerable quantity of destructive beings, was to employ them in clearing away the remains of organized beings deprived of life—thus restoring them sooner to the mass of their component elements, to produce others, and preserve unbroken the eternal circle of creation and destruction.

"For this purpose she has bestowed on the being designed for such operations, tastes, and a mode of life analogous to the functions which they were called upon to fulfil; and in this instance, as in all her works, to produce the necessary degree of perfection, she has gifted these animals with a sense altogether fitted to the manner of their existence.

"It is through the medium which they inhabit, that insects are advertised of the presence of such bodies as may serve for their nutriment. The air, surcharged with odorant emanations, which are continually escaping from it, carries into the respiratory organ all the molecules which it holds in suspension; it thus becomes the invisible guide of the animal, which is seeking to supply its own wants.

"The early observers of nature did not study with suitable attention the mode of decomposition in organized beings. Beholding the almost sudden appearance of destructive insects, and of larvæ, or as they termed them, worms, in carcasses, they regarded them as the product of corruption. It is not two centuries since Redi proved, by conclusive experiments, that the worms were deposited there by flies and other winged insects, and that the latter were attracted by what was named *instinct*, on the bodies in a state of decomposition. This is what is now observed every day.

"It is thus that we behold insects arriving from all quarters, on the residue of aliments, which have been submitted to the digestive action. They raise these substances, perforate them, cause them to present a greater surface to humidity, to dryness, to dissolution, by dispersing or extending them over a greater space—such is the occupation of numerous species, that seem principally summoned to the attack and destruction of organic animal matter deprived of life.

"To these insects the organ of smell has been denied, but they have been supposed to be gifted with a sight so piercing as to supply the place of it. This opinion, however, has been combated, and an opposite one supported with success from experience and observation.

"Certain flowers have a fetid and cadaverous odour, so striking, that at the time when they are full-blown, a great number of insects are observed to seek them, insects which ordinarily live in animal substances in a putrid state. It is thus that the spathes of the snake-root, the corollæ of the variegated stapelia, are often covered or filled with sylphs, beetles, the meat-fly, and other insects, which come not only in the hope of finding their food, but even for the purpose of depositing their eggs. Is it possible to resist this evidence, and not to perceive that, at first, these insects have been deceived by their organs of vision; that, afterwards, those of smell have produced not only the voluntary motion of the insects towards the place where the odorant molecules were floating, but that, moreover, deceived by this illusory sensation, the insect has been induced to deposit its eggs on a substance which its smell alone had indicated as proper for their reception?

"Do we not observe bees, wasps, sphinxes, butterflies, and all insects which live on the juices of vegetables, or nectar of flowers, arrive in great numbers towards the plant which produces them, as soon as they begin to flow, or the petals are opened?

"Perfectly vain is it to attempt to explain this attraction, this movement, by the visual sensation of the insect; for, in spite of all the care of florists in covering their tulips, in spite of those of the grocer, whose honey is concealed in proper vessels, the insect comes, attracted by the odour, and tries all efforts to get at the place from which it directly emanates.

"The insects then, possess the sense of smelling. But in what part of their body is the organ of this perception? It is probable that this sensation is produced in them as in all other animals, through the respiratory organ. But, in insects, respiration takes place through numerous orifices, which correspond to the majority of the rings of the body, except at the head. These apertures are named stigmata, all of which lead to the tracheæ, or vessels with elastic parietes, always filled with the circumambient air, which doubtless arrives there, charged with odorant particles, as happens to other animals. But does the gas thus penetrate into the net-work of the air-vessels, or does it deposit these molecules at the very entrance of the stigmata? This is not easy to decide, as we do not ourselves experience this sort of sensation; for certainly, we should have no idea of the admirable function of the pituitary membrane of animals, if we did not ourselves evidently experience the sensation of odours, and if, under certain appreciable circumstances, we were not deprived of the faculty of smell."

Thus far M. Dumeril. The opinion which he advocates originated with Baster, was adopted by Leckmann, and has received the sanction of that very high authority the Baron Cuvier himself. Notwithstanding, however, the ingenuity and plausibility of this theory, Mr. Spence thinks with apparent reason that the arguments against it are more weighty than those in its favour. The correspondence between that part of the head in insects called the *nose*, and the same part in mammalia, is obvious. Moreover, the close connection between the senses of smell and taste, and the proximity of their organs in all animals where the organ of the former can be clearly ascertained, affords a strong analogy in favour of the seat of this sensation being in the head of insects. The argument of Leckmann that because an insect is smaller than what it feeds upon, it is a matter of indifference whether it smells with its *head* or *tail*, is surely, to say the least of it, somewhat absurd. A flying insect, as Mr. Spence well remarks, would be more readily directed to the object of its pursuit by smelling with the former than the latter, and a feeding insect would be better guided thus in the choice of its food. Besides, to assert that smell is the *necessary* concomitant of the apertures of respiration, is to go beyond what our knowledge of insects will justify. Their other organs of sense are so different in their structure, and often in their mode of being impressed, that there may be an analogous difference in the organ of which we are speaking, and it is possible that insects may smell without inspiring the air. Moreover the same judicious writer observes that we do not smell with our mouths though we breathe with them: thus proving that olfaction and respiration do not invariably accompany each other.

Facts, as well as analogy, are also in opposition to the theory that insects smell with their spiracles. *Bees* shew an especial dislike to the smell of the oil of turpentine. M. P. Huber, anxious to ascertain the seat of smell in these insects, tried the application of this drug with a hair-pencil, and also of oil of marjoram to the abdomen, the head, the trunk, the eyes, and the antennæ, without effect. "But when he pointed the pencil *near the cavity of the mouth, above the insertion of the proboscis*" the little animal exhibited the most extraordinary agitation. Messrs. Kirby and Spence seem to think that the precise organs of smell are discoverable, on dissection, in the common burying beetle, and consist in a pair of circular pulpy cushions found under the nose and the *rhinarium*, or nostril piece. These cushions are covered by a transversely striated membrane. Analogous parts are to be found in other insects. By what channel scents act upon this organ, whether transmitted through the pores of the part representing the nostrils, or received by the mouth, the learned authors just cited do not determine, though they rather incline to the former opinion.

Whatever controversy may have existed respecting hearing and smell, we believe that the sense of taste has never been denied to insects. For a long time it was supposed to reside in the *palpi*, because these parts of the mouth are continually in motion, and applied in all points to the food in proportion as it is divided and bruised by the mandibles and jaws. Observers were led to this notion, because in a great number of species the extremity of the palpi swells, is softened, and becomes, as it were, vesicular. It was in consequence of this peculiarity that some physiologists declared their opinion that the organ of smell might reside in this part. The palpi, however, do not exist in a great number of insects, or at least, are so short as to be utterly inapplicable to this purpose. It is more natural to presume that savours are perceived by insects, as by most animals, in the interior of the digestive canal, and principally at its origin in the mouth. In short, as it is very clearly proved that insects have a tongue, so it is most natural to believe that it is the principal organ of taste. In fact, very distinct nerves repair towards this part. Lyonnet has represented them very well in his Anatomical History of the Caterpillar of the Cossus.

Touch, in all probability, is that sense which in insects is the least developed. Not that these animals are deprived of parts proper to be placed in contact with the different parts of the surface of bodies; but these parts are generally covered with a hard, and often corneous skin, which prevents such an immediate application as is necessary for the full appreciation of the tangible properties of bodies. Not that we are to suppose them altogether destitute of this sense in the hardest parts of their bodies. They must feel sufficiently through the covering of their limbs for all the purposes of motion. The idea of temperature more or less elevated, of softness or solidity, of bulk, or length, breadth or thickness, cannot well be acquired by the insect through the medium of touch. The organs supposed to be designed for this use in insects are, first of all, the antennæ. These are generally allowed to be organs of touch. When elongated and composed of a great number of articulations, they appear to be tentacula, employed by the insect and put continually in motion to explore its path and discover and avoid obstacles. This is observable in sphex, ichneumon, chrysis, and a great number of insects besides. But in others these antennæ are composed of a single hair, or some short rings, as in flies, &c., and though doubtless possessed of this sense, cannot be of any great use as an organ of tact.

In the next place, the palpi are considered as organs of touch. They were considered by Bonsdorf as olfactory in-

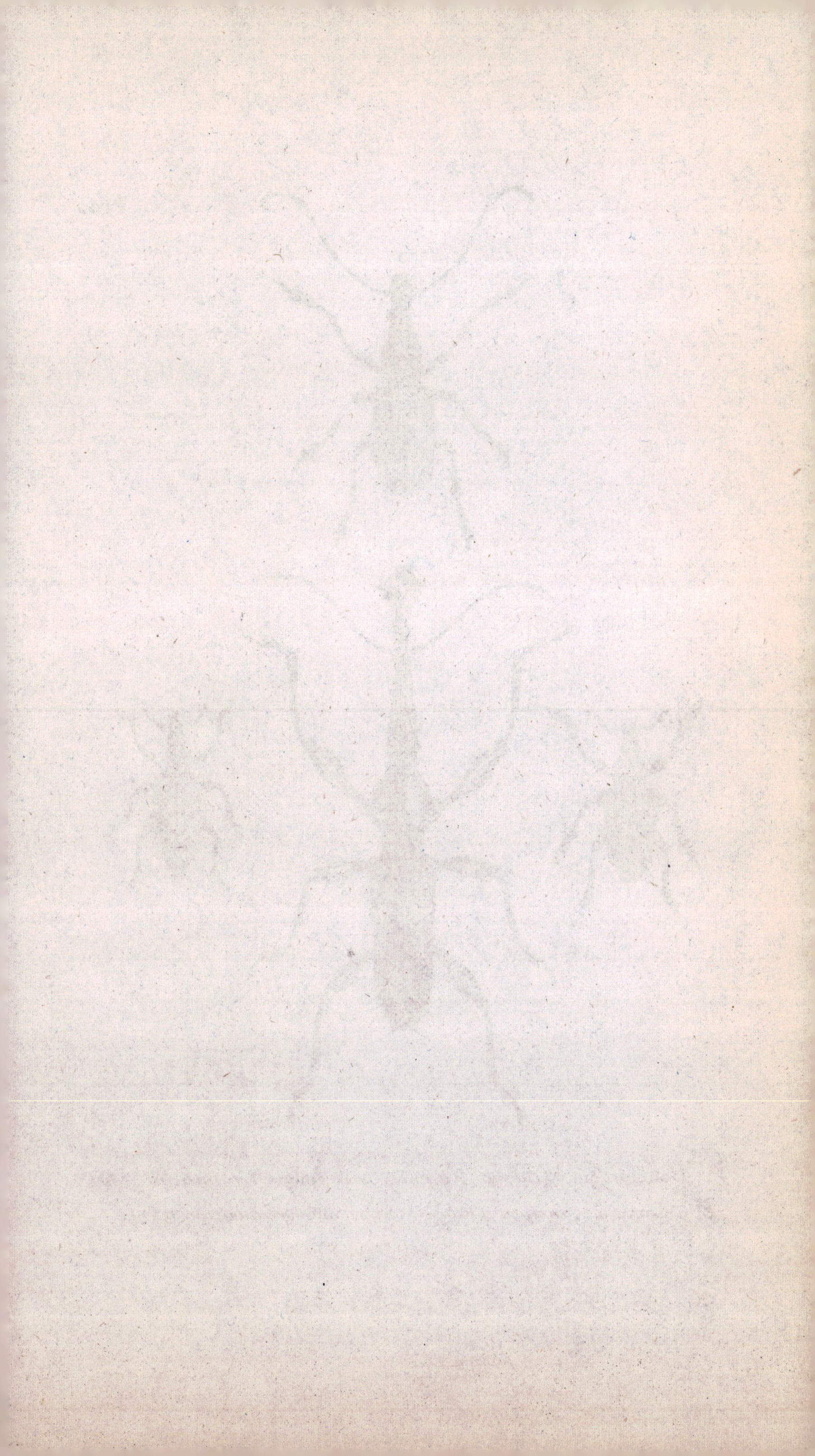

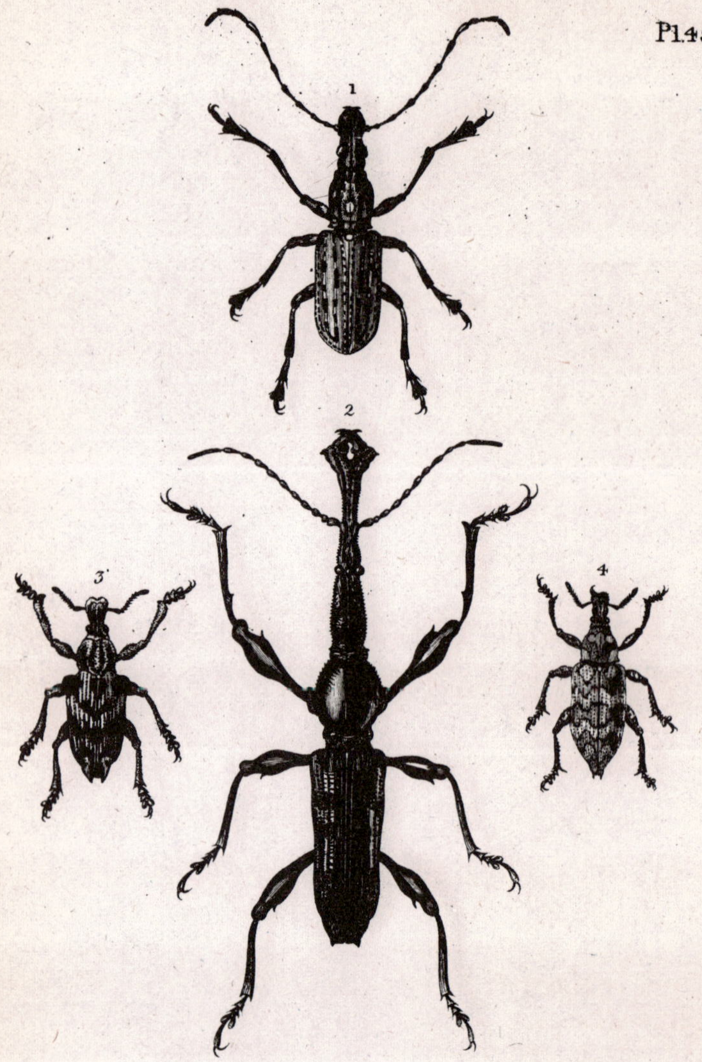

1. Dolicbocera *Childreni. G.R. Gray.* 2. Brenthus *Temminchii. Kuhl.*

3. Chlorima *cyaneipes. G.R. Gray.* 4. Curculio *spectabilis. G.R. Gray.*

struments, and Knoch imagined that the *maxillary* palpi were employed in smell, and the *labial* in taste. But the opinion that they were designed for touch has been embraced by very high authorities, and above all by the Baron himself. M. Dumeril, indeed, observes that the palpi, though very mobile in the masticating insects, are rudimental or altogether nullified in the suctorial, and their form altogether changed.

It would appear that the tarsi are very useful organs of touch in many insects, and the best calculated to give them an idea of the nature of bodies. In the majority they have a broad spongy surface, which in flies, &c. applies with great facility to the superficies of an object. In other insects they are considerably elongated, and composed of very mobile articulations. In many apterous insects these tarsi are obviously the instruments which they employ to explore the solidity and nature of the bodies on which they are about to transport themselves.

Let us now bestow a little attention on the nutritive function in these interesting animals. We have already seen that insects admit of two general divisions founded on the mode in which they take their food, whether by suction or mastication. Like other animated beings they derive the elements of their nutrition from organized bodies, or from substances which have already been borrowed by other living beings from brute or inorganic nature. But the modes of alimentation are extremely varied as we shall now briefly explain.

The kind of food taken by insects varies greatly in one and the same species, at different epochas of its existence. One species is carnassial, or lives on the juices of animals in its first state, and subsequently becomes herbivorous. Another, on the contrary, is first nourished on the debris of vegetables, and afterwards cannot live except on the fluid or solid parts of animal bodies. Some, during a period of their existence, can absorb or suck their aliment in a liquid form, and, consequently, without mastication; while, under other circumstances, the parts of their mouth having changed form, they will attack nothing but solids. This is not the place to quote numerous examples of such modifications; sufficient will be exhibited in the course of this work; but we must mention a few of the more remarkable.

The hydrophili, which in their first state the French call *vers assassins*, which attack and suck the tadpoles of reptiles, small fish, mollusca, and which, in their final form, feed only on aquatic plants, and the half-decomposed leaves of vegetables which fall into the water, furnish us with a clear example of a zoophagus animal becoming phytophagus. It is the same with the anthreni, the telephori, &c. On the other hand, the larvæ of the antlion suck their prey without masticating it, and the perfect insect has a mouth perfectly organized for bruising its aliments. In opposition to this, we find the caterpillars of lepidoptera, as, for instance, the silkworm, gnaw and chew the leaves, while butterflies, bombyces, &c. can only suck the nectar of the flowers with their spiral tongue. These differences of habits and conformation in the parts of the mouth are always connected with important modifications in the other digestive organs.

Each insect knows perfectly well the aliments which are proper for the preservation of its life, and for the growth of its body. It knows how to seek and procure them. There are many, indeed the greater number, which have no occasion to seek their food at any distance. All insects are endowed with an instinct which leads them to deposit their eggs, or the germs of their progeny, where they can easily procure nourishment, or to provide for their subsistence. Notwithstanding the change of food, which we have just remarked, yet insects in the perfect state, notwithstanding that they live so differently from their larvæ, know perfectly well the substances where the latter will find proper nourishment, and deposit their eggs accordingly. Thus the lepidoptera, which live usually on the honey which they extract from flowers, never fail to lay their eggs on or near the plants which are fitted for the support of their caterpillars. The gnats are, in effect at least, perfectly well aware that their larvæ must live and be nourished in the water, and therefore they place their eggs upon its surface. It is the same with many other insects, as the ephemera, libellulinæ, &c.

Among the insects which live in a social state there are some, which (as the *bees*) are obliged to choose a dwelling mutually to assist each other in procuring the necessary food, and to accumulate a certain quantity of it to serve as provision for the winter. Others, like the ants, unite and labour in common to seek food, not so much for themselves as for their larvæ, which are incapable of providing for themselves.

Many insects can subsist only on one species of food, and never vary in their taste. Such are a great number of caterpillars, which live on certain leaves, without being able to touch any others. They die immediately if a supply of these leaves should be wanting.

There are insects which eat often, and have need of almost perpetual nutriment. They cannot fast for any length of time without serious injury. These are the *herbaceous* tribes. Others can remain a long while without taking food, which is the case with *carnassial* insects, and such as subsist on living prey. This is a remarkable analogy between the present and the preceding classes.

Some insects live on the leaves of trees. This is the case with the larvæ of almost all the lepidoptera, and with those of many coleoptera, hymenoptera, &c. Others subsist only on the juice of the leaves and the stalks, as the *cigalæ*, the *tettigoniæ gallinsectæ*, *pucerons*, &c. There are some which live on excrescences of plants and trees named galls, and subsist on these galls. Such are the *cynips*, &c.

For other insects, all these kinds of food seem too gross; they require an aliment of a sweeter and more delicate description, which they find on flowers. The glands of many flowers furnish this honeyed kind of liquid, which modern botanists have decorated with the name of *nectar*. Every one knows that of this nectar the bees compose the substance of honey, after it has undergone a final preparation in their bodies. Ants seek with avidity the saccharine liquid which is ejected by the pucerons, through two peculiar apertures in the abdomen.

Fruits of all descriptions constitute the food of other races. The destruction occasioned by their larvæ in our orchard and garden fruits, is rather too well known. A species of weevil lives in nuts. More precious fruits, such as olives, and also different grains, serve as aliments to caterpillars and larvæ of different species. Green peas, the seeds of thistle and burdock, beans, acorns, and chestnuts, and an immense variety of other grains, serve as food for those little animals.

The insects which feed upon the different species of corn, we know pretty well, to our cost. Many commit great ravages in our barns and store-houses. Such are the *calandra granaria*, *bruchus granarius*, and two species of *tinea* (*granella* and *hordei*). The larvæ of melolonthæ attack plants after another fashion. They gnaw the roots, and thus cause them to perish when they are young. Those of some *cecydomyiæ* and *oscines*, live in the interior of the stems of young cereal plants, principally of barley, and often thus destroy the hopes of agriculture. The larvæ of many insects, and principally those of the coleoptera of the family of *capri-*

cornus, gnaw the interior of the trunks of trees: they pierce the wood or sap, mince it, reduce it into saw-dust, and then eat the particles.

Those of the *tipulæ*, which live under ground, swallow earth, afterwards rejecting all in it that is unfit for nutriment. They prefer the unctuous earth, or soil produced by animal or vegetable decomposition. An immense number of insects live on the excrements of animals; on dead flesh of all kinds; on our butcher's meat, as is well known, where flies deposit their eggs, which soon produce voracious larvæ. The meat which is gnawed by these larvæ rapidly corrupts, which is owing to a kind of fermentation which they occasion, and which accelerates the progress of decomposition. The dried flesh of animals is also attacked by insects, but of genera very different from those that feed on fresh and soft meat. They are usually coleopterous larvæ, and sometimes perfect insects of the same order. They are pretty well known to the amateurs of natural history who make collections of dried specimens. It is scarcely necessary to notice the ravages committed in clothing, furs, &c. by moths and other lepidopterous insects.

Other insects subsist upon the living animal, deriving sustenance both from its fluids and the substance of its flesh. There is especially a singular larva of the genus *œstrus*, which lives on the back and under the skin of horned cattle, where it produces tumours and subsists upon the pus secreted from the wound which it has made. Larvæ of the same genus live in the stomach of horses, round the pylorus, and sometimes in the intestines. It is there alone that they find their subsistence, and their superabundance occasionally proves fatal to these animals. Other œstri are lodged in the frontal sinuses of sheep, where they grow and derive their nutriment from the mucus of the nose.

Even man himself is very far from being exempt from the

daring and troublesome inroads of insect invaders. His self-created rank as lord of the creation affords him no protection from their insolent attacks. They pay no respect to his aristocracy; and the godlike glance, at which the higher animals are said (with what truth, God knows) to quail, cannot scare their formidable minuteness. They invade his dominions without fear, conquer, and colonize. Their expulsion is always difficult, sometimes impossible; and to prevent their irruptions, often calls forth his most strenuous efforts, his utmost skill, and his unwearied vigilance. But he may console himself with the reflection that all other animals suffer in a similar manner—that insects themselves are the prey of other insects, which feed upon their substance, imbibe the streamlets of their life, and occupy the interior cavities of their body, thus illustrating, even in infinite littleness, the inscrutable law of perpetual destruction, and perpetual reproduction.

Though the aliments of insects are, for the most part, in a liquid form—though the majority of species are sustained by the juices or fluids of plants and animals, and thus find both food and drink together, yet some are seen to eat and drink at separate times. The fondness of locusts for drinking, was not unknown to the ancients. They seem to seek, with their antennæ, the drops of dew which hang upon the leaves, and when they have found them, drink them up immediately. Bees, ants, and other insects, exhibit a similar taste.

Of the organs of nutrition, and their various modifications, enough will be found in the subsequent portions of the text, &c., in treating of the different orders and families. We shall only add a word, on the mode in which nutrition probably takes place. There is no doubt of its being performed through the agency of the aliments, which furnish to the different organs not only the means of reparation, but of growth, and of the fulfilment of their functions. But what is the mode of this absorption? This is a question not yet completely answered. M. Cuvier has detailed the reasons which lead us to believe that this absorption, in insects, is operated by a kind of imbibition, because they are destitute of circulatory organs, or of lymphatic, arterial, and venous vessels. It is true that he arrives at this conclusion from negative indications; but his reasonings are so strengthened by analogy, that they acquire the force of proofs.

It is certain that no blood-vessels are to be observed in insects. One vessel, indeed, is known, which is a sort of canal, predominating along the middle part of the back in all its regions (see text). This is very visible in the caterpillars, especially in those which have a smooth body, as the cossus. In this there may be discerned a kind of alternate motion of systole, and diastole, or of dilatation and contraction, which appears to pass from the head to the tail; but neither branches nor roots have been observed, which carry thither any fluid, or proceed from it.

On the other hand, when we know in what manner respiration is performed in insects, we do not find so great a necessity, as in other animals, for the presence of a central agent of circulation, or of canals to convey the fluids towards the place where the air should come especially in contact with them. Animals endowed with circulation, with arterial and venous vessels, which come from the heart and repair thither, have, in consequence of this very circumstance, conglomerated glands, destined for the performance of secretions, such as the liver, the pancreas, the parotid and salivary glands, &c. But in species deprived of a circulating system, the case is very different. There are no glands designed for any such functions. The secretory organs are composed of numerous and distinct filaments, which are immersed in the midst of the nutritive fluid, from which they are to borrow their materials, on which to operate each in its own peculiar mode, for the performance of the secretions.

Such is the mode in which, in the actual state of anatomical and physiological science, we must suppose that nutrition takes place in insects. It is by the porosity of the intestinal tube, that the materials most proper for alimentation are separated from the indigested mass. Their division is such that they then form a sort of vapour, the molecules of which are, probably, absorbed in the form of gas, which would then be contained by the numerous trachæ with which the intestinal tube i scovered.

There are other secretions, however, which take place in insects, but the mode of this separation of humours is very little known. We know, for example, that the acid produced by the ants, and which is analogous to that of vinegar, is disgorged by the insects; that many other animals of the same class, eject or disgorge, at will, certain matters, either fœtid or hurtful, at the moment when they believe themselves in danger.

The secretions of insects, according to Mr. Kirby, are as follow :—*silk* ; *saliva* ; *varnish*, or *gum* ; *jelly* ; *oils* ; *milk* ; *honey*; *wax* ; *poisons* and *acids* ; odorous *fluids* or *vapours*, and *luminous matter*. We must dismiss the consideration of these very briefly here, reserving for the future portions of the work, a more detailed account of them. Many insects discharge from their mouth a fluid resembling the saliva of larger animals. The use of this seems, generally, to be to dilute the food. Several flesh-eating, and herbivorous coleoptera, under the influence of fear, emit a coloured acrid fluid from the mouth. This does not appear to be secreted by the usual salivary vessels, nor must we suppose that every fluid ejected from the mouth of insects is saliva. The eggs of many insects, when excluded from the oviduct, are covered with a sort of varnish or gum, that they may adhere to the substances on which the larvæ are to feed. A kind of jelly, or gluten, serves as a bed for the eggs of some insects, which are committed to the water. Oily substances are secreted by some insects, but their precise nature has not been ascertained, nor whether peculiar organs are appropriated to their secretion. The larva of the *chrysomela populi*, and some other insects, emit a milky fluid from different parts of their bodies. Of silk, honey, and wax we must defer any notice, only remarking that the latter production is not peculiar to bees, and that, in many other insects, it is a *general* transudation from the body; it also appears to form a constituent part of some insects that do not secrete it. A poisonous fluid, producing very unpleasant effects, is secreted by many of these animals. The acid of ants is a very curious phenomenon, as will be seen in the proper place. An acid has been obtained from silkworms, called *bombic acid.*

The powerful scents emitted by insects, whether agreeable or otherwise, are numerous beyond measure. Some are peculiar to certain parts of the body, some exhaled from the whole. Sometimes their vehicle is a fluid secretion, and sometimes gaseous effluvia: of *phosphoric* secretions, we shall speak in treating of the glow-worm.

Fat is found in great abundance in insects in the *larva* state, and more or less in the *imago*. In the former, it occupies the whole of the interior, except the space required for the muscles and internal organs, which it surrounds. It is contained in numerous floating membranes, which sometimes appear like small globules, and sometimes like a thick mucilage. Its colour is white, but sometimes yellow or green.

That insects *perspire*, appears more than probable. A quantity of fluid escapes from them in the pupa state, which has been proved by experiment. The want of a urinary discharge renders it probable that the imago also perspires. But the fact is proved in the case of bees in summer, which when crowded together, perspire most profusely.

The tracheæ, or air-vessels, by which insects respire, have been already mentioned. The larvæ have them as well as the perfect insect, but a very remarkable change in these, as well as in the other parts of the body, takes place at the period of transmutation. Some aquatic insects in particular, respire in a manner totally different, in their three states of larva, nymph, and imago.

We see in a great number of diptera, while they are yet in the larvæ state, the orifices of the tracheæ diversely grouped towards the final rings of the body. In the larvæ of the stratiomys, for instance, is remarked a kind of aigrette, formed of barbed hairs. In the centre of this aigrette, the respiratory organ is observable. In the larvæ and nymphs of the dragon-flies or libellulæ, the mode of respiration is peculiarly singular. The water penetrates into their rectum by a sort of inspiration : it seems probable that it leaves there the oxygen gas, which is combined with it. In the thickness of the parietes of this intestine, are seen a great number of tracheæ, which represent five great leaves or nervures of leaves, composed of small tracheæ, which repair into four principal trunks, two of which, still thicker, are subdivided in all the parts of the body. This mode of respiration in the larvæ of which we are speaking, is rendered sensible to the eye, by letting them stop a few minutes in a coloured water, and then putting them into other water very limpid. At each movement of respiration the coloured water issues from the anus after having washed the tunics of the intestines ; moreover, the animal employs this mode of respiration to facilitate its conveyance or movement in the water, availing itself of the resistance which the jet proceeding from the anus receives upon the liquid mass. The body of the insect thus resting behind, advances on the opposite point, where the head is. In the larvæ of ephemeræ, in those of phryganæ, gnats, &c., genuine gills appear to exist, which are always in motion when the animal respires. This is a kind of anomaly in insects, which merits particular attention, especially in the ephemera, if, as Swammerdam has imagined, these insects have the power of fecundating the eggs, after they are separated from the body of the mother ; this, as we have already seen, constitutes a marked analogy between the fish and some batracian reptiles.

From these anatomical researches, it follows, that as the insects have neither heart nor vessels, it is not the blood which proceeds to seek the air, but the air which visits every part where the fluids exist; the result however is the same, and the two functions are performed, just as two numbers may be multiplied indifferently by each other, and will produce the same quotient.

It has been proved that oxygen is necessary to the respiration of insects ; that this gas is absorbed, that the carbonic acid is disengaged, and that the matter of caloric is developed in this animal operation.

Though it seems probable that the temperature of insects may be pretty nearly the same as that of the atmosphere, yet it must be confessed that we are not positively acquainted with this temperature, for two reasons, first, because it is very difficult to observe and estimate it, and secondly, because the body of the insect is soon placed in equilibrium with the matters which surround it; nevertheless, it has been observed, that insects which live in a social state, and more particularly, in our climates, ants and bees, when they are assembled, develope a temperature almost equal to, if not higher than that of man. A thermometer, placed in winter in the centre of a hive, remains there, constantly elevated from 28° to 30° of Reaumur, and when these insects are excited, their respiration becomes more active, and they suddenly develope a temperature, which rises to two or three degrees above that which they had at first exhibited.

We shall here speak respecting the *voice* of insects, although in reality the sounds produced by these animals cannot always be attributed to the air which proceeds from their bodies. When we come to speak of the *humming* of bees, we shall notice some interesting facts on that subject. Most of the noises produced by insects are owing to rapid vibrations, communicated either to neighbouring bodies, or to certain parts so conformed, as to represent cords or membranes; accordingly it has been said in speaking of insects —*Animalia muta, nisi alio proprio instrumento, sonora.* Some of them move the head against the corslet, or the latter over the elytra, and thus produce a sound; others cause the rings of the abdomen to vibrate against the free extremity of the elytra; some are provided with a pencil of stiff hairs, which operate like a brush rubbing against a solid body—this is the case with the *blaps;* others strike with force the wood which they are gnawing, with some part of the head. The male cigala has a sort of drum or concave scale, under which rolls a cylinder furnished with projecting cords. These last, as well as the males of locusts, crickets, &c., cause some parts of their elytra to resound by crossing them rapidly, or agitating them with the legs, as the cords of certain instruments resound under the bow which is drawn over them. This is a very interesting subject, and one which we shall treat with more detail in the sequel.

It now remains for us to speak a word on the most important function of these animals: that by which they reproduce their like, and the species are preserved.

All insects proceed from other individuals absolutely similar to themselves, and from which they are separated at first in the form of eggs: that is to say, the germ has been deposited with a certain quantity of nutriment, suited to its early age, in a membranous shell of more or less solidity; the configuration of these eggs, and the manner in which they are laid and disposed are most admirable, considered in reference to the future wants of the animal; some are soft, and the shell of others sometimes exhibits an extraordinary degree of solidity. Some are agglomerated, pasted together, united by common or distinct pedicles; they are of various shapes, spherical, oval, cylindrical, flat, depressed, compressed, prismatic, angular, &c. Some are enveloped with protecting substances, proper to keep off the animals that might devour them. These matters consist sometimes of odours, sometimes bristling points, sometimes tight and impenetrable coverings. The female also employs, to protect her progeny, many artful stratagems, even to covering them with her own body, which dries up and guards them like a buckler, as is exemplified in the cochineal insects. The colour of these eggs varies very much according to the species, and the time elapsed from their deposition, because in the latter case, the developed germ communicates its tints to the membranes which enfold it. Some of these eggs exclude in the body of the mother; this is the case with those of the pucerons at certain periods of the year, with those of the blue meat-fly, of the hippoboscus, and in fine, with all the insects which on that account are termed *ovoviparous*.

The sexes are most usually distinct and separated, that is to say existent in different individuals; some are male and others female The number of individuals of both sexes, is in general pretty nearly alike; nevertheless, there are some, which from infancy, are doomed never to have the sexual organs completely developed. When this anomaly occurs, it is the females which are deprived of the (at least apparent) sexual organs; they are then called *neuters* or *mules*. This phenomenon is observable in some genera of hymenoptera, principally in ants, wasps, bees, and in some neuroptera or hemiptera, as the termites and pucerons. It is seldom that insects of one and the same species live in pairs, or in a state of monogamy. The necessity of fecundation alone brings them together for a very short time; shortly after this function is performed the male perishes, but the female survives until after the eggs have been laid. It would seem that the early age of the insect is employed in the elaboration of those juices alone, which are to serve for the work of generation: for as soon as these animals have assumed their final form, and have obtained the faculty of giving or receiving the fluid which transmits life, they couple, lay, and die. The males are in general smaller than the females, they are more elegantly formed, more brilliant and better coloured. The forms of the antennæ, of the wings, and especially of the extremity of the abdomen, where the sexual organs and the instruments destined to place the eggs in situations most favourable to their development, often exhibit notable differences.

Thus the males of ants, of the cochineal insects, of pucerons, and of some herbivorous coleoptera, are exceedingly small if compared with the females. The antennæ of bombyces, of rhipiptera, and others, are much more developed in the males. The wings of some male insects have scarcely any relations with those of the females. Some females are even altogether without wings.

In the majority of insects, the sexual organs are placed at the extremity of the abdomen. They most usually cause an external projection in the males, and sometimes in the females. In some species however, the female organs are differently disposed from those of the male.

The noises produced by some insects are manifestations of the desire of reproduction, and guides of one sex to the other. The phosphoric emissions of others during the obscurity of night, answer the same purpose. Many exhale peculiar odours, by which they are discovered at a distance by the opposite sex.

The configuration of the male and female organs varies too much, not only in the orders, but also in the genera, and even in the species, to enable us to prevent any general notion of it. We shall merely mention that the spermatic vessels are very numerous in the males, and very much inflated at certain times. These vessels, which are twelve or fifteen times the length of the body, are folded and re-folded on themselves, so as to occupy a considerable portion of the cavity of the abdomen. They lead to a common reservoir, or vesiculæ seminales, which have been compared to prostates, to epididymes, to vasa deferentia, &c.

Instruments are often found in the females which facilitate the laying of the eggs, or the various modes in which they ought to be deposited. These are the *ovipositors* (Kirby). The vulva opens into the cloaca; thither the oviducts lead. These are very elongated canals; like the spermatic vessels, but much more bulky. The eggs are distinguished there, and so much the more developed, as they are nearer the common canal which leads into the cloaca.

It is most frequently in this common canal that they receive the glue or viscous humour, which serves to fix or suspend them, by pedicles sometimes very elongated, as may be observed in the eggs of hemerobiæ. There are insects which lay all their eggs at once, like two bunches of grapes, which is the case with the ephemera; but in general, these eggs pass successively one by one, through the orifice of the cloaca. The ovipositors have various forms, of knives, sabres, saws, gimlets, probes, &c.

In this, as in the proper place, we shall briefly notice the general phenomena of insect metamorphoses. The metamorphoses of insects were imperfectly known by the ancients.

We see from many passages of Aristotle, that he knew that several insects, and he particularly names butterflies and bees, come from caterpillars, or *worms*. But these transformations have only been thoroughly known since the researches of Swammerdam, of Redi, and of Gœdært, and the reproduction of insects properly explained.

Besides the notable mutations in form which insects undergo, in the three states which follow their departure from the eggs, they often change their skin, or epidermis, and this epidermis is often altogether of another appearance than that which succeeds it, which imparts quite a different aspect to the insect. This happens to the caterpillar of the silkworm. When it issues from the egg this caterpillar is hairy ; in its subsequent moultings the body is quite smooth, but its tint varies considerably. It is the same with a great number of other larvæ.

Fabricius, in his Entomological Philosophy, has devoted an entire section to the modifications of the metamorphosis in insects. Since his time, the science has made great progress, and is most deeply indebted to the distinguished author of that portion of the " Regne Animal," on which we are now enlarging, M. Latreille.

The larva, or caterpillar state, may be considered as the infancy of the insect. The larvæ, whose forms are only provisional, and whose existence is merely transitory, have received this name from the peculiarity which indicates that they have but a borrowed figure. The Latin word *larva* signified the mask worn by tragedians or comedians, in the representation of characters on the stage.

The larvæ vary in general structure, in manners, and habits, in the different orders. It is, however, under this form, that the insect acquires almost all its growth. For the nymph, to all appearance, and most certainly the perfect insect, increase no more in volume.

In proportion as the larvæ increase, they change their skin. This may be termed their moulting, and one envelope is often succeeded by another of a totally different colour. The time of moulting is a serious crisis for the larvæ, and its periods are hastened or retarded, according to the elevation or depression of temperature, the abundance or deficiency of nutriment. In the larva state, insects are engaged only in their preservation and growth.

The nymphs, or pupæ, to which other names have been given, according to the differences which the species of the various orders, established in the insect class, present in their forms, are individuals passing from the larva state to that of the perfect insect or imago. Under this form, the animal's growth usually ceases. It may still sometimes take nutriment, but it participates much of the form which it is to assume in the sequel. It exhibits as it were the outline, or rough sketch with all its parts, but most generally crowded together, and as it were, swaddled up. Those pupæ which can feed themselves, are, as might be supposed, more or less agile, and pretty nearly conformed in the same manner as the perfect insect. The majority resemble the larvæ, with this difference, that they very frequently exhibit the rudiments of wings. Such is the case with all the orthoptera, and in particular with the locusts. It is also the case with all the hemiptera, as the cigalæ, &c. Similar dispositions are observable in some neuroptera, as the ephemera, and the libellulæ; but in that order there are other species, such as the antlions, the hemerobiæ, &c., which proceed from nymphs altogether different from the larvæ.

It is chiefly according to the modifications which insects undergo at the period when they assume this form of nymphs, that the various modes of their metamorphoses have received their different denominations. Some of these denominations are any thing but happy, yet still it is not perhaps very adviseable to change them for others. We have already had occasion more than once to notice the confusion which reformers in nomenclature have created.

Fabricius has given the name of *complete* metamorphosis to a case in which, in point of fact, the insect undergoes not the slightest change of form, except, perhaps, in the number of feet, and the development of the sexual organs. The majority of the true aptera are in this predicament. They moult, but do not change their forms, and may with propriety be termed *immutable* insects. This is the case with the arachnida, which, however, are now no longer classed with insects, with the scolopendræ, the pediculi, the podura, &c. Others, such as the iuli, wood-lice, &c. acquire some additional limbs. Among all these insects the three separate states of larva, nymph or pupa, and imago do not exist. It is proper, however, to observe that many apterous insects undergo real transformations.

It is to the *semi*-metamorphosis that Fabricius refers the series of changes which insects undergo whose forms remain pretty nearly the same, or, in other words, whose larvæ do not differ from the nymphs, except in size and dimensions of parts, or in the absence, the rudiments, or the complete development of the wings. These insects preserve in their three states the same habits and mode of subsistence. The orthoptera, hemiptera, and some neuroptera are, as we have already hinted, in this predicament, the insect preserving the form of the species during its entire existence, though the pupa and larva are distinct.

The third mode of metamorphosis is that exhibited by the insects which, as the coleoptera and most part of the hymenoptera, proceed from larvæ, which are more or less active, according as they are destined to feed themselves, as they are nourished in advance, or daily by their parents until the period when, after divers moultings which the growth of their body required, they pass into that state termed by Fabricius *incomplete* nymph. This is when the larva at its last moulting changes form suddenly, and allows the perfect insect to appear, but at first in a state of extreme softness. It gradually assumes greater solidity, and the animal appears with all its members, with its six feet, with its wings, but bent, folded back upon themselves, and in a state of almost absolute paralysis. From this sort of nymph state the insect does not issue, but by losing the external coating of skin which retained all its parts in a state of forced immobility. With some slight modifications this is the predicament of the flea among the aptera, and many larvæ of the tipulæ among the diptera, as well as certain neuroptera, such as the antlions, hemerobiæ, &c.

The fourth principal mode of transformation is exhibited by the butterflies and other lepidoptera whose caterpillars are changed into chrysalids. This is the sort of nymph which Fabricius names *obtected*, and which is also termed *pupa* and *aurelia*. At the moment in which the insect quits for the last time the skin of the caterpillar, it appears under quite another form than that which it is to assume in the sequel. It is an undivided body, variable in form, most usually conical towards one of its extremities, and presenting on one of the faces of the opposite extremity some prominent traits which mark some parts of the perfect insect, particularly the antennæ, the feet, and wings, but very closely contracted together. Some of these nymphs or chrysalids, which are almost always condemned to an immoveable state, undergo this change in the open air and naked. Such are the chrysalids of the diurnal butterflies. Others come from caterpillars which are sheltered in a cocoon of silk, which they spin around their body, or like the tineæ, or some *pyrales*, they undergo transformation in the case itself which they inhabited. In fine, some chrysalids when about to assume their last form advance out of their shell by the assistance of the stiff points with which the segments of their bodies are furnished.

The fifth and last mode of metamorphosis is exhibited by the pupa of the majority of two-winged insects. This is the sort of nymph which Fabricius terms *coarcta*. The larvæ of these insects, which are very improperly called the *worms* of flies, are indeed devoid of feet. They move, however, by the aid of some peculiar organs, and with more or less agility. Most of them are developed in very humid situations, or substances, and even in liquids. They change their skin several times, but at their last moulting they totally lose their primitive forms. Their body grows shorter, contracts itself so as to present the appearance of an egg-shell, or elongated ball, the envelope of which, at first soft and whitish, afterwards grows hard and brown, leaving externally to be distinguished no trace, no lineament, no appearance of any kind of the insect which it contains. It is, in fact, a sort of corneous shell independent of the animal which it. protects. When it is opened, an insect is found in the nymph state, analogous to that of the coleoptera in the contraction of its members. When this pupa has acquired a sufficient degree of consistence, it begins to work upon the walls of its prison, which tear constantly and circularly so as to allow the body of the insect to be disclosed, which comes forth quite moist, and the wings but little developed. These, however, soon become sufficiently extended to sustain the insect in the atmospheric air, in which its new mode of locomotion is to be exerted.

Such are the principal metamorphoses of insects. There are some which participate in many of the modes of transformation we have now described. The study of this period of insects is certainly one of the most curious and interesting that can engage the attention of the naturalist. In some species in particular, the change of the nymph into the perfect insect takes place with the most extreme rapidity, and the observer can either accelerate or retard this operation, so as to see it performed at will under his immediate inspection, in the duration of a minute. Interesting, however, as this subject is, we must dismiss it for the present, reserving any minuter details thereon for the supplements to the separate orders of this class.

A subject of perhaps not less interest is the determination of the climates which are proper to the different races of insects, or, in other words, their geographical distribution. Such distribution is in close relation with the nutrition of these animals; for since the Author of Nature has extended living beings over every point of the surface of the globe, capable of sustaining them, and since these beings vary with climate, it must happen that the alimentary substances by which they are sustained, must also differ according to the places in which they pass their lives, and be subjected, like the animals themselves, to the same geographical circumscription.

Independently of this consideration, the temperature which is suitable to the development of one species, is not always equally proper for that of another. Thus the extent of the countries occupied by certain species, has its necessary limits, beyond which they cannot pass, at least suddenly, without ceasing to exist.

Another consequence follows from these principles: where the empire of Flora ends, there likewise terminates the domain of Zoology. The animals which feed on vegetables, could not exist in regions of complete sterility, and those which are carnivorous would be equally deprived of their prey.

From observation we find that the countries most abounding in animals with articulated feet, and especially in insects, are those whose vegetation is the richest and the most speedily renewed. Such are the effects of a powerful and permanent heat, of moderate humidity, and of the variety of the soil. The more, on the contrary, that we approach the boundary where begins the reign of everlasting ice and snows, whether we proceed towards the poles, or ascend to the heights of lofty mountains, where the diminution of caloric produces similar phenomena, the greater is the decrease of plants and of insects. Thus Otho Fabricius, who has published a good Fauna of Greenland, enumerates but 468 species of animals altogether, and the number of those of insects, including, according to the Linnæan system, the crustacea and arachnida, amounts only to 110. The catalogue of this naturalist may perhaps be not complete; but even so, we are justified in believing that the number of insects in that country must be extremely limited. But when we come to the desolate domains of eternal winter, life totally disappears, and the productive force of nature is at an end. The plains in the neighbourhood of the poles, the very elevated regions of the mountains of the temperate and torid zones, covered with perpetual snow, exhibit, in this respect, the same state of inertia. These mountains, considered with regard to the animals and vegetables which are peculiar to them, form gradually, and by superposition, particular climates, whose temperature and productions are similar to those of the plains of the most northern countries. Thus we find that the Alps are the habitation of many species of insects which are again to be found only in the north of Europe. The *prionus depsarius*, which had long been supposed to belong to no other country but Sweden, has been discovered in the mountains of Switzerland M. Latreille has found at Cantal the *lycus minutus*, which used to be considered peculiar to the most northern provinces of Europe. The butterfly named *Apollo* by Linnæus, very common in the fields and gardens of Upsal, as well as in other parts of Sweden, is found in France only on mountains of more than three or four thousand feet above the level of the sea.

Similar observations might be made with regard to other insects, which in southern countries, such as the south of France, Italy, or Spain, establish their domicile on alpine or sub-alpine mountains, while in higher latitudes they rise but little above the level of the soil. These animals find in both situations a similar temperature and similar food. The elevation above the sea, and the mean temperature of the places where insects are taken, will always be carefully noted by the enlightened entomologist.

Thus, as well as geographers, naturalists have divided the surface of the earth into different climates. The former have taken as their bases the progressive differences in the longest duration of the natural day; but the latter have founded their divisions on the mean temperature of the regions proper to animals and vegetables. In the entomological philosophy of Fabricius, the acceptation of the word *climate* is general, and embraces the entire of the habitations of insects, or rather, of all animals with articulated feet. He divides the climates into eight stations, or into as many particular sub-climates. These are termed the *Indian*, *Egyptian*, the *Austral*, the *Mediterranean*, the *Boreal*, the *Oriental*, the *Occidental*, and the *Alpine*. But it is not difficult, as M. Latreille observes, to see, from the enumeration of the countries which he refers to each of them, that these divisions are not always established on positive proofs, and that if the principle on which they are founded be pushed to its full extent, which principle is the mean temperature, some of them must be suppressed. The sub-climate which he calls *mediterranean*, comprehends the countries adjacent to the Mediterranean sea, along with Media and Armenia. The *boreal*, or northern, extends from Paris to Lapland. The *oriental* consists of the north of Asia, of Siberia, and the cold or mountainous districts of Syria. The *occidental* contains Canada, the United States, Japan, and China.

This simple view is sufficient to prove that these divisions are very arbitrary. Many of these countries may have, and in fact have, an identical mean temperature: they cannot, however, be considered as possessing the same climate. But independently of these divisions being of comparatively little use to science, since of some places where the temperature is the same, the animals are different, it is impossible, in the present state of our knowledge, to obtain a solid basis for these divisions of climate. The different elevations of the soil above the level of the sea, its mineralogical composition, the varying quantity of its waters, the changes produced on temperature by the extent, height, and direction of the mountains, by the magnitude of forests, and even by the influence of neighbouring climates, occasion uncertainty and confusion in our calculations from the difficulty of appreciating the just effect of such causes, either singly or combined.

M. Latreille would consider climates under another point of view. This is presented by the genera of arachnida and insects which are exclusively proper to certain determined spaces of the surface of the earth. With the imperfection of catalogues, especially in foreign species, it would be difficult to pursue a different plan. Even the entomology of Europe cannot be considered in this point of view as any thing but very remote from perfection. Many, even scientific travellers, may be reproached with having neglected the precise indication of the places where they have found the objects with which cabinets of natural history are enriched. Their inattention, therefore, to the physical and mineralogical characters of the soil in such situations, can excite little surprise: yet such particulars constitute an essential part of the history of animals. The *licini*, *papilio cleopatra*, some *lamiæ*, &c. are only found in calcareous soils. The *pimelia bipunctata* has been observed to move to no great distance from the sea-coast. They are numerous near Marseilles, where M. Latreille made this observation. Other species of the same genus are found in the interior of some countries, such as Barbary, Syria, and Egypt; but in places where the soil is impregnated with saline particles, or abounds in plants of the genus *salsola*. Thus they always inhabit localities similar to that of the first-mentioned species. The insects of those countries which border on the Mediterranean, the Caspian, and the Black Seas, exhibit very great inter-relations, and remain for the most part on the ground, or on plants of no great height. Those countries seem to be the chief habitat of certain coleoptera: and though the Cape of Good Hope is so far distant from them, many of its insects exhibit family traits with the others. These facts prove that many characters of affinity exist between the soils and vegetable productions of those widely distant regions.

The entire, or nearly so, of arachnida and insects which inhabit countries whose temperature and soil are the same, but which are widely distant, is composed in general of different species, and this even when those countries are under the same parallel of latitude. All the insects and arachnida brought from the most eastern parts of Asia, such as China, are distinct from those of Europe and Africa, whatever may be the latitudes and temperatures of those Asiatic countries. The majority of these same animals also differ specifically when the countries which they inhabit, identical in soil and temperature, are separated from each other (the difference of latitude making no difference) by natural barriers, interrupting the communication of these animals, or rendering it very difficult, such as seas, very high mountain-chains, or vast deserts. Thus we find that the arachnida and insects, and even many other animals, of America and

New Holland cannot be confounded with those of the ancient world of the same classes. The insects of the United States, though often very nearly approximating to ours, are yet removed from them by some characters. The insects of New Granada and Peru, countries neighbouring to Guiana, and alike equinoxial, yet differ in a great measure from those of the latter, because those regions are separated by the barrier of the Cordilleras. A similar sudden change is observable in passing from Piedmont into France by the defile of Tende. These rules may admit of some exceptions with regard to aquatic species. On the other hand, we find insects whose habitat is very much extended. The *papilio cardui*, so common with us, and even in Sweden, is to be found at the Cape of Good Hope, and a species nearly related to it in New Holland. The *sphinx nerion* and the *sphinx celerio* are found in France and in the Isle of France, the former seeming to be their northern, and the latter their southern boundary. The *dytiscus griseus*, which inhabits the waters of Provence and Piedmont, is no stranger in Bengal.

Many *genera* of insects, and particularly those which feed on vegetables, are spread over a great number of points of the principal divisions of the globe. Some others are exclusively peculiar to a certain extent of country, whether in the Old or New World.

Many species in their native country exclusively affect certain localities, either in low situations, or in those which are elevated, and at an equally constant height. Some Alpine butterflies are continually confined to the regions of perpetual snow. In ascending mountains, when we come to a height where the temperature, vegetation and soil are the same as those of a much more northern country, we discover there many species which are peculiar to the latter, and which would be sought for in vain in the plains and vallies which

1. Phrenapates *Bennettii*. *Kirby*. 2. Antimachus *cornutus*. *Gebl*. 3. Toxicum *Richesiacum*. *Latr*. 4. Ryſsocheton *politus* *G.R.Gray*. 5. Zopherus *Mexicana*. *Hope*. 6. Nyctelia *Hopei*. *G.R.Gray*.

are at the foot of those monntains. If, in the same country, the temperature of its low-lying situations, or even of those upon a level with the horizon, be modified by certain local circumstances, we may observe a corresponding modification in the species of insects there to be found. For instance, if the mean temperature is lowered, the species are the same as those to be found a little more northwards ; if raised, we shall find the same species as exist a little more to the south.

M. Latreille would divide the old and new continents into zones, extended successively in the direction of the meridians, and the breadth of which is measured by a portion of a circle, parallel to the equator. The species proper to one of these zones disappear gradually, and give place to those of the following zone, so that from interval to interval, the dominant, or even the entire of the species, shall cease to be the same. These changes he compares to the series of horizons discovered by the traveller in proportion as he recedes from the point whence he originally set out.

Sweden has many insects peculiar to itself, some of which are located in its most northern provinces, as in Lapland. In the southern, as in Scania, are found, though in small number, many insects of Germany. France, even towards the forty-fourth and forty-fifth degrees of latitude, has many which are to be found in the former countries. But it would appear that the Rhine and its eastern mountains form, with regard to some other species, a frontier which they have not passed. The first of those which are proper to the warm countries of western Europe begin to shew themselves towards the lower course of the Seine, precisely at the point where the vine begins to flourish in the open plain without the assistance of any local circumstances. This is still more remarkable at Fontainbleau, and in the neighbourhood of Orleans.

But these insects are only the forerunners of those which

are peculiar to the truly southern countries. The domain of the latter is recognized by the appearance of certain species of *Cigala mantes, termes, &c.*, but especially by the presence of the *European scorpion*, and the *ateuchus sacer*. The culture of the olive—the spontaneous growth of the arbutus, speak still more sensibly to the observation. This change becomes extremely remarkable on proceeding from Paris to Marseilles, about the territory of Montelimart. The shores of the Mediterranean are something warmer; there we find the *Mygalæ*, the *Cebriones*, the *Onitis*, the *Scarites, &c.* appearing for the first time. If we pursue our way into Spain, and visit the beautiful districts of the eastern part of that country, where the orange and the palm-tree flourish in the open plains, we are saluted by a new order of arachnida and insects, intermingled with some already observed in the south of France. We find there *nemoptera*, and various other insects analogous to those of Barbary and the Levant. Having examined these species, the entomology of the Atlantic countries of Africa, or of those which are situated on the Mediterranean, as far as Atlas, will not occasion much surprise. There we also discover certain genera of insects which are peculiarly predominant in the countries comprised between the tropics, such as the *graphiptera, &c.*

Respecting the insects of the south-east of Europe, our knowledge is yet somewhat imperfect. It may, however, be observed, that the *papilio chrysippus* of Linnæus, which is common in Egypt and the East Indies, is also found in the kingdom of Naples. Most of the Egyptian species are strangers to Europe. Its southern extremity, bordering on Nubia, presents several individuals of the *Midas*, which like the *bucephalus antenor*, and *gigas*, inhabit only the hottest climates of the old continent approximating to the equinoctial line. From the banks of the Senegal and onward, in the interior countries more to the south, we no longer find any European insect. It is from those burning regions that the largest species of the genus *Goliath* of Lamarck proceeds; the others come from South America and Java. The colony of the Cape of Good Hope, especially, abounds in species of the genus *Anthia* and *Brachycera*. In the same quarter of the globe, are found numerous lepidoptera, which constitute peculiar sections.

All the successions of species take place gradually from west to east, and *vice versa*. Many of those which are found in Normandy and Britanny, are also to be met with in the southern parts of England. The departments situated on the left bank of the Rhine to the north, are in this respect similar to the neighbouring provinces of Germany, but only as to a small portion of species. Some insects of the Levant appear to have travelled westward, and established themselves in the neighbourhood of Vienna and throughout the Austrian territory; such are the *cantharis orientalis*, &c. M. Latreille is of opinion, that the insects of Asia Minor, Syria, Persia, &c., though closely allied to those of the south of Europe, are nevertheless, for the most part, specifically distinct. He also thinks similarly respecting those of southern Russia, and of the Crimea. The arachnida and insects of the coast of Coromandel, of Bengal, of southern China, and even of Thibet, exhibit very great inter-relations, but they are absolutely distinct from those of Europe, though for the most part they may be classed in the same genera, and in some of the African genera. No genera which nature appears to have peculiarly allocated in the southern and western parts of the ancient continent, such as the *graphiptera*, &c., are to be found there; some species of *brachycera*, according to Fabricius, are to be found in the East Indies; but this assertion M. Latreille does not appear to consider as well-founded. The genus *anthia* is found in Bengal, and is replaced in New Holland by that of *helluo*.

The island of Madagascar approximates in some respects, as to the natural families of insects, to Africa, but its species are very distinct, and even many of them have no analogues. The isles of France and Bourbon, also present vestiges of similar affinities; the insects of these colonies seem in general to partake more of the character of those of the East Indies. Their number is very limited.

Although the entomology of New Holland forms a special type, it is nevertheless composed in a great measure of species analogous to those of the moluccas and south-eastern India. New Holland is to be sure less abundant in its insect population, being, at least in the parts with which we are acquainted, more dry and less wooded, and analogous to that of the Cape of Good Hope. The genus of the *Mylabris*, the species of which are so abundant in the south of Europe, in Africa, and in Asia, does not appear to proceed beyond the Isle of Timor. New Holland, would seem, in this respect to have traits of similitude with America. There we find certain of the *passalus*, a genus whose species peculiarly inhabit the New World. It is probable, that considered under the relation of generic groups, the productions of the western hemisphere have more affinity with those of the east of Asia than with ours. We know, for instance, that the pouched animals are confined to the eastern extremities of the old continent. Many similar examples might be alleged, and the class crustacea furnishes not a few.

The insects of New Zealand, of New Caledonia, and probably those of the neighbouring islands, have much affinity with the insects of New Holland. It may be presumed that the same holds good of some other Archipelagoes of the great Austral Ocean. Those islands, composed for the most part of aggregations of polyparia, form a chain which unites them on the west to the preceding islands, and have possibly received from them their productions. A similar communication, in consequence of the want of similar means, cannot exist from the coast of America. Thus many of these islands though American in their geographical position, may yet be considered Asiatic, as to the animal and vegetable productions.

The New Continent presents a similar progressive march, in the changes of species relative to the notable differences of latitude and longitude. M. Bosc found in Carolina many species not to be found in Pennsylvania, and still less in the province of New York. In Georgia, some species of lepidoptera are found, whose principal habitat is the West Indies. The banks of the Missouri, about twenty degrees west of Philadelphia, are the habitation of many peculiar species. Other changes of this kind have been observed in Louisiana. The entomology of the Antilles, with a few exceptions, forms an absolute contrast to that of the United States. The isle of Trinity, in ten degrees of north latitude, has some equatorial species, as for instance, butterflies of those divisions named *menalaus*, *teucer*, &c., which are not to be seen in St. Domingo. In the former island are also found quadrupeds unknown in the latter. There are species in Brazil, which are likewise to be found in Cayenne; but it has also a multitude of others peculiar to itself.

Nevertheless, in comparing the parallels of the Old and New World, under the relation of temperature suitable to the different species of insects, we shall find that these parallels do not correspond in this respect. The southern insects of the western hemisphere are not found so high as in ours. With us, they commence to appear between 48° and 49° N. L.; there, they scarcely appear higher than 43°. When we reflect on the physical constitution of America, and consider that its soil is very much watered, considerably mountainous, covered with large forests, and that its atmosphere is extremely humid, we may easily conceive that certain genera of insects of the old continent, such as anthia, pemeliæ, &c., which frequent dry grounds, sandy and very warm situations, could not live in the unctuous, moist, and umbrageous soil of the New World. Accordingly in proportion, we find much fewer of the carnivorous coleoptera in America, than in the older quarters of the globe. The size, too, of such insects there, is often less than with us. The scorpions of Cayenne, and of the other equinoxial countries of America, are hardly larger than that one of the South of Europe, which is named *occitanus*. They are far from equalling the size of the African scorpion, which is as large as our river crab. But, with regard to the species which feed on vegetables, America does not yield to any country, even the most fertile of the ancient continent. This is particularly true of *lepidoptera*, in general, of *scarabeidæ*, of *chrysomelina*, *cerambycinæ*, &c., and particularly wasps, ants, *orthoptera*, and *araneidæ;* nevertheless, Southern China, and the Moluccas, appear to preserve a sort of superiority in giving birth to such lepidoptera as the *papilio priamus*, the *bombyx atlas*, &c., whose dimensions surpass those of the American lepidoptera. A remarkable fact is, that Europe, Africa, and Western Asia, have scarcely any insects of the genus *phasma*, or *spectrum*, and that the species found there are small, while the Moluccas, and South America, present us with some of a remarkable size. The habitual atmospheric humidity of the new continent, its narrow and elongated form, the vast extent of ocean with which it is surrounded on all parts, and the nature of its soil, furnish a sufficient explanation of the discordance which is observed between its climates, and those of our hemisphere, considered under the same parallels. The New World is to the Ancient Continent, what England is to a great part of Europe.

Normandy, and Britanny, compared to the provinces of France, situated to the east, exhibit similar approximations.

M. Latreille, has, as we before hinted, given a new and ingenious division of the earth, into climates or zones, the limits of which circumscribe in an approximating manner, the places of habitation exclusively proper to the different races of these animals. These geographical sections are founded on the following observations:—first, the northern extremities of Greenland and Spitzbergen appear to be, in our hemisphere, the final limit of vegetation. It is arrested, towards the south pole, at the Land of Sandwich, the *ne plus ultra* of geographical discoveries in the Austral hemisphere. The eighty-fourth degree of north latitude, and the sixtieth of south latitude, will thus form the two extremities of that part of our globe, which serves as a habitat for plants and insects. Second, the entomology of the new continent, beginning at least with the north of the United States, and proceeding southwards, differs even as to species, from the entomology of the old continent. Third, that portion of Greenland, with the zoology of which we have been made acquainted by Otho Fabricius, presents many insects, and even other animals, which may be found again in the most northern and western extremities of Europe. We may thus consider Greenland as forming to the north, and under this point of view, the limit of the two worlds. Fourth, the insects of eastern Asia, proceeding from the countries whose longitude is about 62 degrees more east than the meridian of Paris, the insects of New Holland, those of that part of Africa which extends from Atlas and the tropic of Cancer, to the southern extremity of that peninsula, differ from the animals of the same class which inhabit the other countries of the old continent. Fifth, a space of latitude measured by the arch of a circle of 12 degrees, produces, abstracting some local variations, a sensible change in the mass of species; it is even almost total, if this arch be doubled, or consist of 24 degrees, as from the north of Sweden to the north of Spain.

Resting on these observations and some others less general, M. Latreille separates, to the west, the two hemispheres, by a meridian which, proceeding from Greenland, and following a mean direction between the Canary Islands, those of Cape Verd, Madeira, and the most advanced point to the east of South America, Cape St. Roch, ends near Sandwich Land. Its longitude is 34 degrees west of the meridian of Paris. Another meridian more east by 62 degrees than that of that city, detaches the oriental part of Asia from its western countries, from Europe, and from Africa. The difference of longitude of these two meridians is thus 96 degrees. A third more eastern, and of the same quantity, will determine the limits of the Old World and the New to the eastward, traversing the great ocean.

The other 144 degrees will complete the circle of the equator, and will be in longitude, the extent of the grand zone proper to the insects of America. M. Latreille divides it by means of a fourth meridian into two equal portions, having each 72 degrees in longitude.

These four great zones will be arctic, or antarctic, according to their situation within, or beyond the equinoxial line. M. Latreille divides each of them into climates to which he gives twelve degrees of latitude. That which is comprised between 84 north latitude, and 72 degrees, is called the *tropical.* Then, continuing the duodecimal division, come the following climates: the *sub-polar, superior, intermediate, sur-tropical, tropical, equatorial.* The antarctic zones, divided in the same manner, have two climates less, the *polar*, and *sub-polar*, since they terminate at the 60th degree of south latitude. These zones considered in each hemisphere are distinguished by M. Latreille, into *occidental* and *oriental.*

Such is an outline of the system of M. Latreille, from which he conceives that geography may derive sound advantages. It may thereby be determined to which of the two hemispheres naturally belong many islands situated on their limits. Such considerations, he also thinks, may be useful to mineralogists and geologists. We see, for instance, that the insects, and even the plants of those countries by which the basin of the Mediterranean is circumscribed, and also those of the Black and Caspian Seas, singularly resemble each other as to the genera and families in which those productions are grouped. Such relations seem to favour the opinion of naturalists, that the latest uncovered countries have been those which were formerly bathed by the waters of the ocean.

The increasing progression of the intensity and duration of caloric, has a very great influence on the volume and the development of the mucous tissue of insects and arachnida. The more, in general, that we advance towards the equinoxial regions, the more we find of species remarkable for their size, for the inequalities and prominences of their bodies, and for the variety of their colours. The increase of light tends to convert yellow into red and orange. The diurnal butterflies of our mountains have usually the ground of their wings white, or of a brown more or less deep. Such observations, properly followed up, would greatly assist in the distinction of species and varieties.

Having considered the general structure, the functions, and the geographical distribution of insects, we shall indulge in a few remarks on that most interesting of all subjects, the character of their intellectual faculties. In this point of view, although the labours of insects appear to announce a degree of industry of which the superior classes of animals presents us with few examples, yet there can be no comparison between them and the latter as to extent of intellect.

We cannot possibly attribute to them such combinations of ideas, and such judgment, as a more perfect, or, at least, a more complicated organization allows to other animals. The insects seem to possess, when born, all the knowledge which their destination requires, and which is composed of a certain number of ideas relative to their wants and the employment of their organs. The circle of their actions is traced out: they cannot pass beyond it. This natural disposition, which fits them for exercising, in a determined and constant manner, all that is necessary for the preservation of their lives and the propagation of their race, is what is usually termed instinct. They can have no better guide. Too transitory on the scene of nature, they have neither time to deliberate nor to profit by the lessons of experience. Their races would soon become extinct, if there were no better agent employed in their preservation than the acquired knowledge of experience.

The bee is scarcely more than born when it begins to apply itself to labour, and to exhibit the talents of the most accomplished artist, to execute in the most exact and regular proportions, without any model, and without the least hesitation, a work which pre-supposes calculations of the higher geometry, and which a skilful mechanician could not perform but after long trials, and with instruments of which the bee is deprived. Even were this insect provided with them, it would be impossible for it to construct, beforehand, its alveoli in suitable proportions to the number of the future population, which it cannot foresee, and to give to those which are destined to contain the male and female eggs the requisite size for those individuals not yet in existence. But Nature has been the preceptor of the bee, and has formed her a geometrician. We even see among men certain individuals born with happy dispositions for the mechanical arts, and excelling in them without having had the advantage of a master. The most just and ingenious ideas, which are, in general, the result of meditation, study, and instruction, present themselves spontaneously to such minds, without effort, and in the liveliest manner. The most perfect instinct of insects is clearly analogous to this accidental gift of nature, converted into a necessary and persevering habit, and perpetuated from generation to generation.

To the habitual and stimulating causes which call this faculty into action, such as the impression excited on the senses by external objects, hunger, and the desire of reproduction, we must add, a predominating sentiment, that of the preservation of posterity. Wherefore does the neuter bee, which can have no feeling of maternity, from the time she is placed from the moment of her birth in a new hive, immediately set to work on the construction of its radii? If this was merely for her own proper subsistence would it be necessary that she devote herself to such long and painful labours? and wherefore should she suffer herself to die of hunger when deprived of the queen which is to propagate her race? What can stimulate her to such minute, such attentive cares? Why do the females of insects, which live in a solitary and isolated manner, deposit their eggs with studious care previously to terminating their own existence? Doubtless all this is the result of an internal impulse, or maternal sentiment, which they are compelled to obey.

The earlier naturalists, inspired with a sort of respect for the industrious society of bees, and also considering its utility, accorded to it the first place in their methodical classifications of insects. But on this point internal organization is the essential character which ought to guide us. The possession of an instinct more or less extended would be an indication of order but little to be depended on. Among the bees themselves several species are to be found, which although very much approximating in organization to the common bee, yet living in a solitary manner, are much inferior on the score of instinct. The perfection, therefore, of this quality seems to be an appendage to the organization of the animal and not the result. The beaver, though more industrious than the carnivorous mammalia, is yet much below them as to organization. But, in truth, all the attempts which have been made to discover or conjecture the *cause* of instinct have been utterly vain and unsatisfactory. To notice the many absurd and even unintelligible theories which have been advanced by otherwise clever men on this subject, would be to waste the time and weary the patience of our readers. It is sufficient to observe that they are all mere assumptions, without a shadow of proof, or like that one which refers the instincts of insects to their *sensations*, little better than identical propositions. As to the doctrine that instinct is the result of the immediate inspiration of the Deity, that is sufficiently refuted by the fact that it is sometimes found to be at fault. That it is identical with reason is a proposition equally untenable, for it is exercised with the same facility and perfection by the insect just disclosed from the pupa as by the oldest. But the opinion of Mr. Spence on this subject so completely coincides with our own, and is expressed so much better than we could hope to express it, that we shall give it in his own language.

"Instinct then is *not* the result of a plastic nature; of a system of machinery; of diseased bodily action; of models impressed upon the brain; nor of organic *shootings-out*:—it is not the effect of the habitual determination for ages of the nervous fluid to certain organs; nor is it either the impulse of the Deity or reason. Without pretending to give a logical definition of it, which while we are ignorant of the essence of reason is impossible, we may call the instincts of animals those unknown faculties implanted in their constitution by the Creator, by which, independent of instruction, observation, or experience, and without a knowledge of the end in view, they are impelled to the performance of certain actions tending to the well-being of the individual and the preservation of the species; and with this description, which is, in fact, merely a confession of ignorance, we must, in the present state of metaphysical science, content ourselves.

"I here say nothing of that supposed connection of the instinct of animals with their *sensations,* which has been introduced into many definitions of this mysterious power, for two reasons. In the first place, this definition merely sets the world upon the tortoise; for what do we know more than before about the nature of instinct when we have called it, with Brown, a predisposition to certain actions when certain sensations exist, or with Tucker have ascribed it to the operation of the senses, or that internal feeling called appetite? But secondly, this connection of instinct with bodily sensation, though probable enough in some instances, is by no means generally evident. We may explain in this way the instincts connected with hunger, and the sexual passion, and some other particular facts, such as the laying of the eggs of the flesh-fly in the flowers of *Stapelia hirsuta* instead of in carrion, their proper nidus, and those of the common house-fly in snuff instead of dung; for in these instances the smell seems so clearly the guide that it leads into error. But what connection between sensation and instinct do we see in the conduct of the working bees which fabricate some of the cells in a comb larger than others expressly to contain the eggs and future grubs of drones, though those eggs are not laid by themselves, and are still in the ovaries of the queen? So we may plausibly enough conjecture that the fury with which, in ordinary circumstances, at a certain period of the year the working bees are inspired towards the drones, is the effect of some disagreeable smell or emanation proceeding from them at that particular time; but how can we explain on similar grounds the fact that in a hive deprived of a queen no massacre of the drones takes place? Lastly, to omit a hundred other instances, as many of them will be subsequently adverted to, if we may, with some shew of reason, suppose that it is the sensation of heat which causes bees to swarm; yet what possible conception can we form of its being bodily sensation that leads bees to send out scouts in search of a hive suitable for the new colony, several days before *swarming*."

We confess that nothing appears to us to be more conclusive and satisfactory than the above observations.

There are some very material peculiarities which distinguish the instincts of insects from those of other animals. These are reduced by Mr. Spence under the heads of *exquisiteness*, *number*, and extraordinary development. The instincts of other animals are for the most part of a very simple kind, having reference to the selection of food, propagation, defence of themselves and their young, and expression of their feelings by vocal sounds. Some of the larger animals have more developed instincts, which extend to laying up food for winter, and constructing habitations for the young, &c.

Insects have all these instincts, but many of them exhibit an *exquisiteness* of instinct unknown to other animals. No animal, of the preceding classes, catches its prey with such exquisite contrivance as the spider. No carnivorous beast digs a pit-fall for its intended victim, concealing itself below until the prey is caught, like the ant-lion. It is only necessary to hint at the dwellings of bees, ants, &c. in this place. Even the exquisite construction of some birds'-nests, and the buildings of the beavers, are not to be compared with them.

A very remarkable character in the instinct of insects, is its accommodation to circumstances. In the larger animals, the modifications of instinct are but slight. When a birds'-nest is removed, though it may remain close by the tree, it never thinks of replacing it, but sets about constructing another. But insects often exhibit the most ingenious resources in certain circumstances. We sometimes find a honey-comb of a form altogether peculiar, with differently shaped cells, and differently placed, and all adapted to some new circumstance, which did not modify the construction of the other combs. This is a matter of great surprise, and nothing similar is observable in the phenomena of instinct in other animals. The examples of this accommodation to circumstances are very numerous among the insect tribes, but we must defer any detail of them, until we come to speak of the orders separately. We shall find occasion to give many curious instances of this, when we come to treat of bees in particular.

This peculiarity of adaptation to circumstances, looks, on first consideration, more like reason than instinct. On this point, the observations of Mr. Spence are peculiarly excellent. He remarks, in the first place, that these variations are limited in extent. Insects have a certain number of resources, but cannot increase them. Bees, for instance, cement their combs to the top of the hive, when they are becoming heavy, with *mitys* instead of wax, and there is no reason to doubt that they did the same in the time of Aristotle and Pliny. They also occasionally vary their procedures, securing the combs with wax, or with propolis, only added to the upper range of cells, or arranged in braces and ties to the adjoining combs. But they never employ other methods of strengthening the combs, which might yet be considered obvious enough. They do not use mud, for instance, instead of wax or propolis, or mortar instead of mitys. Any instance of this description, would be decidedly in favour of the hypothesis of reason. But, considering that bees are confined to the same limits in this respect that they were two thousand years ago, and since the new-born perform all their operations as scientifically as the old, Mr. Spence justly considers the actions of which we are speaking, as merely variations of instinct.

Besides, no degree of reason, that can with any probability be attributed to these insects, could be adequate to the production of such complicated labours. If considered to be the result of reason, they must pre-suppose most various and extensive knowledge. A man could not perform such operations in a year, even with the best natural mechanical genius, and a competent knowledge of geometry, without which he could not perform it at all. It is not possible, therefore, that such problems could be solved by animals, perhaps not two days old, which supposition is involved in according to them the principle of reason.

The greater number of instincts, in the animals we are considering, is another distinctive mark, worthy of attention. In the higher animals, this number is considerably limited. Insects, on the contrary, possessing all the instincts which are to be found in superior animals, have, at the same time, many in addition to those. The bees, for instance, have one instinct of sending out scouts, before they swarm, in search of a proper place of settlement. Another, of following the queen wherever she goes. A third, of cleansing their new abode from all dirt, &c. &c. But we cannot afford to do more in this place, than hint at this most interesting subject, reserving its details for a future occasion. Mr. Spence has enumerated thirty distinct instincts in the nurses among the working bees, and the number might be very considerably increased.

The extraordinary development of instinct in insects, is another remarkable circumstance. In the larger animals the development of some insects depends upon age, and the state of the organs. To this, the different instincts exhibited by insects under their different metamorphoses are quite analogous. But a development of instinct, not dependent on such causes, but on external circumstances, with individuals of the same species and structure, is a singular phenomenon in the insect world. Thus as we have already mentioned, the working bees destroy the drones every autumn, but do not do so when they have lost their queen. They are certain to be destroyed without mercy, if there is a properly impregnated queen ; but if she be lost, or her impregnation retarded, the drones are equally sure to live out the winter. In a hive where this is the case, not a single drone perishes, though they are massacred with unrelenting ferocity in all the neighbouring hives. To this last proceeding there can be no doubt that the workers are impelled by an imperious instinct. But how account for the change ? Surely it cannot be attributed to reason, to a foresight equal at least to human, and a degree of self-command superior. This would, indeed, be going too far. Such conduct can be only accounted for by the development of a new instinct, suited to the occasion, or at all events, by the extinction of the old. But a new and peculiar instinct is unquestionably developed in some positive operations subsequent to the loss of a queen. But to these we can only now advert.

The question that remains to be noticed, is whether insects are guided solely by instinct, and have no share whatever of reason. Mr. Spence is inclined to think that they have no inconsiderable portion of the latter.

To larger animals, few philosophers deny some share of reason. But it is not so generally accorded to insects. This arises, in all probability, from the opinion that instinct and reason co-exist in an inverse ratio in animals. This is true, in some instances, and particularly of man, in whom instinct is weak, and reason pre-eminent. But it is far from being invariably the case. Many animals whose instincts are not particularly acute, are far from possessing a supply of reason to atone for the deficiency.

With most of the actions of insects, which are determinate and perfect, and performed without instruction and experience, reason has evidently no connection. A young bee, as we have before observed, builds as skilfully as the oldest; it has no knowledge of the design of any of its operations, or of the effects which will follow. It is thus destitute of the materials of reasoning. In the deviations of instinct which we have noticed, we cannot suppose insects to be influenced by reason. "They are still limited in number," says Mr. Spence, "and involve acts far too complex and recondite to spring from any process of ratiocination in an animal whose term of life does not exceed two years." But the same gentleman thinks, that reason may have a part "in *inducing* some of these last-mentioned actions, though the actions themselves are purely instinctive." This distinction is illustrated by a case cited from Huber, in which some bees bent a comb at right angles, to escape a slip of glass; the consequent variation in the form of the cells, are referred to instinct, by Mr. S. He then goes on to say, "Yet the original determination to avoid the glass, seems, as Huber himself observes, to indicate something more than instinct, since glass is not a substance against which nature can be supposed to have forewarned bees, there being nothing in hollow trees, their natural abode, at all resembling it in substance or polish; and what was most striking in their operations was, that they did not wait until they had reached the surface of the glass before changing the direction of the comb, but adopted this variation at a considerable distance, as though they foresaw the inconveniences which might result from another mode of construction. However difficult it may be to form a clear conception of this union of instinct and reason in the same operation, or to define precisely the limits of each, instances of these *mixed* actions are sufficiently common among animals to leave little doubt of the fact. It is instinct which leads a greyhound to pursue a hare, but it must be reason that directs an old greyhound to trust the more fatiguing part of the chase to the younger, and to place himself so as to meet the hare in her doubles."

With due submission, the two cases are not exactly analogous, unless it can be proved that the bees which took the above-mentioned method, were old bees, and had availed themselves, like the greyhound, of the lights of observation and experience. This is not stated to have been the fact. Besides it is questionable, whether the duration of a bee's life, would admit of the degree of experience necessary, even for a calculation like this.

It must, however, be confessed, that this question is involved in no common difficulty. It is altogether a question to be decided by facts, the number and authenticity of which must materially affect its negative or affirmative. The correctness of the observations made, must be taken into consideration, and not less so, the possibility of systematic prejudices in the observer. It is impossible, within our present limits, to do it justice; there are, however, many very strong facts brought forward by Mr. Spence, in favour of his view of the question. Two of these we shall find room for here.

"A bee which Huber watched, while soldering the angles of a cell with propolis, detached a thread of this material with which she entered the cell. Instinct would have taught her to separate it of the exact length required; but after applying it to the angle of the cell, she found it too long, and cut off a portion so as to fit it to her purpose."

The second instance is cited from Dr. Darwin, who informs us "that, walking one day in his garden, he perceived a wasp upon the gravel-walk, with a large fly nearly as big as itself, which it had caught. Kneeling down, he distinctly saw it cut off the head and abdomen, and then taking up with its feet the trunk or middle portion of the body, to which the wings remained attached, fly away; but a breeze of wind acting upon the wings of the fly turned round the wasp with its burthen, and impeded its progress. Upon this it alighted again upon the gravel walk, sawed off first one wing and then the other, and having thus removed the cause of its embarrassment, flew off with its booty. Could any process of ratiocination be more perfect? 'Something acts upon the wings of this fly and impedes my flight; if I wish to reach my nest quickly, I must get rid of them—to effect which the shortest way will be to alight again and cut them off.' These reflections or others of similar import, must be supposed to have passed through the mind of the wasp, or its actions are altogether inexplicable. Instinct might have taught it to cut off the wings of *all* flies, previously to flying away with them; but here it first attempted to fly with the wings on—was impeded by a certain cause—discovered what this cause was, and alighted to remove it. The chain of evidence seems perfect in proof, that nothing but reason could have been its prompter."

To these latter observations nothing can be replied, but that their value entirely depends on the exact truth of the fact brought forward to support them. Now, though we would not willingly impeach the high authority on which it rests, it must be remembered that Dr. Darwin was an advocate of the hypothesis, that all the actions of insects are referable to a reasoning faculty.

One of the strongest points in favour of Mr. Spence, is the fact so well ascertained by a number of proofs, that insects receive and communicate information, which we can hardly suppose could be done if they were wholly devoid of reason. Dr. Franklin relates a story to this effect which is very remarkable. "Having placed a pot containing treacle in a closet infested with ants, these insects found their way into it, and were feasting very heartily when he discovered them. He then shook them out, and suspended the pot by a string from the ceiling. By chance one ant remained, which, after eating its fill, with some difficulty found its way up the string, and thence reaching the ceiling, escaped by the wall to its nest. In less than half an hour, a great company of ants sallied out of their hole, climbed the ceiling, crept along the string into the pot, and began to eat again. This they continued until the treacle was all consumed, one swarm running up the string, while another passed down. It seems indisputable that the one ant had in this instance conveyed news of the booty to his comrades, who would not otherwise have at once directed their steps in a body to the only accessible route."

Now we confess, that this does not appear to us so very indisputable. Is it not possible that the ants, which are certainly not destitute of the sense of smelling, might have been attracted by the smell of the treacle?

The last story we shall quote on this subject for the amusement of our readers, is from the same valuable work of Messrs. Kirby and Spence.

" A German artist, a man of strict veracity, states, that in his journey through Italy, he was an eye-witness to the following occurrence. He observed a species of scarabæus, *(Ateuchus pilularius?)* busily employed in making for the reception of its egg, a pellet of dung, which when finished, it rolled to the summit of a small hillock, and repeatedly suffered to tumble down its side, apparently for the sake of consolidating it by the earth, which each time adhered to it. During this process the pellet unluckily fell into an adjoining hole, out of which all the efforts of the beetle to extricate it were in vain. After several ineffectual trials the insect repaired to an adjoining heap of dung, and soon returned with three of his companions. All four now applied their united strength to the pellet, and at length succeeded in pushing it out; which being done the three assistant beetles left the spot, and returned to their own quarters."

That insects possess memory, seems quite certain. Bees after hours of wandering return to their own particular hives, which they clearly distinguish from all others in the same neighbourhood. This fact is well expressed by Mr. Rogers, in his poem of the "Pleasures of Memory," though he is mistaken in supposing the bee to be reconducted to its hive by the scent of the various flowers which it has visited, for bees fly *straight* to their hives from great distances.

> "Hark! the bee winds her small but mellow horn,
> Blithe to salute the sunny smile of morn.
> O'er thymy downs she bends her busy course,
> And many a stream allures her to its source.
> 'Tis noon, 'tis night. That eye so finely wrought,
> Beyond the search of sense, the soar of thought,
> Now vainly asks the scenes she left behind;
> Its orb so full, its vision so confined!
> Who guides the patient pilgrim to her cell?
> Who bids her soul with conscious triumph swell?
> With conscious truth retrace the mazy clue
> Of varied scents that charmed her as she flew?
> Hail! MEMORY, hail! thy universal reign
> Guards the least link of Being's glorious chain."

In concluding this imperfect survey of the intellectual faculties of insects, we may sum up by saying, that the great majority of their actions are guided by a principle of instinct, toally distinct from reason; that however, they seem capable, though to a very limited extent, of forming judgments from their immediate perceptions, and of acting thereupon. At least that they perform actions which will hardly admit of any other solution than this. We may also conclude that they have some mode of inter-communication with each other, and that they possess memory seems fully certain. We may repeat, however, that the subject is surrounded with difficulties and obscurities, which, we fear, will not be speedily removed and dissipated. We cannot, in a case of this kind, be too cautious in the admission of anecdotes, or too strict in the investigation of facts on which to found our theories.

We often find insects assembled in great numbers in one and the same place: but if their individual preservation is the sole motive of these re-unions—if they are only attracted to a particular spot by the abundance of aliments which they find there, common to them all, or a shelter where they may be less exposed to the intemperature of the weather, or the attacks of their enemies, such accidental assemblages cannot be considered in the light of societies, properly so called. Certain caterpillars, which have been designated under the title of common, processionary, &c. already in approximation when in the egg state, weave in concert a web, which, like a hammock, or tent, serves as a habitation for them until their final metamorphosis. But such labours have reference solely to their own proper existence. They are occupied only with themselves. They have no family to bring up, no pains or cares respecting generations yet unborn. "There reigns among them," says Bonnet, "the most perfect equality; there is no distinction of sex, and almost no distinction of size. They all resemble, all have the same share in the labours. All compose, properly speaking, but one family, issuing from the same mother." This temporary association is dissolved the moment the caterpillars pass into the chrysalis state. All becomes inertness, and absolute isolation.

It is not thus with the perfect societies of insects, and which alone properly deserve the name. They are eminently distinguished from the preceding, not only by reason of the remarkable differences observed in the external forms of the individuals which compose them, but still more by the institutions by which they are governed. Their principal end is the education of the young, and even those which, under the form of nymphæ, have no further need of nourishment, find active and vigilant sentinels to prevent or avert the dangers by which they may be threatened.

At the period when this education is finished, these associations present us with three sorts of perfect individuals in the full enjoyment of all their faculties, males, females, and individuals of the latter sex, but unfit for reproduction. These last have been designated as neuters, mules, workers, and even soldiers, as among the termites. The denomination of workers is frequently equivocal, since the females, in some species, are equally industrious. Neuter appears to be preferable.

These societies are either temporary or continuous. When temporary, they owe their origin to a female, who, without assistants, abandoned to her own resources, lays the foundations of the colony, and soon finds auxiliaries in the neuters which she begins to bring into the world. Such are the societies of wasps and humble bees. But those which are continuous, have neuters at all periods. Sometime these neuters, as among the ants and bees, are charged exclusively with all the labours and cares of the family. Sometimes they have no other functions than to watch for the defence of the community, and perhaps for the preservation of the germs of the race, as among the termites.

The countries situated between the tropics are those where nature has most energy, and where these assemblages of insects are most multiplied and most formidable. The action exercised on animal and vegetable substances by insects, which live in an isolated or solitary state, is usually slow, and its effects are not sensible but in the course of a time,

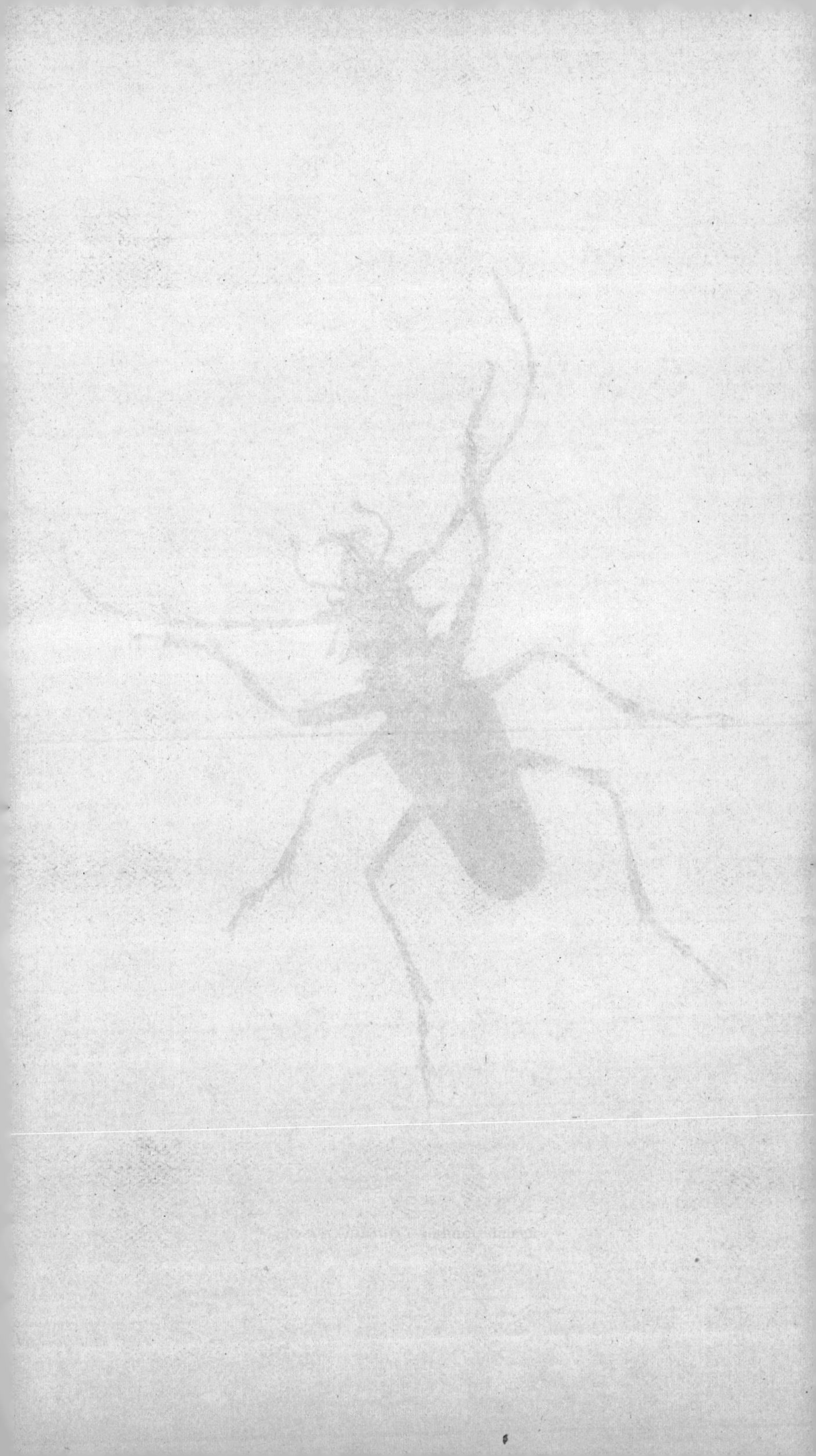

Pl. 6.

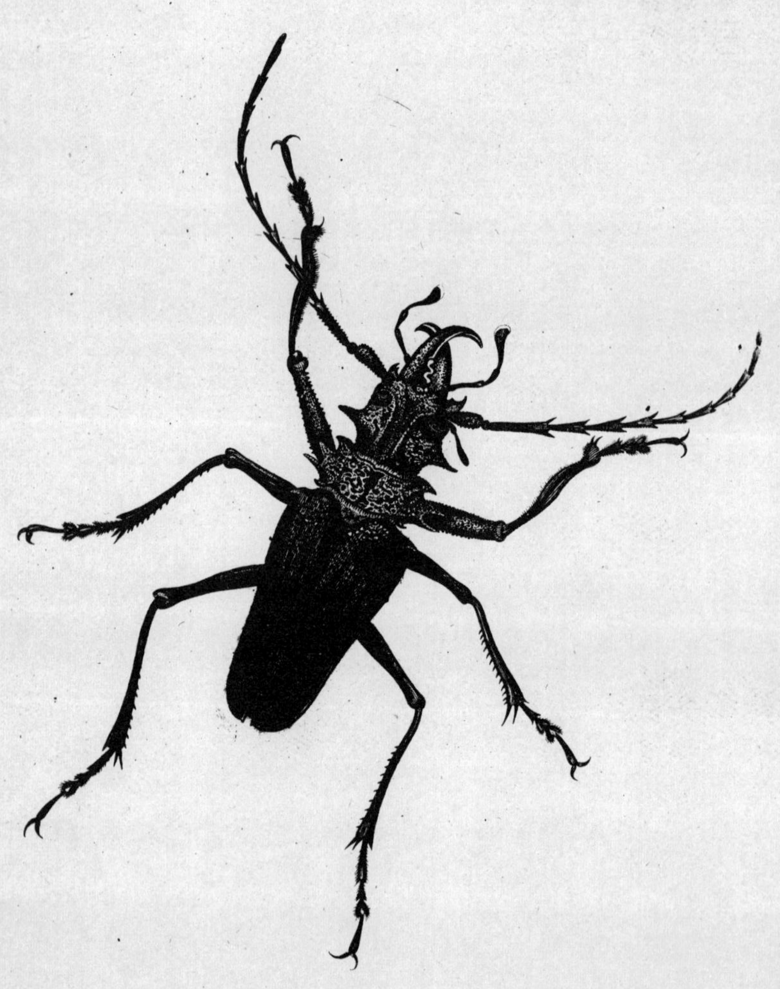

Psalidognathus *Friendii.* G. R. Gray.

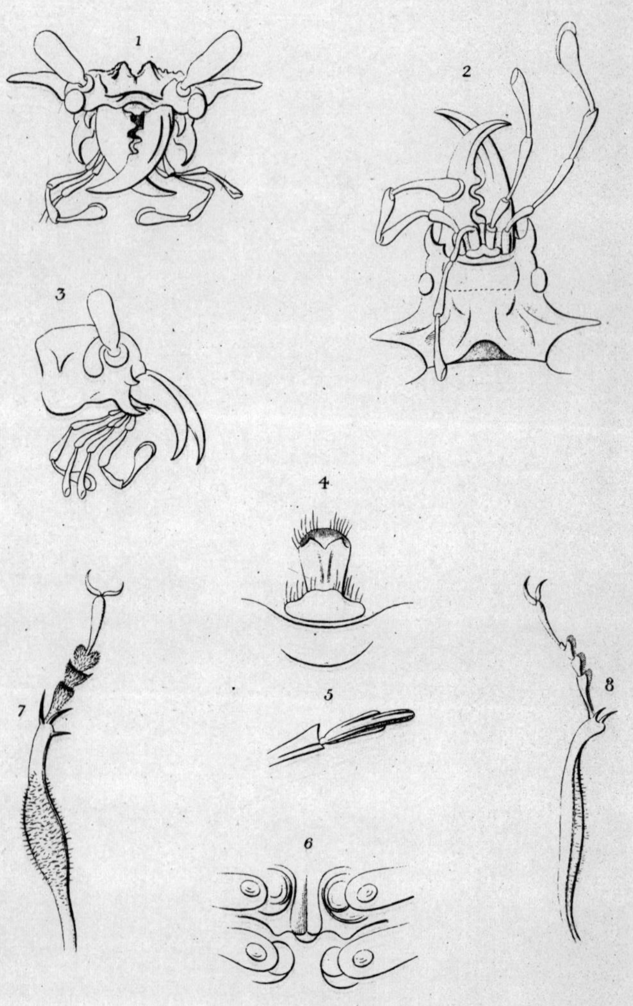

Details of Psalidognathus *Friendii*.

and sometimes one of sufficient length. But when these animals are assembled in the same place in large bodies, when they form, like the termites and ants, innumerable legions, soon, in spite of their littleness, they will devour and cause to disappear all the organized bodies which they find deprived of life. The end of the Author of Nature in establishing such societies of insects, appears to have been to augment the energy of this active and re-active force, which maintains equilibrium among beings, and which, by continual creations and destructions, perpetually regenerates organized matter on the surface of our globe. If the regions neighbouring to the equator develope more numerous productions, the number of destructive agents, by a just compensation, is also much more considerable there. Millions of ants and termites work incessantly, in purging the surface of the soil from carcasses by which the air would be speedily corrupted, and so great are their voracity and their numbers, that often in a single day, these armies of insect-vultures devour the flesh of a colossal quadruped. In their turn they become the food of an infinity of birds, reptiles, quadrupeds, not to mention the enemies opposed to them, by the class of animals to which they themselves belong.

The females of the social insects are distinguished by their prodigious fecundity. Reaumer estimates at twelve millions, the number of eggs laid by the domestic bee in spring, within the space of twenty days. But this fecundity is much inferior to that of the termites of the same sex. At the time of laying, their belly is so distended by the number of eggs with which it is filled, that this part is then, according to Smeathman, five hundred or two thousand times more bulky than the rest of the body. Its volume is twenty or thirty thousand times larger than that of the belly of the neuter. In fine, the number of eggs which the female can lay in the space of one day, amounts to eighty thousand. Now, this exceeding fecundity of insects which live in society, seems to establish the necessity of a third division of individuals, such as the neuters, which shall possess the affections of maternity, without the reproductive faculty.

All these insects, with the exception of the termites, are of the number of those which undergo complete metamorphoses, and their larvæ, unlike the caterpillars, are quite unable, from feebleness, absence of feet, and the extreme smallness of the mouth to provide for themselves. Besides, they would seek in vain to procure their food, which consists in vegetable or animal matter, which has undergone preparatively a digestive process. In this state of things almost daily assistance is necessary for them. This the mothers, had they been alone, could by no possibility have afforded. They never could have found strength or time to collect magazines of provisions for so numerous a family, nor could the provisions have been preserved in a proper state, up to the time that they were wanted. If the existence of the mothers were prolonged beyond the time of the disclosing of the young, and the bringing up and the education of the latter were entrusted to them, their difficulties would still increase. They could not find every day the quantity of aliments required, especially in rainy weather, and even supposing that they could procure them, how could they distribute them to each individual larva? How could they watch over, and preserve them from the infinite number of perils by which they are menaced? It is very different with solitary insects. Their family, few in number, isolated, concealed, and occupying but a small space, can easily be withdrawn from the investigations of its enemies. But insects, united in great numbers in the same nest, have more unfavourable risks to run. The careful attention of the ants to their progeny affords an apt illustration of this point.

The preservation therefore of such animals, and the prosperity of their community, could only be assured by the establishment of a particular and numerous order of individuals, who should supply the functions of mothers, and who should possess nothing of the character but its sentiments and affections. Nature in forming, in this instance, neuter individuals, was obliged *to* depart from her usual laws, to ensure the subsistence of her work, and her foresight has modified her resources according to the circumstances in which the beings were to be placed. For example, she has followed a different plan with regard to the termites, the young individuals of which have not this feeble infancy, and differ from those which are adult only by a smaller size, the absence or shortness of the wings, and some other peculiarities of no great importance. Then the neuters, justly called soldiers, have a large head, strong jaws, which act like pincers, and they scarcely compose more than the hundredth part of the population. They simply act the part of videts and defenders. The other individuals, up to the moment in which their organs are entirely developed, remain exclusively charged with all the internal works. Still delicate and without defence, they have only need to be preserved from the strong impression of the heat, and from the attacks of such enemies as might introduce themselves into their habitation. In working under cover, and in subterranean galleries, they avoid the first of these dangers, and the armed neuters preserve them against the second, while the society is maintained by this reciprocity of service. An activity common to all the members of the society, is thus the distinguishing characteristic of the termites, which are one of the most terrible agents of destruction in the equatorial countries. As they work only in their infancy, and are at this age deprived of wings, or only possess the rudiments of them, they then very much resemble ants in their habits. But their numbers being far greater, they construct vaster and more solid habitations, and as their wants are greater, their destructive force is more powerful. Their invasions are less easily opposed, as they operate in darkness, and thus escape the notice of man and his vengeance.

The historian of these insects, Smeathman, was not acquainted with their nymphæ. The individuals which he seemed to consider as such are the neuters, which defend the habitation, and the individuals which he calls workers, are only the termites in their first age, or in the larva state. These insects do not undergo their last metamorphosis at the same period. The individuals less advanced collect the females which have been fecundated, and take care of the eggs. The termites form then, under all relations, a society quite distinct from those of the ants, wasps, humble-bees, and bees, insects, all of which undergo a perfect metamorphosis. These last-mentioned societies, according to the consideration of the organs of locomotion, are established upon three models.

In one, such as that of the ants, the neuters are deprived of wings, and for the completion of their labours have no other instruments than the parts of the mouth.

All the individuals of the other societies have wings, but the wasps are not better provided than the ants in regard of the means for the execution of their work. It is not so with the humble-bees and bees. The legs and tarsi of their hinder limbs have a peculiar form, which allows them to collect the pollen of flowers. These insects also have organs solely destined to elaborate and secrete the honey and the wax. As well as among the wasps, the female constitutes an integral part of the society, as long as it subsists. The females of wasps and humble-bees even commence the establishment, and are foundresses and queens at the same time.

These organic differences have a great influence on the instinct of these insects, for the perfection of their work is proportioned to their means.

Being destitute of wings, the neuter ants live on the ground, or establish themselves in the clefts of walls and trees, at but a small elevation above the soil. Those which construct habitations employ a considerable time in bringing together the materials which are to compose it. All their works are of a rustic and very simple construction. The wasps, humble-bees, and bees, to which the organs of flight afford a facility of removing rapidly and to great distances from their dwellings, and to return thither with equal celerity, after having collected the materials of their choice, are the most favoured in their works. Their productions are known, and are the objects of our admiration. But the following observation seems to have escaped the historians of these animals. Of all the substances which it is possible for them to employ, those which they prefer are the most proper for the construction of an edifice which, suspended in the air, should be, in a given volume, the best arranged for the number of the population, the least weighty, and the most solid in relation to the duration of the society. Thus the nests of wasps consist of a sort of pasteboard or very thick paper, in the construction of which the ligneous substance predominates. The bee knows how to gather and prepare a resinous substance, capable, from its ductility, of being reduced into very slender laminæ, of being fashioned at the will of the animal, namely, wax, a substance equally light and resisting, of which the bee is the sole fabricator. The entrance of the pieces which compose the edifice is sometimes vertical, sometimes horizontal, but always at the under part, which places its inhabitants in shelter from the rain, even when the solid walls do not protect them.

The bee is, of all these insects, the one whose instinct is the most perfect, the only one whose habits are not at all carnivorous, and whose existence is one of the blessings of nature. The others are born for destruction; she, on the contrary, appears to have been made to secure the fecundation of vegetables, by transporting from one to the other the pollen of their flowers, which the winds alone could not as certainly have propagated. She alone has a brush and basket to collect this pollen, a species of syphon to draw in the honey, and special internal organs where it is received, elaborated, and converted into wax. The radii which she constructs are disposed upon a vertical plane, and furnished on both sides with alveoli; while those of the wasps are always horizontal, and present but a single range of cells. The society of wasps is temporary; that of bees, whose regime is monarchical, is durable, and only ceases from accidental circumstances. Our domestic bee can live in every climate: it can brave the wintry breath of Siberia, and the heats of the torrid zone.

The societies of which we have been speaking hitherto are all composed of individuals of the same species; but two sorts of ants, commonly called the red and the sanguine, present us, in this respect, with a very singular fact, for the observation of which we are indebted to the younger Huber. The societies of these insects are mixed. There are found amongst them, besides the three sorts of ordinary individuals, some neuters proceeding from one or even from two other species of ants, carried away from their habitations in the larva or nymph state. The neuters of the reddish species compose a population of warriors, from which circumstance they are termed Amazons and Legionaries by M. Huber. Towards the moment in which the heat of the day begins to decline, if the weather be favourable, and regularly at the same hour, at least for many consecutive days, these ants quit their nests, advance in a serried column, more or less numerous according to the population, proceed towards the ant-hill which they intend to invade, enter in spite of the resistance of the proprietors, seize with their jaws the larvæ or nymphs of the neuter ants in the habitation, and transport them, still pursuing the same order, to their own proper domicile. Other neuter ants of the conquered species, born among these warriors, and formerly kidnapped in their larva state, take care of the newly-imported larvæ, as well as of the posterity of their ravishers.

We might enter here into many further details concerning the societies of insects, only that our limits forbid it, and that they will come with more propriety under some future heads. What we have done in this way was necessary to render our general observations intelligible. From every thing respecting such societies, this consequence may be deduced—the laws which govern the societies of insects, those even which appear to us the most anomalous, constitute a system combined with the most profound wisdom, and established from the beginning, not resulting from accident. Such a consideration should elevate our thoughts with a religious respect towards that Eternal Intelligence, which, in giving existence to so many various beings, has thought proper to perpetuate their generations by means sure and invariable in their operation, concealed from our feeble reason, but ever to be regarded with the most intense admiration.

Notwithstanding that we have, perhaps, transgressed the limits to which the character of our Work necessarily confines us, in this introductory essay, we yet feel that much of the greatest interest has been left unnoticed, and that the sketch is, upon the whole, meagre, feeble and imperfect. But the candid reader will overlook this, when he considers the vastness of the subject, even in the most general way of viewing it, and the impossibility of entering at large into details, which might yet be deemed indispensably necessary, to illustrate our observations. If Messrs. Kirby and Spence have given to their magnificent work, in four large volumes, the modest title of an *Introduction*, how was it possible for us to comprise within our scanty limits, all the general points of interest with which the science of entomology is so eminently pregnant. If we have given such a view as may prove sufficient for the comprehension of the ensuing scientific details of our respected author, and if, though at the humblest distance, we have contributed, not to improve the science, but to spread a taste for a pursuit so highly calculated to expand and elevate the mind, and impress it with the truest principles of philosophical religion, we have attained our object and are satisfied. We shall now in imitation of Messrs. Kirby and Spence, terminate this feeble sketch with the same quotation with which they have closed their Introduction:

" Happy, if full of days—but happier far,
If, ere we yet discern life's evening star,
Sick of the service of a world that feeds
Its patient drudges with dry chaff and weeds,
We can escape from custom's idiot sway
To serve the Sovereign we were born t' obey.
Then sweet to muse upon his skill display'd
(Infinite skill) in all that he has made!
To trace, in Nature's most minute design,
The signature and stamp of power divine—
Contrivance intricate, expressed with ease,
Where unassisted sight no beauty sees;
The shapely limb, and lubricated joint,
Within the small dimensions of a point;
Muscle and nerve miraculously spun,
His mighty work, who speaks and it is done,
Th' invisible in things, scarce seen reveal'd,
To whom an atom is an ample field;
To wonder at a thousand insect forms,
These hatch'd, and those resuscitated worms,
New life ordained and brighter scenes to share,
Once prone on earth, now buoyant upon air,
Whose shape would make them, had they bulk and size,
More hideous foes than fancy can devise;

With helmet-heads and dragon scales adorn'd,
The mighty myriads, now securely scorn'd,
Would mock the majesty of man's high birth,
Despise his bulwarks and unpeople earth:
Then with a glance of fancy to survey,
Far as the faculty can stretch away,
Ten thousand rivers pour'd at his command,
From urns that never fail through every land;
These like a deluge with impetuous force,
Those winding modestly a silent course;
The cloud-surmounting Alps, the fruitful vales;
Seas on which every nation spreads her sails;
The sun, a world whence other worlds drink light;
The crescent moon, the diadem of night;
Stars countless, each in his appointed place,
Fast anchor'd in the deep abyss of space—
At such a sight to catch the poet's flame,
And with a rapture like his own exclaim:
These are Thy glorious works, thou Source of Good,
How dimly seen, how faintly understood!
Thine, and upheld by thy paternal care,
This universal frame, thus wondrous fair.
Thy power divine, and bounty beyond thought,
Adored and prais'd in all that thou hast wrought,
Absorb'd in that immensity I see,
I shrink abash'd, and yet aspire to thee;
Instruct me, guide me to that heavenly day,
Thy words, more clearly than thy works display,
That while thy truths my grosser thoughts refine,
I may resemble thee, and call thee mine."

COWPER.

THE FIRST ORDER OF INSECTS.

The Myriapods (Myriapoda—*Mitosata*, Fab.),

Vulgarly named *millipedes*, are the only animals of this class, which have more than six feet in their perfect state, and whose abdomen is not distinct from the trunk. Their body, deprived of wings, is composed of a considerable series of rings, most frequently equal, each of them, with the exception of the first, having two pairs of feet, most usually terminated by a single crotchet: whether these rings be undivided, or divided into two half segments, each has a pair of these organs, and one of them only exhibits two stigmata.*

The myriapods, for the most part, resemble small serpents, or nereïdes, having feet very much approximated to each other, in the entire length of the body; the form of these organs extends even to the parts of the mouth. The mandibles are biarticulated, and immediately followed by a piece

* The rings of the body of insects have in general two stigmata. If we consider under this point of view the rings of the body of the scolopendræ, particularly of the large species, those which have twenty-one pairs of feet, we shall see that they are alternately provided with, or deprived of, two stigmata; and thus, comparatively to those last animals, these are really only semi-rings. Each complete segment has two pairs of feet, one of which is supernumerary, since we find, in other insects, that the rings provided with feet have but two feet.

formed like a lip, quadrified, with articulated divisions, or resembling little feet, and which by its situation, corresponds to the largest of the crustacea. Then come two pairs of little feet, of which the second, in the form of large crotchets, in many species, appear to replace the four jaws, or at least the two, as well as the lower lip of insects; these may be considered as species of buccal feet. The antennæ, two in number, are short, a little thicker towards the end, or almost filiform, with seven articulations in some species, a great number in others, and setaceous; their eyes are usually formed of an union of smooth eyes, and if, in others, they have a cornea with facets, these lenses are, nevertheless, proportionally larger, rounder, and more distinct than those of the eyes of insects. Their stigmata are often very small, and their quantity, in proportion to that of the rings, is usually more considerable than in other insects, in which, at the most, it amounts to no more than eighteen or twenty. The number of these rings and that of the feet, augment with age, a character, which strongly distinguishes the myriapods from other insects, the latter being always born with the number of segments proper to them, and all their feet being developed at once, either at the period of birth or when they pass into the nymph state. M. Savi, the son, professor of Mineralogy at Pisa, who has made a peculiar study of the iuli, has observed, that when they issue from the egg, they are devoid of these organs: these animals, therefore, undergo a true metamorphosis. In some, the male sexual organs are always placed after the seventh pair of feet, on the sixth or seventh segment of the body, and those of the female, near the origin of the second feet; in the others, these sorts of organs are situated in the usual place, near the posterior extremity of the body. The position of the masculine parts of the first, compared with that which they have in the crustacea and arachnida, would seem to indicate the separation of the trunk and abdomen. With regard to the other myriapods in which the sexual organs are posterior, we remark that in an analogous portion of the body of certain species (*Scolopendra morsitans*), an inversion takes place in the successive order of the stigmata, which would seem to announce the same distinction.

The myriapods live and grow a longer time than the other insects, and according to M. Savi, two years, at least, are necessary to some of them, (the iuli,) for the genital organs to become apparent.

From this assemblage of facts, we may conclude, that these animals, on the one side, approximate to the crustacea and arachnida, and on the other, to the insects; but considered with regard to the presence, form, and direction of the tracheæ, they certainly appertain to the last class.

We shall divide them into two families, perfectly distinct, both by reason of their organization and habits, and composing, in Linnæus, two generic sections.

The first family of the Myriapods, that of

The Chilognatha, Latr., or the genus of the Iuli of Linnæus,

Has the body generally crustaceous, and often cylindrical; the antennæ a little thicker towards the end, or almost of an equal thickness, and composed of seven articulations; two thick mandibles without palpi, very distinctly divided into two portions, by a medial articulation, with imbricated teeth, and implanted in a concavity of its upper extremity; a sort of lip (lanquet), situated immediately below them, covering them, crustaceous, plane, divided at its exterior surface, by longitudinal sutures and emarginations, into four principal areas, tuburculated at the upper edge, and of which the two intermediate ones are more narrow and short, situated at the upper extremity of another area, serving them as a common basis; the feet very short, and terminated by a single crotchet; four feet situated immediately under the preceding piece, of the form of those which follow, but approximating more to their basis, with the radical articulation proportionately longer, and the majority of the others attached in double pairs to a single ring. The male genital organs are situated immediately after the seventh pair of feet, and those of the other sex behind the second. The stigmata are placed alternately, outside the origin of each pair of feet, and very small.

The chilognatha walk very slowly, or slide along, as it were, and roll themselves up into a spiral or ball. The first segment of the body (and in some the following one) is the largest, and presents the form of a corslet or small buckle. It is seldom until the fourth in some, and the fifth or sixth in others, that the duplicature of the pairs of feet commences; the first two or four feet are even entirely free as far as their origin, when they adhere to their respective segments, only by a medial or sternal line. The last two or three rings are apodal; on each side of the body is seen a series of pores, which have been taken for stigmata, but which, according M. Savi, are destined for the issue of a sort of acid liquor, and of a disagreeable odour, which seems to serve as a defence against enemies. The apertures proper to respiration, the discovery of which is owing to him, are placed on the sternal piece of each segment, and communicate interiorly, with a double series of pneumatic pouches, arranged like chaplets all along the body, and from which proceed the tracheal branches, which spread themselves over the other organs. According to the observation of M. Straus, the pouches or vesicular tracheæ, are not connected one with the other in the ordinary way by a principal trachea.

In the environs of Pisa, where M. Savi collected the foregoing observations, the season of reproduction with the common iulus commences towards the end of December, and finishes towards the middle of May. The copulative organs of the male are placed, in this species, under the sixth segment, but they are not apparent, as genital parts, until the individual has attained about a third of its ordinary size. Until then, this place is occupied by a pair of feet, the fifteenth, and which always exists in the females. In them, the orifice of the sexual parts, is placed between the first and second segment. In certain glomeris and female iuli, I have observed behind the origin of the second pair of feet, two small convex nipples, which appear to characterize the sex. That of the males also consists of two nipples, but each terminated by a scaly crochet. In copulation, these insects are upright, and apply face to face, the anterior extremities of the body, entwining below. The body of individuals just born, is kidney-formed, perfectly smooth, and without appendages. Eighteen days after, it undergoes a first moulting, and they then assume the form of the adults. But, still, they have only twenty-two segments, and the total number of their feet is twenty-six pairs. M. Savi appears to contradict the assertion of De Geer, who says that he counted but three pairs, and eight rings in the young individuals. But is it very certain the moulting of which M. Savi speaks is really the first? and should we not, on the contrary, presume that these young individuals do not pass suddenly from a state in which they exhibit no locomotive appendage, to that in which they have twenty-six pairs of feet? or that, in a word, some other changes of the skin, but which may have escaped M. Savi, have taken place, and have successively developed this number of feet? Do not the observations of Reaumur, the Swedish naturalist, confirm these gradual transitions? Be this, however, as it may, according to M. Savi, the first eighteen pairs of feet only serve for locomotion. At the second moulting, the animal has thirty-six

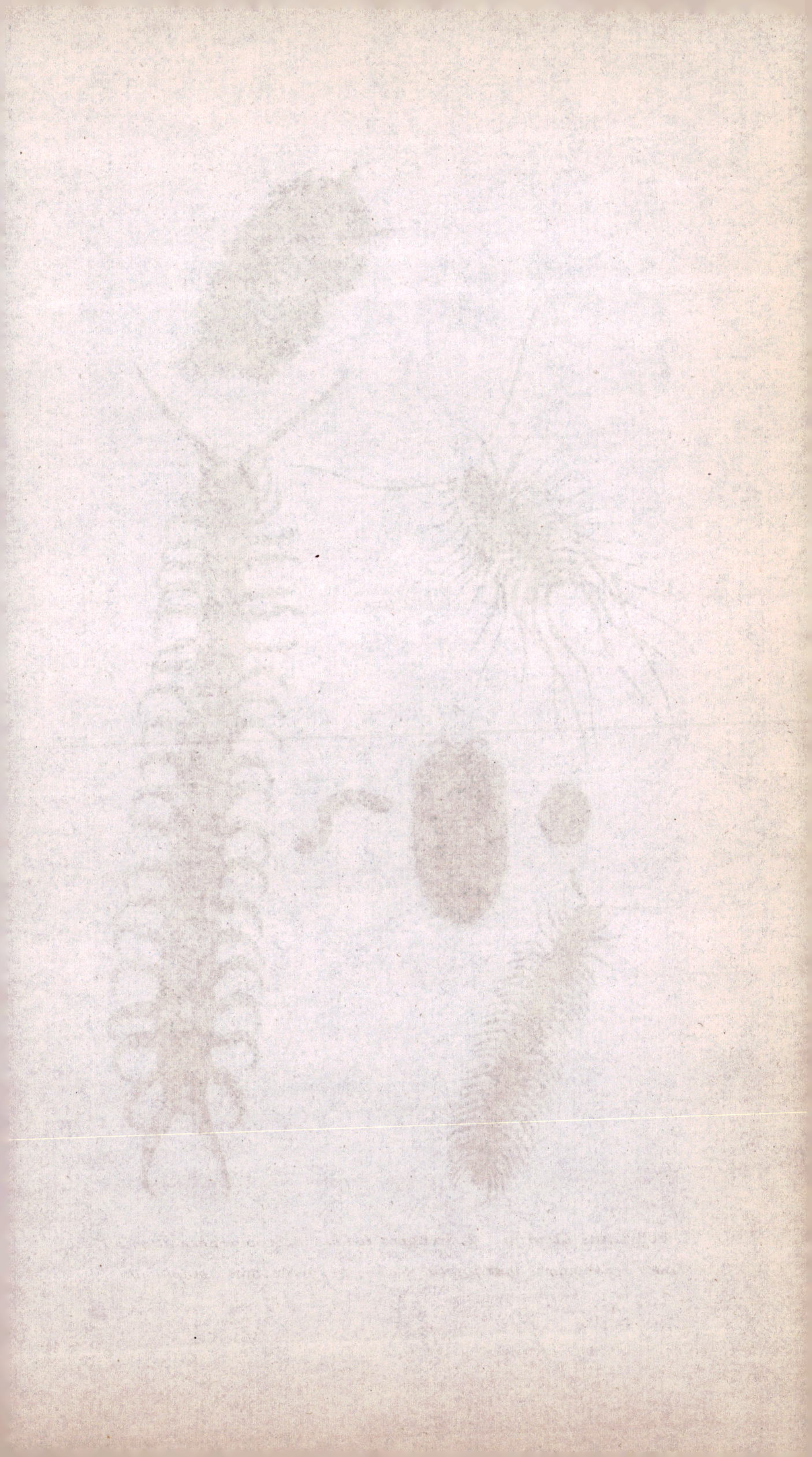

Pl. 1.

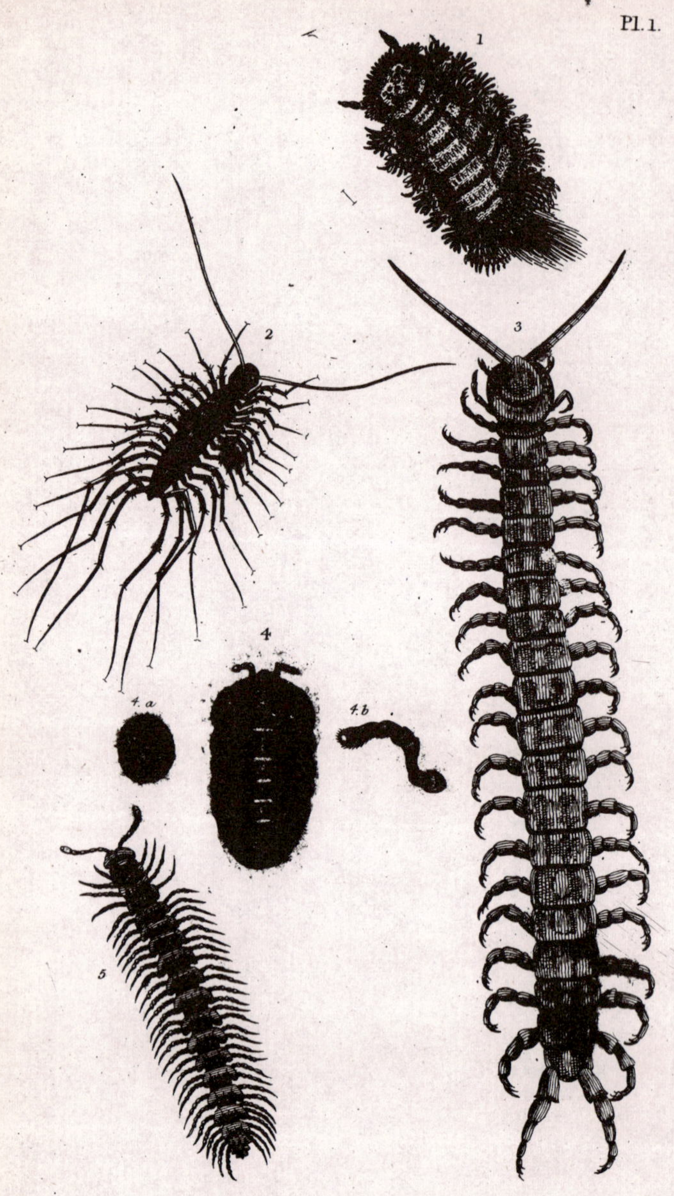

1. Pollyxenus *lagurus*. 2. Scutigera *livida*. 3. Scolopendra *alternans*. *Leach*. 4. Glomeris *marginata*. *Leach*. 5. Polydesmus *complanatus*.

pair, and at the third forty-three. The body then, is composed of thirty segments. Finally, in the adult state, the male has thirty-nine, and the female sixty-four. Two years after, they moult again, and it is only then, that the organs of generation appear. From the birth, which takes place in March to November, the time in which M. Savi ceased his observations, these changes of the skin are renewed from month to month very nearly. In what the insect casts off, we can discover even the membrane which lines the alimentary canal, and the tracheæ. The organs of the mouth are the only parts which M. Savi could not find.

These insects feed either on animal or vegetable substances, but in a state of death and decomposition, and they lay in the earth a great number of eggs.

They form in Linnæus, but one genus—

THE IULI (IULUS, L.),

Which we divide as follows :—

Some have the body crustaceous, without appendages to the end, and the antennæ swelled towards the summit.

THE GLOMERIS (GLOMERIS, Latr.),

Similar to wood-lice, oval, and rolling themselves up in a ball.

Their body, convex above, and concave underneath, has, along each of its lower sides, a range of small scales, analogous to the lateral divisions of the trilobites. It is composed, without reckoning the head, of but twelve segments or tablets, the first of which, more narrow, forms a sort of collar, in a transverse semi-circle. The following segment and the last are the largest of all. The number of feet in the females is thirty-four, and thirty-two in the males, their sexual organs replacing the pair which is wanting. These animals are terrestrial and live under stones, in mountainous places.*

THE IULI, proper (IULUS, LINN.),

Which have the body cylindrical, and very long, rolling themselves spirally, and without any projection, or trenchant edge on the sides of the rings.

The largest species live on the ground, particularly in sandy places and woods, and shed a disagreeable odour. The smallest feed on fruits, roots, or the leaves of pot-herbs. Some others are found under the bark of trees, in moss, &c.

The very large Iulus (*I. Maximus*, Lin.), Marcg. Bras. p. 255. Proper to South America, and is about seven inches long.

The Iulus of the sands (*I. Sabulosus*, Lin.), Scheff. Elem. Entom. lxxiii. *I. fasciatus*. De G. Insect. VII. xxxvi. 9, 10. Leach, Zool. Misc. cxxxiii. About sixteen lines in length, of a blackish brown, with two reddish lines along the back; fifty-four segments, of which the last but one is terminated by a strong point, hairy and corneous at the end. In Europe.

The terrestrial Iulus (*I. terrestris*, Linn.), Geoff. Insect. II. xxii. 5. About one-fourth smaller, bluish ash, intersected with clear yellowish, forty-two to forty-seven segments. With the preceding.†

* *Iulus Ovalis*, Lin. Gronov. Zooph. pl. xvii. 4, 5.—*Oniscus Zonatus* Panz. Faun Insect. Germ. IX. xxiii.—*Glomeris Marginata*, Leach, Zool. Misc. cxxxii.—*Oniscus Pustulatus*, Fab. Panz. ibid. xxii.

† See the two before-cited memoirs of M. Savi. and the Zool. Misc. of Dr. Leach, vol. 3. with respect to these two species and some other English ones. Add, *Iulus Indus*, Lin. De G. VII. xliii. 7; Seb. Mus. II. xxiv. 4, 5; Seb. Mus. I. lxxxi. 5; Schrœt Abhandl. I. iii. 7.

THE POLYDESMI (POLYDESMUS, Latr.),

Similar to the Iuli in the linear form of their body, and the habit of rolling themselves spirally, but having the segments compressed on the lower sides, with a crest-formed projection above.

They are found on stones, and most frequently in humid places.*

The species, which have apparent eyes, form the genus *Craspedosome*, of Dr. Leach.†

The others have the body membranous, very soft, and terminated by pencils of small scales. Their antennæ are of the same thickness. Such are,

THE POLYXENI (POLYXENUS, Latr.),

Which comprehend as yet but one species, classed with the Scolopendra (*Sc. lagura*, L.), by Linnæus, Geoffroy, and Fabricius.

It is the *Iulus with pencilled tail* of De Geer, Insect. VII. xxxvi. 1, 2, 3; Zool. Misc. cxxxv. B. *Polyxenus lagurus*. This insect is very small, oblong, with aigrettes of small scales on the sides, and a white pencil at the posterior extremity of the body. It has twelve pairs of feet, placed on as many semi-rings.

It sojourns in the clefts of walls, and under the bark of old trees.

The second family of the MYRAPODS,

The CHILOPODA (CHILOPODA, Latr.), or the genus of the SCOLOPENDRÆ (SCOLOPENDRA), of Linnæus,

Have the antennæ more slender towards their extremity,

* The *Iulus Complanatus*, (Zool. Misc. cxxxv. A.;) *depressus*, *stigma*, *tridentatus* of Fabricius; his scolopendra? *dorsalis clypeata*.

† The species, unknown before the researches of Dr. Leach, appear peculiar to Great Britain.—See the plate cxxxiv. of his Zoological Miscellany, Vol. III.

of fourteen articulations and upwards; a mouth composed of two mandibles, provided with a small appendage, in the form of a palpus, presenting in their centre the appearance of a synostosis, and terminated like the bowl of a spoon indented on the edges; one quadrifid lip,* the two lateral divisions of which are larger, annulated transversely, similar to the membranous feet of caterpillars; two palpi, or little feet united at their base, unguiculated at the end; and a second lip† formed by a second pair of feet, dilated and joined at their origin, and terminated by a strong crotchet, mobile, and pierced under its extremity with a hole, for the issue of a poisonous fluid.

The body is depressed and membranous. Each of its rings is covered with a coriaceous and cartilaginous plate, and has, for the most part, but one pair of feet. In this case, however, they are but semi-rings. The last pair is usually thrown backwards, and elongated in the form of a tail. The

* A piece analogous to the lower lip of the chylognatha, representing, in my opinion, the tongue of the crustacea, but also fit to perform the office of jaws. This is what M. Savigny names first auxiliary lip.

† The second auxiliary lip of M. Savigny. It is not joined with the head, but with the anterior extremity of the first semi-segment. The two feet with crotchets form, by their union and dilatation of their first articulation, a plate in the form of a chin and lip. The same semi-segment bears the two first common feet. In the scolopendræ proper of Dr. Leach, the first two stigmata are situated on the third semi-segment. The second and the following, will compose the first complete ring, and then the first two stigmata will be placed, as in the other insects, in a space corresponding to the prothorax. This second auxiliary lip may thus represent the lower lip of the grinding hexapode insects; but here the pharynx is situated in front of this lip, whereas, in the myriapods, it is placed in front of the first auxiliary lip. It is from these relations, and many others furnished by the entomostracea and arachnida, that I consider the feet of the hexapod insects as the analogues of the six jaw-feet of the decapode crustacea.

organs of respiration are composed either wholly or in part of tubular tracheæ.

These animals run very fast, are carnassial, avoid light, and conceal themselves under stones, old beams of wood, bark of trees, in earth, dung, &c. The inhabitants of warm climates dread them greatly, the species which are found there being very large, and their poison probably more active. The *scolopendra morsitans* is designated in the Antilles, by the epithet *maleficent.* Some are known to possess a phosphoric property.

The sexual organs are interior, and situated at the posterior extremity of the body, as in the majority of the following insects. The stigmata are more sensible than in the preceding family, and are either lateral or dorsal.

This family, which in the method of Dr. Leach forms his order *sygnatha*, may, according to these last mentioned characters, the nature of the respiratory organs, and the feet, be thus divided. Some have but fifteen pairs of feet,* and their body seen above, exhibits fewer segments than below—

THE SCUTIGERA (SCUTIGERA, Lam. *Cermatia*, Il.),

Which have the body covered with eight plates in the form of shields, under each of which M. Marcel de Serres, has observed two pneumatic pouches, or vesicular tracheæ, receiving the air, and communicating with the lateral and lower tubular tracheæ. The under part of the body is divided into fifteen semi-rings, having each a pair of feet terminated by a very long tarsus, narrow, and very much articulated. The last pairs are more elongated. The eyes are large, and with facets.

They have slender and tolerably long antennæ. The two

* Dr. Leach counts two pair more, because he comprehends in this number the palpi, and crotchet-formed feet of the head.

palpi are salient, and furnished with small spines. The body is shorter than in the other genera of the same family, with the articulations of the feet proportionally longer.

The scutigera, which, according to these characters, form the passage from the preceding family to this, are very agile and often lose a portion of the feet when taken hold of.

The species of our country conceals itself between the beams or rafters of the scaffold-work of houses.*

THE LITHOBII (LITHOBIUS, Leach),

Which have the stigmata lateral, the body divided, both above and underneath into an equal number of segments, each having a pair of feet. The upper plates are alternately longer and shorter, as far as the posterior extremity.

The forked Lithobius (*Scolopendra forficata*, Lin.), Fab. De G. Geoff. Hist. des insect. II. xxii. 3. Panz. Faun. insect. Germ. L. xiii. Leach, Zool. Misc. cxxxvii.†

The others have at least twenty-one pairs of feet, and the segments are, as well above as below, of equal size, and the same number.

THE SCOLOPENDRÆ proper (SCOLOPENDRA, Linn.),

Are those, which reckoning from the two feet coming immediately after the two crotchets forming the external lip, exhibit only twenty-one pairs, and the antennæ of which have seventeen articulations. They compose the genera *Scolo-*

* The *scoloptendra with twenty-eight feet,* of Geoffroy, which appears to differ from the *S. Coleoptrata* of Panzer, Faun. Insect. Germ. L. xii., and from that of Linnæus; *Iulus Araneoides*, Pall. Spic. Zool. IX. iv. 16.; *Scolopendra Longicornis*, Fab. of Tranquebor.—See also Leach Zool. Misc., *Cermatia Livida*, cxxxvi., and the 14th vol. of the Linnæan Transactions.

† *L. Variegatus Lævilabrum*, Leach, Trans. Linn. Soc. XI.—See also the third vol. of his Zoological Miscellany.

pendra, and *Crytops* of Dr. Leach. The eyes are distinct, eight in number, four on each side, in the first, and that which comprehends the largest species; they amount to nothing, or are scarcely visible in the second.

The most southern departments of France, and of other countries of the South of Europe, present us with one species (*Scolopendra, cingulata*, Latr. *Sc. Morsitans,* Vill. Entom. tom. IV. xi. 17, 18,) almost as large sometimes as the ordinary species of the Antilles, but having the body more flatted.*

The crytops have their antennæ more grained than the scolopendræ, and the two hinder feet more slender. Dr. Leach mentions two species found in the neighbourhood of London.†

In the scolopendræ composing the genus *Geophilus* of the same writer, the number of feet is above forty-two, and often very considerable. The antennæ have but fourteen articulations, and their extremity is less slender. The body is proportionally more narrow and long. The eyes are not very distinct. Some species are electrical.‡

* *Scolopendra Morsitans*, Linn. De G. Insect.VII. xliii. 1.—See, for the other Species, the third volume of the Zoological Miscellany, by Dr. Leach. The *Scolopendra Gigantea* of Linnæus (Brown, Iam. xlii. 4.), and other large species, but incompletely described.

† *Crytops Hortensis*, Zool. Misc. cxxxix. ejusd. ibid.; *Crytops Lavignii.*

‡ *S. Electrica*, Lin. Frisch. Insect. XI. viii. 1.; *S. occidentalis*, Linn. List. ibid, vi. *S. Phosphorea*, Lin. Fallen from the sky on a vessel, at one hundred miles distant from land. See the third volume of the Zool. Miscell Dr. Leach's *Geophilus maritimus*, CXL. 1, 2.—*G. longicornis*. Tab. eadem, 6, and some other species.

SUPPLEMENT

TO THE

ORDER MYRIAPODA.

The characters of this order have been amply detailed in the text, and our supplementary notice of it must be slight indeed. There is little of any interest to be found, at least in a popular point of view, in the insects which compose it.

The myriapods live and grow for a longer time than the other animals of this class, as has been mentioned in the text. They can scarcely be considered as undergoing metamorphosis, unless we regard the development of additional feet and rings, in proportion to age, as such. This is what M. Latreille calls "*ebauchée*," or a mere imperfect sketch of transformation. The myriapods may thus be considered as forming a passage from the insects to the crustacea. In external forms they approximate to the latter, but their internal organization, the only proper basis of methodical divisions, associates them to the former. It is just thus that the trachean arachnida resemble in exterior the pulmonary arachnida, but are nevertheless closer to the insects in all the relations of internal structure.

The myriapods make their habitation in the earth under the different bodies placed upon its surface, under the bark of trees, moss, and between the leaves of some vegetables cultivated in our gardens, and are greatly attached to concealment and obscurity.

Certain very singular fossil animals, the analogues of which have not been discovered, and many of which form the mineralogical composition of the strata in which they have been found, appear to belong to races totally annihilated by the ancient revolutions of the globe: the trilobites seem to fill up the void between the myriapoda and the crustacea.

We shall content ourselves here with a slight notice of the genus IULUS, in the family *chilognatha* of this order. The body is composed of a great number of short rings, of a hard smooth substance, somewhat calcareous. The number of these rings varies according to the species. With the exception of two or three at each extremity, they are equal, and each of them have, underneath, two pairs of feet, contiguous or closely approximating at their origin. The head of the iuli is of the same breadth as the body, flat underneath, convex and rounded above in the hinder part, a little more narrow and almost squared afterwards from the eyes. The anterior edge is emarginated in the middle. The eyes are buried in the surface of the head. They are oval, level, and formed of small grains of an irregularly hexagonal figure. Near their internal side are inserted the two antennæ, which are but little longer than the head, rather thick, and with seven articulations. The mouth is composed of two large mandibles, and a crustaceous piece or sort of under lip, covering transversely the under part of the head. The mandibles have some relations with those of the woodlice, and a structure altogether peculiar, to which nothing analogous is found out of the order myriapoda. They are composed of a scaly stem, at the extremity of which is an articulation equally scaly, and surmounted by a piece in which are implanted transversely some small corneous parts, which are trenchant, and may be considered as so many teeth. The back of each mandible is, moreover, emboxed externally in a scaly capsule, which is large, articulated at its base, angular, and formed, as it were, of two levels, the extremity of each of which is emarginated.

The under lip, which M. Savigny considers as two pair of jaws united, is divided by many sutures or imprinted lines. Below and in the middle is seen a piece, the edges of which are angular, above which two narrow pieces rise parallely contiguous at their internal edge, and the extremity of which is obtusely edged. On each side, in the direction of the preceding, rises a scaly piece of the same figure as the two middle ones, but larger, somewhat widened, and rounded on the external side at the top. Towards its internal angle are two tubercles, which might be taken for two palpi. The general piece is flat, and resembles, being very slender, a membranous leaf.

The first two rings of the body do not form an entire circle. They are open below. Also, the first two pair of feet have a peculiar membranous support which fills the intervals. These first two feet, and even the second, seem to be attached immediately under the mouth, and are thus analogous to the jaw-feet of the crustacea. The first ring is very open, and may be considered as a sort of corslet, being plate-formed, and longer than each of the others. The third, though nearly forming an entire round, is nevertheless open, and has but a single pair of feet, inverted like the preceding. The fourth segment is more closed, but has only a single pair of feet. It is only at the fifth that the duplication of feet commences. Thus, supposing that the first segment, which may be considered the corslet, has no feet, the first pair of these organs of motion will correspond to the second segment, the second to the third, the third to the fourth, fourth to the fifth, fifth to the sixth. The duplication afterwards is continued without interruption in the females, but in the males there is but a single pair at the seventh segment, in consequence of the situation of the sexual organs in that part. These first segments thus appear to represent the trunk of other insects.

The determination of the species of iuli, having been established on the number of feet, it was natural to expect some error in the specific characters, as authors in general believed that each ring had two pair of feet. The two last rings are totally without them. The penult has the middle of its hinder edge advanced into a point, it partly receives the terminal segment, which is composed of two valves rounded at the internal edge, applied against each other, and opening in a passage for the excrements and eggs.

The feet are very small, disposed in two series closely approximating in a horizontal direction at their base, and subsequently forming a crotchet; they are composed of six small articulations, and have a conical and corneous point.

The iuli, notwithstanding the great number of their feet, are by no means agile, on the contrary they walk very slowly and seem to glide along like earth-worms. They cause their feet to act, one after the other, in slow and regular succession. Each range forms a sort of undulation; they move at the same time their antennæ, appearing to employ them to feel the ground; and the body on which they are walking. In a state of repose, these insects have their body rolled up in a circular or spiral figure, the head being in the middle: they might be taken for little serpents.

The iuli are found under stones, bark of trees, moss, &c. They are generally fond of places somewhat humid and obscure. There is a southern species, however, which is rather large, and remains in the open air, being found in great quantities in calcareous soils. Degeer has observed an iulus gnawing the larva of a fly, and partly eating it. It would therefore seem probable that these insects are of a carnivorous nature; nevertheless, the most common opinion is, that they feed on earth. Several species, but very small ones, also gnaw fruits, the leaves of potage plants, and especiallv those which fall to the ground and are decomposing.

The iuli are oviparous. On opening the gravid females, the ovaries are found to be full of white eggs, tolerably bulky. Degeer found on the young, at the moment in which they were disclosed, only six feet, which were attached in pairs to the first three rings. The total number of rings at this time, is but seven or eight; but, in four days time, four other pairs of feet, and some additional rings at the posterior extremity shoot forth. The antennæ, which at first had but four apparent articulations, now have six. The naturalist just mentioned did not observe any vestige of spoils near these insects, but it seems probable, notwitstanding, that they change their skin to acquire the development of these new parts. The exactitude of the researches of Degeer, can hardly permit us to doubt of the truth of this observation. The young iuli may then be considered as undergoing a metamorphosis, since the number of their organs of motion, and of the segments of the body, increases with age, and doubtless by the means of successive moultings.

This genus is very numerous in species, and one belonging to America is very remarkable for its size (*Iulus maximus*).

The family of the *Chilopoda* are distinguished from the preceding, not only by the characters given in the text, but by the swiftness of their movements. They are carnivorous, and conceal themselves under stones, old beams, the bark of trees, in the earth, in dunghills, &c. They feed on living insects, which they prick with the crotchets of their second lip, just as the araneida do with the talons of their mandibles, and distil into the wound a very active poison, active at least in its operation on those little animals, for they perish instantly on its application. According to the ingenious notion of M. Savigny, the mandibles of the araneida represent the second pair of the auxiliary jaws of the crustacea. Now, in the chilopoda, the second lip also corresponds to those same jaw-feet, and answers similar purposes.

The inhabitants of warm climates are much afraid of these insects, the species found there being very large, and their poison, in all probability more active. There is one, even in the southern provinces of France, the bite of which sometimes occasions very serious accidents.

The Scolopendræ, are the only insects of this family worth detaining the attention of the reader in this place. They vary considerably in their dimensions. The largest of those found in Europe, are little more than two inches long. Those of India, are as much as eight. They are sometimes termed millipedes. They will bite, or rather pinch with considerable force. They live in the earth, in old rotten wood, under stones, and in other humid places; they feed on earth-worms and living insects. Some species of them emit a phosphoric light.

These insects for a long time have been reputed venomous, because when they are seized, they separate their crotchets with which they attempt to bite, and in the place where they have bitten, a painful inflammation ensues; but according to the report of travellers, though the pain occasioned by the bite of the large scolopendræ of India, is much more violent than that produced by the sting of the scorpion, yet the consequences are not fatal. Leeuwenhoek, who has examined the crotchets of these insects, has found near their point an aperture, communicating with a cavity, which extends as far as their extremities, this he conceives to be the issue for the acrid fluid which the scolopendræ introduces into the wound, and which causes the severe pain which is felt after the bite. This aperture constitutes a relation between these insects and the araneïda.

The scolopendræ are very lively, and run with great agility. M. Veiss compares the gait of the forked scolopendra, which is one of the genus Lithobium of M. Latreille, to that of an Helix, and supposes that the mechanism of their movements takes place in nearly the same manner. According to him, there is this difference, that the scolopendra, instead of walking, moves in succession a great number of feet. One set of these feet operate in the line of position, while the others are raised up; the latter are quickly put to the ground, while the last ones of each division are raised. All these various movements, which follow the body, from the head to the hinder extremity, produce kinds of undulations. The insect varies its movements and their degree of force according to necessity; each foot resting on the line in which it walks, transports, in the same way as do the muscles of the snail, the body, to the distance in which the limb can act.

The ancients believed that the scolopendræ were reproduced in the same way as the tape-worms. But though this opinion is inadmissible, it is no less certain that the mode of generation in those insects yet remains a mystery. The sexual organs would seem to be situated at the extremity of the body. It is known that those insects moult and quit their skin, in nearly the same manner that the wood-lice do. Several species are found in foreign countries.

THE SECOND ORDER OF INSECTS.

The THYSANOURA

comprehends the apterous insects, borne solely on six feet, without metamorphosis, and having, moreover, either on the sides, or at the extremity of the abdomen, peculiar organs of locomotion.

The first family of the *Thysanoura*, that of

The Lepismenæ, *Lat.*

has the antennæ in the form of bristles, and divided from their origin into a great number of small articulations; palpi very distinct, and projecting at the mouth; the abdomen provided on each side, underneath, with a range of mobile appendages, in the form of false feet, and terminated by articulated bristles, three of which are most remarkable, and the body is always furnished with shining scales.

It comprehends but a single genus of Linnæus—

Lepisma.

Their body is elongated, and covered with small scales, often silvery and brilliant, which circumstance has caused the most common species to be compared to a little fish. The antennæ are in the form of bristles, and usually very long. The mouth is composed of a lip (*labrum*), of two mandibles, almost membranous, of two jaws with two divisions, with a palpus of from five to six articulations, and a lip with four sections, and bearing two palpi with four articulations. The thorax is of three pieces; the abdomen, which is contracted by little and little towards its posterior extremity, has, along each side of the belly, a range of small appendages supported on a short articulation, and terminating in silken points. The last are the longest. From the anus issues a sort of scaly stylet, compressed, and consisting of two pieces. Then come the three articulated bristles, which are prolonged beyond the body. The feet are short, and often have very large haunches, strongly compressed, and in the form of scales.

Many species conceal themselves in the clefts of window-frames which are seldom opened, or always closed; under planks which are somewhat moist; in presses, &c. Others live retired under stones.

These insects run very fast: some of them leap by means of the threads of their tail.

Two subgenera have been formed of them—

MACHILIS, *Lat.*, *Petrobius*, Leach,

whose eyes are very much composed, almost contiguous, and occupy the major portion of the head. They have the body convex, and arched above, and the abdomen terminated by small threads, proper for leaping, and of which the middle one, placed above the other two, is by far the longest.

The maxillary palpi are very large, and in the form of small feet; the thorax is contracted centrally, and its first segment is smaller than the second, and vaulted.

These insects leap remarkably well, and frequent stony

and covered places. All the known species belong to Europe.*

LEPISMA, *Lin.*, *Forbicina*, Geoff. Leach,

which have the eyes very small, very much separated, and composed of but a small number of grains; the body flatted, and terminated by three threads of the same length, inserted on the same line, but employed in leaping.

Their haunches are very large. Most part of the species are found in the interior of houses.

The *Sugar Lepisme* (*L. Saccharina*, Lin.), *Forbicini, Plate*, Geoff. Insect. II. xx. 3. Schœff. Elem. Entom. lxxv., four lines long, of a silvery and somewhat leaden colour, without spots; is, as is said, originally of America, but has become common in our houses.

With it, and in the same places, is often found the *Banded Lepisme* (*Vittata*, Fabr.), which has the body ashen, punctuated with four stripes of the latter colour along the back of the abdomen. Other species are to be found under stones.

The second family of the THYSANOURA is that of the

PODURELLÆ, Lat.,

whose antennæ consist of four pieces, and the mouth presents no distinct and prominent palpi. The abdomen is terminated by a forked tail, which, in a state of repose, is applied under the belly. It is employed in leaping. In Linnæus this family constitutes but a single genus—

PODURA.

These insects are very small, very soft, and elongated, with the head oval, and the two eyes composed each of eight

* *Lepisma polypoda*, Lin.; *L Saccharina*, Vill. Entom. Lin. IV. xi. 1.; Roem. Gen. Insect. xxix. 1.; *Forbicine Cylindrique*, Geoff.; *Lepisma Thezeana*, Fab.; *Petrobius Maritimus*, Leach. Zoolog. Miscellan. CXLV.

small grains. Their feet have only four distinct articulations. The tail is soft, flexible, and composed of a lower piece, mobile at the base, at the extremity of which two stems are articulated, susceptible of approximation, separation, or crossings, and which constitute the teeth of the fork. These insects can erect their tail, push it with force against the plane of their position, as if they let go a spring, and thus elevate themselves into the air, and leap in the same manner that fleas do, but to a less height. They usually fall down on their backs, the tail being extended behind. The middle of their belly presents a raised portion, which is oval, and divided by a fissure.

Some remain on trees, on plants, under the bark, or under stones; others on the surface of dormant waters, and sometimes even on the snow during the thaw. Many unite in numerous societies on the ground, or on sandy paths, and resemble at a distance a small heap of gunpowder. The multiplication of some species appears to take place in winter.

The PODURA proper (PODURA, Lat.),

have the antennæ of the same thickness, and without rings or small articulations at the final piece. Their body is almost linear or cylindrical, with the trunk distinctly articulated, and the abdomen narrow and oblong.*

SMYNTHURUS, Latreille,

have the antennæ more slender towards their extremity, and terminated by an annulated piece, or composed of small arti-

* *Podura Arborea,* Lin.; De Geer. Insect. VII ii. 1—7.; *P. Alvalis,* Lin. De G. *ibid.* 8—10.; *P. Aquatica,* Lin. De G. *ibid.* ii. 17.; *P. Plumbea,* Lin. De Geer, *ibid.* iii. 1—4.; *P. Ambulans,* Lin De G. *ibid.* 5, 6; *P. Aquatica Grisea,* De G. *ibid.* ii. 18, 21.

The Poduræ *vaga, villosa, cincta, annulata, pusilla, lignorum, fimetaria* of Fabricius.

culations. The trunk and abdomen are united in a globular or oval mass.*

N.B. There is so little which can be added on this order, taken separately, that we shall proceed with the translation of the text, and reserve our supplementary observations to the end of the two following orders.

* *Podura atra,* Lin. De Geer. *ibid.* iii 7—14.; *poduræ viridis, polypoda, minuta, signata* of Fabricius.

THE THIRD ORDER OF INSECTS.

The Parasita, Lat., Anoplura, Leach,

Thus named from their habits, (see lower down,) have but six feet, and are apterous, like the thysanoura; but their abdomen has no articulated and mobile appendages. For organs of sight they have but four or two small simple eyes. Their mouth is in a great measure interior, and presents externally either a muzzle or mammella, enclosing a retractile sucker, or two membranous and approximating lips, with two mandibles crotchet-formed. They form, in Linnæus, only the genus of

The Lice, (Pediculus,) Lin.

Their body is flatted, almost transparent, divided into twelve or eleven distinct segments, of which three are for the trunk, each having a pair of feet. The first of these segments often forms a sort of corslet. The stigmata are very distinct. The antennæ are short, of the same thickness, composed of five articulations, and often inserted in an emargination. Each side of the head presents one or two simple eyes. The feet are short, and terminated by a very strong claw, or by two crochets directed one towards the other. These animals thus crook themselves easily, either to the hairs of quadrupeds, or the feathers of birds, the blood of which they suck, and on the body of which they pass their lives and multiply. They attach their eggs to these cutaneous appendages. Their generations are numerous and succeed each other with great rapidity. Some particular causes, which are unknown to us, favour them in an extraordinary manner, and this is exemplified in regard to *man*, in the pedicular malady, or *phtiriasis*, and even in our infancy. These insects constantly live on the same quadrupeds, and the same birds, or at all events, on animals of those classes, which have analogous characters and habits. One bird frequently breeds two sorts. Their motion, in general, is rather slow.

Some (*Pediculea*, Leach,) such as,

The Lice proper, Pediculus, Deg.

Have for mouth a very small nipple, tubular, situated at the anterior extremity of the head, in the form of a muzzle, and inclosing in the state of inaction, a sucker. Their tarsi are composed of an articulation, whose thickness almost equals that of the leg, terminated by a very strong claw, folding back on a tooth-formed prominence of the leg, and performing with this point the office of a pincers. Those which I have observed, have but two smooth eyes, one on each side.

There are three sorts of them peculiar to man. Their eggs are known under the name of *nits*.

In the two following species, the thorax is very distinct from the abdomen, in its breadth and medial length. They form the genus *pediculus*, properly so called of Dr. Leach.

The human body-louse, (*P. humanus corporis*, De G. Insect, VII. 1. 7). Of a dirty white without spots, with the sections of the abdomen less projecting than in the following. It comes solely on the body of man, and breeds to a terrific extent in the pedicular disease.

The human head-louse (*P. humanus capitis*, De G. Insect, VII. 1. 6). Ash-coloured, with spaces wherein are situated brown or blackish stigmata; lobes or sections of the abdomen rounded; on the head of man and particularly of children.

The males of this species and the preceding have, at the posterior extremity of their abdomen, a small scaly and conical piece, formed like a sting, probably the sexual organ.

The Hottentots, the negroes, and different Simiæ, eat lice,

or are *phtirophagi*. Oviedo pretends to have observed that these vermin, at the elevation of the tropics, abandon the Spanish sailors that are going to the Indies, and attack them again at the same point, on their return to Europe. It is also reported, that in India, however dirty one may be, these animals are never to be found except in the head.

There was a time when physicians employed the human louse in suppressions of urine, introducing it into the canal of the urethra.

Dr. Leach forms a proper genus, *phtirus*, of the *louse of the pubis* (*P. pubis*, Lin.), Redi. Expt. xix. 1.; which has the body rounded and broad, the thorax very short, being almost confounded with the abdomen, and the four hinder feet very strong. Its vulgar designation is *crab-louse* (*Morpion*, Fr.). It attaches itself to the hairs of certain parts, and to those of the eyebrows. Its bite is extremely strong.

Consult, for the species which live on man, the excellent treatise on cutaneous diseases by Dr. Alberti, physician to the king.

Redi has figured, but rudely, many other species which are found on various quadrupeds. That which lives on the pig, has the thorax very narrow, and abdomen very broad. It is the type of the genus Hæmatopinus of Dr. Leach.* The *louse of the buffalo*, figured by De Geer, presents more important characters.

The others, (*Nirmidea*, Leach,) such as the

RICINUS, De Geer—*Nirmus*, Herm. Leach,

Have the lower mouth, and composed at the exterior of two lips, and two crotchet-formed mandibles. Their tarsi are very distinct, articulated, and terminated by two equal crotchets.

* Zool. Misc. CXLVI. *P. Suis*, Panz. Faun. Insect. Germ. LI. xvi. The louse of the stag. Panz. *ibid.* xv. belongs to the genus *Melophagus*, of the order of diptera.

With the exception of a single species, that of the dog, all the others are exclusively found on birds. Their head is generally large, sometimes triangular, sometimes crescented, and often has angular projections. It differs sometimes in the two sexes, as well as the antennæ. I have observed in many, two simple eyes, approximating on each side of the head. According to the observations communicated to me by M. Savigny, these insects have jaws, with a very small palpus on each of them, and are concealed by the under lip which has also two organs of the same sort. They have, besides, a species of tongue.

M. Leclerc de Laval, has told me, that he has seen in their stomach, particles of the feathers of birds, and believes that this constitutes their only nutriment; Degeer, however, assures us, that he found the stomach of the *ricinus fringillæ* filled with blood, with which the insect had just gorged itself. It is also known that these insects cannot continue to exist any length of time upon dead birds; they are then seen to crawl uneasily over the feathers, particularly over those of the head and the neighbourhood of the bill.

Redi has also represented a great number of their species.

Some have the mouth situated near the anterior extremity of the head. The antennæ are inserted at the side, far from the eyes, and are very small.*

In the others the mouth is almost central. The antennæ are situated very near the eyes, and their length equals almost the half of that of the head.†

* *Pediculus stermæ hirundinis.* Linn. D. G. Insect, VII. iv. 12;—*pediculus corvi coracis.* Linn. D. G. *ibid.* 11;—*Ricinus fringillæ*, (D. G.) *ibid.* 5, 6, 7;—*pediculus tinnunculi.* Panz. *ibid.* XVII.

† *Ricinus gallinæ*, D. G. *ibid.* 15; on the hen partridge and pheasant;—*R. emberizæ.* D. G. *ibid.* 9; *R. Mergi*, D. G. *ibid.* 13, 14;—R. Canis, D. G. *ibid.* 16;—*Pediculus pavonis*, Panz. *ibid.*, xix. Lat. Hist. Nat. des Fourm. 389, xii. 5. See again, Panzer pl. xx., xxiv. His *pediculus ardeæ*, xviii., appears to be the same as the *ricinus of the diver* of Degeer, iv. 13.

A celebrated German naturalist, Dr. Nitzsch, professor at Halle, has made a very profound study of the organization, both internal and external of these animals, as is proved by his memoir on the epizoic insects, inserted in the entomological magazine of M. Geunar. The genus *pediculus* properly so called, or that whose species are provided with a sucker, is ranged by him with the epizoic hemiptera. The ricini of Degeer, and others, or the nirmi of Hermann, that is, the species provided with mandibles and jaws, are referred to the order orthoptera, and designated, collectively, by the name of *mallophagi.* Two genera of this division approach the preceding, inasmuch as these animals also live on the mammifera: such are the *trichodectes* and *gyropus.* In the first, the maxillary palpi are wanting, or indistinct, and the antennæ are filiform and of three articulations. Their species are found on the dog, the badger, the weasel, the ferret, &c. In the second, the maxillary palpi are apparent, the antennæ are thicker towards the end, and have four articulations; its mandibles have no teeth, and the labial palpi are nullified, and the four posterior tarsi have but a single crotchet at the end. These latter characters distinguish it from another genus, having maxillary palpi visible, antennæ with four articulations, and thicker at the end, and the anterior mouth, as of the *liotheum.* Here the mandibles are bidenticulated, the labial palpi are distinct, and all the tarsi are terminated by two crotchets; the species are found on divers birds, whereas the gyropi live on the quadrupeds vulgarly called *guinea-pigs.* A fourth and last genus, that is that of *philopterus*, whose species are exclusively proper to birds. The antennæ have five articulations, the third of which, in the males, often presents a branch forming a pincers with the first; these organs are filiform. The maxillary palpi are invisible. The tarsi have two crotchets at their extremity, but not divergent like those of the liothea. Here, moreover, the males have six testicles, three on each side, and their four biliary vessels are thickened towards the middle of their length. Those of the trichodictes and philopteri do not exhibit this sort of swelling, and their testicles are only four in number, two on each side. In these two genera, besides, there are ten ovaries, five on each side. In the female liothea, as far as the above-mentioned philosopher was able to observe, there are but six, three on each side. There is no positive knowledge concerning the number of those of the female gyropi, nor of that of the testicles in the males. In all those genera the thorax is bi-parted, that is to say, the prothorax and mesothorax compose the apparent trunk; and its third division, or the metathorax is united and confounded with the abdomen. Mr. Kirby was the first, as I believe, who thus designated this segment; but M. Nitzsch appears to me to have also employed the first of the other two denominations. See the generalitics of the class of insects. The limits of this work interdict us from the exposition of the sub-genera which he has established. We shall only remark, that that which he has named *goniodes*, the fourth of the genus philopterus, is solely peculiar to the gallinacea. In the collection of memoirs which terminates our history of the ants, we have described in detail a species of ricinus—philoptère.

M. Leon Dufour has formed, of the *pediculus melittæ* of Mr. Kirby, already well observed by Degeer, who took it for the larvæ of the melöe proscarabeus, as well as by this celebrated English entomologist, a new genus, *(Triongulin des andrenettes,)* of which he has published and represented the characters in the thirteenth volume (19, B.) of the Annals of Natural Science. If this insect was not the larva of this meloë, as Mr. Kirby thinks, there is no doubt that it should form in the order of the parasites a proper sub-genus; but according to the researches of M. M. Lepeletur and Servile, the opinion of Degeer is confirmed.

THE FOURTH ORDER OF INSECTS.

THE SUCTORIA De G.—SIPHONAPTERA, Leach,

Which composes the last of the apterous insects, have for mouth, a sucker of three pieces,* enclosed between two articulated laminæ, forming, when united, a proboscis or bill, either cylindrical or conical, and the basis of which is covered by two scales. These characters exclusively distinguish this order from all others, and even from the hemiptera, to which it approximates the most under these relations, and in which Fabricius has placed the present insects. The suctoria, moreover, undergo true metamorphoses, analogous to those of many insects with two wings, as for instance, the *tipularia*.

This order is composed but of a single genus, that of

THE FLEAS (PULEX L.),

Their body is oval, compressed, clothed with a skin of sufficient firmness, and divided into twelve segments, four of

* Rœsel represents but two; but M. M. Kirby and Straus have observed an additional one. According to the latter, the two scales covering the basis of the bill are palpi.

which comprehend the trunk, which is short, and the others the abdomen. The head is small, very much compressed, rounded above, truncated, and ciliated in front. It has, on each side, a small and rounded eye, behind which is a fosset, in which a small mobile body is discovered, furnished with small spines. At the anterior edge, near the origin of the bill, the pieces are inverted, which are taken for antennæ, which are scarcely of the length of the head, and composed of four articulations almost cylindrical. The sheath or bill is divided into three articulations. The abdomen is very large, and each of its rings is divided into two, or formed of two laminæ, one superior and the other inferior. The feet are strong, particularly the hinder, proper for leaping, spiny, with large haunches and thighs, and the tarsi composed of five articulations, the last of which is terminated by two elongated crotchets. The two anterior feet are almost inserted under the head, and the bill is placed between them.

The male, in copulation, is placed under the female, so that their heads face each other. The female lays a dozen of eggs, white, and a little viscous. From them come forth little larvæ without feet, very much elongated, like little worms, exceedingly lively, rolling themselves in a circle or spiral, and serpentine in their gait. They are at first white, and afterwards reddish. Their body is composed of a scaly head, without eyes, having two very small antennæ, and of thirteen segments, having little tufts of hairs, with two sorts of crotchets at the end of the last. Their mouth presents some small moveable pieces, of which these larvæ make use in pushing themselves forward. After having remained a dozen days under this form, the larvæ enclose themselves in a small silken shell, where they become nymphs, and from which they issue in the perfect state, at the end of a space of the same duration.

Every one knows the *common flea* (*pulex irritans*), Ræs· Ins. II. ii. iv., which lives upon the blood of man, dog and cat. Its larva inhabits in ordure, under the nails of filthy men, in the nests of birds, especially of pigeons, attaching itself to the neck of their young ones, and sucking them to such a degree as to become altogether red.

The *penetrating flea (pul. penetrans*), Catesby Carol, III. x. 3. and of which M. Dumeril has given an excellent figure in his work entitled " General considerations on the Class of Insects," and in the Dictionary of Natural Sciences, may probably form a peculiar genus. Its bill is of the length of the body. It is known in America, under the name of *chigoe.* It introduces itself under the nails of the feet, and under the skin of the heel, and there very speedily acquires the volume of a small pea, by the prompt increase of the eggs, which it carries in a membranous sac under the belly.

The numerous family to which it gives birth, occasion, by their remaining in the wound, a malignant ulcer, difficult to be destroyed and sometimes mortal. People are but little exposed to this disagreeable inconvenience, if they are careful to wash frequently, and especially if they rub the feet with the leaves of tobacco bruised with *roucou*, and other acrid and bitter plants. The negroes are very dexterous in extracting the animal from the part in which it has established itself.

Divers quadrupeds and birds nourish fleas, which seem to differ specifically from the preceding.

SUPPLEMENT TO THE ORDERS,

THYSANOURA, PARASITA AND SUCTORIA.

NOTWITHSTANDING that we have thus grouped these orders together, in consequence of the scantiness of the materials which they present for our supplementary observations, we shall not find a great deal to say in addition to the excellent descriptions in the text.

As to the first, we must pass it over altogether, for we could say nothing concerning it, except in relation to the quick movements, and the localities of the animals which compose it, which have been already mentioned in the text. We shall pass on, therefore, at once, to the consideration of those parasitic animals with which, unfortunately, our own species is but too closely connected.

All LICE live on blood, some on that of man, others on that of quadrupeds. They suck it with their proboscis, which is hardly ever perceived, unless it be in action. There is no quadruped which has not its particular louse, and some nourish several. Man, as we have already seen, is attacked by three species.

Swammerdam, who has given us the anatomy of the human louse, was unable to discover any male among those which he examined. He always found in them an ovary; which occasioned him to suspect that they were hermaphrodites. But the observations of Leeuwenhock, differ much from those of that author. He has observed individuals among these insects provided with all the parts which characterize the male sex, and he has given the figures of those parts. The same author has also discovered in those which he regards as males, a recurved sort of sting, situated under the abdomen, with which, according to him, they can prickle. He believes that the great itching which they occasion proceeds from the pricking of this sting, having remarked that the introduction of their proboscis into the flesh produces scarcely any sensation, unless perchance that it touch on any of the nerves. Degeer tells us, that he has seen a similar sting situated at the end of the abdomen in several human lice as well those of the body, as of the head. The individuals, which according to the opinion of Leeuwenhoëk are the males, have, according to Degeer, the end of the abdomen rounded, whereas the females, or those which have no sting, have it emarginated. M. Latreille has also observed, very distinctly in a great number of individuals, the sting, or at least the conical and scaly point of which the aforesaid authors have made mention.

The multiplication of these insects is unfortunately far too great, but they are not quite so prolific as some have represented. Experiments have proved that in six days a louse can lay fifty eggs, and that some will still remain in the belly. The young ones soon issue from the eggs, (at the end of six days,) and change their skin several times, after which, they are in a state to reproduce. This occurs about eighteen days after, when they begin to lay in their turn. According to these observations, and the calculations to which they have given rise, two female lice may have eighteen thousand descendants in the course of two months.

Linnæus considered the louse, which remains constantly on the human head, as a variety of the common louse. It is, however, from its characters, as described in the text, and also from its undeviating locality, proved to be a distinct species. This same great naturalist also informs us, that the largest lice he ever met with, are to be found in the warm caverns of Fahlun, in Sweden.

Children, and persons who totally neglect their hair, and are otherwise of filthy habits, are generally infested by the head-lice. When powder was fashionable, and preposterously esteemed an embellishment to the hair, it frequently proved a productive cause of these vermin. Those who left too long upon their head the dandriff produced by the use of powder, or who employed adulterated powder for the ornament of their hair, were very liable to the intrusion of these amiable guests. The other species, which is found on all parts of the body, excepting those of generation, comes generally in consequence of the neglect of personal cleanliness, and the not changing the linen and internal garments sufficiently often. It is from this same species that those persons suffer who are afflicted with that most dreadful and disgusting malady *phthiriasis*. It may be observed that their appearance, though in fewer numbers, is sometimes indicative of the approach, or even of the crisis of other diseases, and that where there is no neglect of personal cleanliness.

As for the third species, we believe, that its appearance is invariably the result of indiscriminate debauchery, accompanied of course with personal neglect.

Dirt attracts these insects, and prepares for them a nidus favourable to the reproduction of their posterity. This is the vice which must be first eradicated if we are desirous of being protected from their inroads, or if already attacked by them, of rendering more efficacious any other means employed for their destruction.

The remedies employed against these vermin act in two ways. Some of them, such as oily and fatty substances, or such as contain azotic gas, close up the stigmata of these insects, or the apertures destined for the admission of air, and smother them. Others, such as the seeds of *staphis agria*, of larks-spur, tobacco, &c. reduced to a powder, produce the effect of a violent poison, and exercise their influence on the general organization of the animals. Mercurial preparations are, of all others, the most certain and speedy agents of their destruction. They may also be employed for the extinction of the species which infest our domestic animals.

It is said that lice, by piercing the skin, often produce pustules which may be converted into itch or tinea. Their multiplication, in certain subjects, is carried to such an extent as to produce the mortal malady which has been already noticed, namely, *morbus pedicularis* or *phthiriasis*. History has afforded us many examples of this; Pheretima, mentioned by Herodotus; Sylla, Antiochus Epiphanes, the two Herods, Maximin, and Philip the Second, perished of this disease, or something very analogous to it. Mr. Kirby is prone to think that it has fallen particularly as a judgment from God on the oppressors of mankind and the persecutors of religion. But this and all such notions ought to be admitted with caution, or rather rejected altogether. Until it is shewn that the disease in question has fallen upon such persons alone, and on no others, we must discard the idea of its being an instrument of Divine vengeance. But this is so far from being the case, that the reverend entomologist himself informs us that " this most loathsome of all maladies, or one equally disgusting, has been the inheritance of the rich, the wise, the noble, and the mighty; and in the list of those who have fallen victims to it, you will find poets, philosophers, princes, kings, and emperors." It would be somewhat hard to believe that among all these there was not one less bloody than the first, or less profligate than the second Herod, less oppressive than Maximin, or less bigotted than Philip. The fact is, diseases of all kinds, like the rain of heaven, fall upon the just and the unjust. They follow to be sure, as a pretty certain consequence, the indulgence in some vices. Intemperance of all kinds will entail them on the practiser as a necessary, but a *natural* punishment. But the man who unites a sound constitution to a prudent care of his health, may be at once supremely wicked and remarkably healthy. He may cheat and rob, and oppress and murder, but while he avoids the vices which are ruinous to health, he is more likely to find the reward of his villanies here from the indignation of mankind than from any heaven-descended visitation of disease.

Mr. Kirby seems doubtful whether all the cases recorded as of *morbus pedicularis* are referrable to the same specific cause. He believes there is sufficient reason for thinking that three different kinds of insects are concerned in the production of diseases which have all been confounded under the name of *phthiriasis*. Besides those produced by *pediculi*, this eminent naturalist refers many cases to the agency of *acari*, or mites, and *larvæ* in general.

It has never been proved, in any well authenticated way, that the species of the genus pediculus burrow beneath the skin, or are *subcutaneous*. This remark is equally applicable to man, and the lower animals, as far as we know anything of the habits of the genus in question. For this we have the highest medical authority, as far as man is concerned. Dr. Mead informs us that "the louse feeds on the surface of the skin," and Dr. Willan, in a work on cutaneous diseases, states, respecting the body-louse, that "the nits or eggs, are deposited on the small hairs of the skin, and the animals are found on the skin, or on the linen, and not under the cuticle, as some authors have represented." The same writer informs us, "that many marvellous stories are related by Forestus, Schenkius and others, respecting lice bred under the skin, and discharged in swarms from abscesses, strumous ulcers, and vesications. The mode in which pediculi are generated being now so well ascertained, no credit can be given to these accounts." He thinks, however, that those authors have mistaken some other insects for lice, as some such animals may sometimes be found in putrid ulcers.

It appears, therefore, that cases of disease from animals residing under the cuticle, cannot be referred to pediculi. The poet Alcman, and Pherecydes Syrius, the philosopher, are mentioned by Aristotle, as having died of some complaint of this kind. But it could not have been a true phthiriasis, as that great philosopher, who attributed it to lice, imagined; for he tells us, that "they are produced in the flesh, in small pustules, like tumours, which have no pus, and from which, when punctured, they issue." Dr. Heberden, in his Commentaries, informs us of a similar case, which he terms *morbus pedicularis*, but which could not be so, for the same reason. He represents the insect as inhabiting tumours, from which, when opened, they issue. He also tells us, that in all respects they resemble the common louse, but in being whiter. But an observer not accurately skilled in entomology, might, as Mr. Kirby remarks, easily mistake an *acarus* for a *pediculus*.

Dr. Willan has cited two other cases, which he seems to think may with propriety be referred to true *phthiriasis*. In one of these cases, it is stated that the pediculi so abounded, that two black servants had no other employment than that of carrying baskets full of these insects, and throwing them into the sea. This, as Mr. Kirby observes, appears to be somewhat exaggerated and hyperbolical. We shall conclude this part of the subject, by observing, that phthiriasis must vary in its types, according to the species of pediculus by which it is produced.

The facts mentioned in the text, that lice abandon the Spanish sailors, in a certain degree of latitude when going to the Indies, and revisit them again on their return, and that body lice are unknown in India, are observations that have need of being corroborated by more certain testimonies than we are yet in possession of. But, if true, there would be nothing in the fact very surprising. A degree of considerable heat, and a more abundant transpiration, might prove unfavourable to the propagation of the *pediculi corporis*. As their skin is more tender, the influence of the air might prove detrimental to them in those burning climates.

The disgusting fact, mentioned in the text, of these vermin being eaten, is not confined to the Hottentots, the Negroes of Western Africa, and the Simiæ. It has been observed to prevail among some of the American tribes, and is not uncommon even in Europe, as amongst the beggars in Spain and Portugal. Some authors have pointed out the courses which should be adopted to protect or free the person from these disgusting insects. The best of all, in ordinary cases, is cleanliness. Of the medical uses to which these animals have been applied, it is unnecessary to dilate here. No one, we fancy, of the least degree of intelligence, gives any credit to such remedies at present. It was imagined that their introduction into the urethra of new-born infants, troubled with suppression of urine, might, by the titillation which they caused in that canal, force the sphincter to relax, and give passage to the urine. Farriers used to employ the same remedy with horses in similar cases.

The *pediculi* proper, are confined to men and quadrupeds. The *ricini*, or bird-lice, to the feathered race.

It appears that the ancients designated under the name of *ricinus*, those acarides, or acari, vulgarly called ticks, which attach themselves to the skin of dogs, oxen, &c. Degeer might, therefore, have better employed another denomination for the present genus, and which preceding naturalists confounded with that of lice. Accordingly, Dr. Leach has adopted, from Herman, the denomination of *nirmus*.

It was by no means surprising, that the earlier naturalists did not distinguish these animals from the lice; their external physiognomy is almost the same, but their organization, as appears from the text, is different in many essential points. We shall not repeat that description here, but merely remark that it is evident from it, that these two genera approximate in a natural series. From the consideration of their resemblances and differences, and from some other facts, furnished by the trachean arachnida, and the branchiopoda, we may see that nature, in preserving always the same type of general forms, is pleased to modify, and sometimes rather abruptly, that of the manducatory apparatus, or that she easily converts organs adapted for grinding, into sucking organs, and *vice versâ*. Her views in this respect are subordinate to the model on which the body of the animal is at first formed. This consideration caused M. Latreille to reject, as a primary character, the division of insects into two lines—one composed of the grinding, and the other of the sucking insects.

The ricini live exclusively on animals of the class Aves. Degeer, it is true, makes mention of a ricinus found on the body of a dog; but this species M. Latreille refers to pediculus, and could discover no mandibles on it.

These insects remain by preference under the wings, about the axilla, and on the head of birds; they attach themselves there very strongly, by means of the two robust and equal crotchets which terminate their tarsi. They multiply there sometimes, in such quantities, that the birds grow considerably thinner, and may even die in consequence. Care should be taken to examine domestic birds, supposed to be infested by these vermin, and often to clean out the places in which they are kept, and where they are accustomed to rest. By such means they may also be protected from a species of mite, which multiplies prodigiously in such places, and by which these domestic animals are seriously incommoded.

The genus of the ricini is very numerous; there is no bird without one or two species. Redi has figured a great number of them, and though his figures are rude, one may easily see how many varieties the forms of those insects present. Their characters, manners, &c., are, with the exceptions already stated, the same as those of the lice. There is one singularity in the *ricinus pavonis*, and that is, that the antennæ are forked.

We now arrive at that well known genus, which constitutes the order SUCTORIA, the PULEX, or fleas. In dividing, as M. de Lamarck has done, the insects which undergo metamorphosis, into two grand sections, those which have mandibles and jaws, and those in which these organs are transformed into suckers, the order Suctoria seems to be intermediate between the Hemiptera and the Diptera.

If a certain number of female fleas be shut up in a vessel, at the time in which they begin to appear, some one among them will immediately commence to lay. They lay almost a dozen of eggs; these eggs are tolerably bulky, ellipsoïd, white, and somewhat clammy. Rœsel pretends that the mother lets them drop at hazard: but it is probable that she attaches or glues them to different bodies. When the season is favourable, these eggs disclose the young at the end of five or six days. On issuing from its envelope, the larvæ is white and transparent, but when a little more aged it becomes reddish—a fact which might surprise us, did we not know that nature employs so much sagacity and address in her means of preserving the posterity of insects : thus the difficulty we experience in finding this larva in our apartments. No question but that they are very numerous in such places, and yet, if we examine the sweepings of our bed-rooms, with ever so much attention, we rarely succeed in finding a single one. It seems probable that they remain concealed in the folds of such pieces as compose our beds, or in some other place which completely shelters them from our pursuit. It is more easy to find them in the nests of birds, and especially of pigeons.

These larvæ are without feet, though some authors have thought proper to advance the contrary. They look like little worms, and are exceedingly lively after changing into nymphs. Rœsel thinks that the individuals which are more clear in colour are the males, but M. Latreille is of a different opinion, as this difference of tint is not observable in the perfect state, at least as an indication of sex. Eleven or twelve days after this larva has been enveloped in its tomb, the nymph is stripped of a pellicle which enveloped its limbs, becomes the perfect insect. and appears under the form which it is destined always to preserve. It begins to leap at the first moment of its new existence. Those larvæ which are born at the close of summer, pass the winter under the same form.

All the world knows, the fleas are parasitical insects; they prefer the delicate skin of women and children to that of other persons. They nestle in the fur of hares, dogs, and cats, which are very much tormented by them, especially in summer and autumn. Many birds are annoyed by them, particularly pigeons, hens, and swallows.

According to the testimony of Ovington, the Hindoos, conformably with their belief in the metempsychosis, lavish upon these animals, as well as on all species of vermin which suck the human blood, the most extravagant cares. An hospital has been established for them near Surat. Their sustenance is purchased at the expence of some fool, who is delivered during the night to the voracity of these animals.

The fleas have afforded materials to the ingenuity of man, and have been made to produce surprising effects of skill. A flea of middling size has been seen to draw a silver cannon supported on two little wheels, weighing eighty times as much as itself, which was charged with powder and let off, without the insect discovering the slightest symptom of alarm. Mouffet relates, that another flea drew with facility a gold chain of the length of a finger, with a padlock shutting with a key, and which with the animal scarcely weighed a grain An English artist, according to Hook, constructed an ivory coach for six horses, holding four persons, having two lackeys behind, and a coachman on the box, with a dog between his legs, which was drawn by a single flea! "What fineness and delicacy of labour!" cries M. Latreille, "but why not devote it to objects of greater utility?" This observation, though under some circumstances just, we cannot give an unqualified assent to. If every work of man was to be measured by its *direct* utility, some of the noblest productions of art and genius might be undervalued and despised. The exercise of human ingenuity is in itself laudable, and though it be employed on an object of no direct utility, it may, and not unfrequently does, lead to the most useful discoveries and inventions. The man employed in such a task as we have described, was at all events improving himself in his art, and increasing the delicacy of his tact, and the accuracy of his vision. Better to be so employed than in fabricating engines of destruction, or mingling in scenes of dissipation, vice, and folly.

In studying so small an animal as the flea, many subjects of admiration present themselves to our mind. What prodigious force of muscle must not that be which enables this insect to raise itself thirty times its height! How singular the structure of that tube with which it sucks our blood! Nature, with her usual wisdom and foresight, has given this animal a compressed form, which enables it to insinuate itself with more facility between the hairs of animals, and conceal itself there. She has encased its body in a sort of armour, by enveloping it in a firm and elastic skin, capable of resisting the pressure of our fingers.

It is not necessary to enter into any detail here, of all the means which have been prescribed for the destruction of these troublesome insects. Some recommend the placing in apartments plants of a powerful and penetrating odour, such as savory, wormwood, &c., or acrid plants, such as *persicaria*, or vegetables with glutinous leaves, and branches of the alder. Others have recourse to a mercurial unguent, to boiling water, into which simply mercury has been put, and which is scattered throughout the chamber. Some prescribe the vapour of sulphur. The inhabitants of Dalecarlia, place in their habitations a hare's skin. These insects take refuge there, and are then easily destroyed by fire or water.

We often murmur against Nature, and consider fleas and other vermin as a spot which soils the beautiful picture which she presents to our eyes. But let us be reasonable, and admire the wisdom of her designs, in having chosen the sensation of pain as a sentinel to give us warning of the consequences of our vices, or the irregularity of our habits. We should conform ourselves to her views. Cleanliness without fastidiousness should be observed in our dwellings.

If, towards the end of autumn, and the commencement of spring, the different articles of furniture that we use, were exposed to a heat of sufficient strength, the sources of our inconveniences would be speedily destroyed, and at all events we should cease to calumniate Nature, even if we had not sufficient gratitude to study and admire her. But a small number of the species of the flea is as yet known; but it is probable that if the fleas of different animals were examined with a little more attention, that several others might be discovered.

THE FIFTH ORDER OF INSECTS.

The Coleoptera, *Eleutherata*, Fab.,

Have four wings, the upper two crustaceous, in the form of scales, horizontal, and uniting together, at the internal edge, in a right line; mandibles and jaws; the lower wings merely folded crosswise, and covered by two others which form a sort of cases, and which are designated either under that name or that of elytra.*

They are, of all insects, the most numerous and the best known. The singular forms, the brilliant and agreeable colours by which some of their species are characterised, the volume of their body, the more solid consistence of their integuments, which renders their preservation more easy, and the numerous advantages derived from the study of the variety of forms in their external organs, have attracted the particular attention of naturalists to this order. Their head presents two antennæ of different forms, and of which the number is almost always eleven. They have two eyes with

* See, for the anatomical characters of the insects of this order, the Annals of Natural Science, vol viii. p. 56, where M. Dufour presents a summary of them.

facets, but no simple eyes. It is true that in some brachelytra, two small yellowish points have been observed, which have been taken for simple eyes, but, in my opinion, without sufficient examination; for the forficulæ, a genus of the orthoptera, the nearest to the coleoptera, exhibit none. The mouth is composed of a labrum, of two mandibles, most frequently of a scaly consistence; of two jaws, each having one or two palpi, and a labium formed of two pieces, the mentum and the lingua, and accompanied with two palpi, usually inserted on this latter piece. Those of the jaws or their externals, when there are two, have never more than four articulations: those of the labium have generally but three.

The anterior segment of the trunk, or that which is in front of the wings, and which is usually called the *corslet*, has the first pair of feet, and considerably exceeds in extent the two other segments. The interior membrane presents, on each side, behind, a stigma, a character which, I believe, had hitherto escaped notice, but the existence of which was presumable. The other two segments unite closely with the base of the abdomen, and their lower part, or the *postpectus*, serves as an attachment to another pair of feet. The second, on which is placed the scutellum, grows narrow in front, and forms a short pedicle, which is inclosed in the interior cavity of the first, and serves for a pivot in its movements.

The elytra and the wings originate on the lateral and upper edges of the posterior trunk. The elytra are crustaceous, and, in a state of repose, are applied, one against the other, in a right line along the length of their internal edge, or at the suture, and always in an horizontal position. Almost always they conceal the wings, which are broad and folded transversely. Many species are apterous; but the elytra always exist. The abdomen is sessile, or united to the trunk in its greatest breadth. It is composed, at the external part, of six or seven rings, membranous above, or of a consistence less solid below. The number of articulations in the tarsi varies from three to five. To judge by analogy, the coleoptera called monomera, have probably three articulations in the tarsi; but the first two are invisible. This section and that of the dimera, have been suppressed.

The coleoptera undergo a complete metamorphosis. The larva resembles a worm, having a scaly head, a mouth analogous, in the number and functions of its parts, to that of the perfect insect, and usually six feet. Some species, few in number, are without feet, or have but simple mammellæ.

The mesothorax is always short and narrow, and the metathorax, often spacious, is furrowed longitudinally in the middle.

The nymph is inactive, and takes no nutriment. The habitat, the mode of life, and the other habits of these insects, either in their first or last stage of existence, vary very considerably.

I divide this order into four sections, according to the number of articulations in the tarsi.

The first comprehends the *pentamera*, or those of which all the tarsi have five articulations; and it is composed of six families, of which the first two are distinguished from the others by the existence of a double excrementitious apparatus. According to M. Dufour, the *silpha*, a genus of the fourth family, also exhibits one, but single, or one side alone.

The first family of the PENTAMEROUS COLEOPTERA is

THE CARNIVORA, *Cuv.* (*Adephages*, Clair.),

with two palpi on each jaw, or six in all. The antennæ are almost always thread or silk-formed, and simple.

The jaws are terminated by a scaly piece, talon-formed, or

crooked, and the interior part is furnished with hairs or small spines. The ligula is enchased in an emargination of the mentum. The two anterior feet are inserted on the sides of a compressed sternum, and borne on a large *rotula*. The posterior two have a very strong trochanter at their origin. The first articulation is large, and appears to be confounded with the postpectus, and has the form of a curvilinear triangle, with the external side excavated.

These insects give chase to others, and devour them. Many of them have no wings under their elytra. The anterior tarsi of most part of the males are dilated or enlarged.

The larvæ of this family are also extremely carnivorous. They have in general a cylindrical, elongated body, composed of twelve rings. The head, which is not comprised in this number, is large, scaly, armed with two powerful mandibles curved at their points, and presents two short and conical antennæ, two jaws divided into two branches (one of which is formed by a palpus), a ligula bearing two palpi, shorter than the preceding, and six small simple eyes on each side. The first ring is covered by a scaly plate; the others are soft, or but little closed. The first three have each a pair of feet, the extremity of which is curved in front.

These larvæ differ according to the genera. Those of the cincindelæ, for example, have the upper part of the head sunk in the centre in a basket-form, while its lower part is gibbous. They have on each side two small simple eggs, much more bulky and similar to those of the lycosæ, or *wolf-spiders*. The upper plate of the first segment is large, and in the form of a semi-circular buckler. The eighth ring has on the back two mammellæ with crotchets. The last has no remarkable appendages.

In the other larvæ of this family which are known to us,

with the exception of that of the omophronta, the head is less strong and more equal. The simple eyes are very small and more similar to each other. The scaly piece of the first ring is squared and does not edge beyond the body ; the eighth has no mammellæ, and the last is terminated by two conical appendages besides a membranous tube, formed by the elongation of that part of the body in which the anus is situated. These appendages are corneous and dentated in the larvæ of the calosomi and the carabi. They are fleshy, articulated, and longer in those of the harpali and licinæ. The body of the last but one is a little shorter, with the head a little more bulky. The form of the mandibles of the one and the other approaches to that which they have in the perfect insect. The larva of one species of omophron, according to the observations of M. Desmarest, has a conical form, a large head, with two very strong mandibles, and presents but two eyes. The posterior extremity of the body, which is narrowed by degrees, is terminated by an appendage of four articulations. I have counted but two in those of the larvæ of the licini and harpali.

This family has always the first stomach short and fleshy ; a second elongated, and, as it were, hairy at the exterior, in consequence of the numerous small vessels with which it is furnished, and an intestine short and narrow. The hepatic vessels, four in number, are inserted near the pylorus.

There are terrestrial and aquatic insects of this order.

The terrestrial have feet solely proper for running, the four hinder ones of which are inserted at equal distances, the mandibles entirely uncovered : the piece terminating the jaws is straight underneath and covered only at its extremity, and most frequently the body is oblong, with the eyes projecting. All their tracheæ are tubular or elastic. Their intestine is terminated by a widened cloaca, furnished with two small sacs which secrete an acrid humour.*

They are divided into two tribes.

* M. Leon Dufour has presented in the Annals of Natural Science, Vol. viii. p. 36, the following recapitulation of the anatomical characters of the insects of this division.

" The carabici are hunters and carnivora. The length of their digestive tube does not exceed that of twice the length of their body. The œsophagus is short. It is followed by a membrano-muscular crop, very well developed and extremely dilatable. Then comes a gizzard, oval or rounded, with cellular and elastic parieties, armed internally with mobile corneous pieces, proper for trituration, and provided with a valvule at its two orifices. The *chylific ventricle* which succeeds it is of a soft and expansible texture, constantly bristling with papillæ more or less prominent, and narrowed in the hinder part. The *narrow* intestine is rather short. The cœcum has the form of a crop. The rectum is short in both sexes. The *hepatic vessels* are but two in number, in the form of an arc variously folded, and are implanted in four isolated insertions around the terminations around the chlyific ventricle. The testicles are formed each by the agglomerated circumvolutions of a single spermatic vessel; sometimes they are almost naked, sometimes clothed with an adipose bed, of a sort of *tunica vaginalis.* The *vasa deferentia* are often refolded like an epididymis. The *vesiculæ seminales,* two in number only, are filiform. The *ejaculatiorius* is short, the penis narrow and elongated, and the *armatura copulatrix* more or less complicated. The ovaries have but from seven to twelve ovigerous sheaths to each, multilocular and united in a cone-formed bundle. The oviduct is short. The *sebaceous gland* is composed of a secretory vessel, sometimes filiform, sometimes swelled at the extremity, and of a reservoir. The *vulva* is accompanied by two retractile crochets. The eggs are oblong-oval. The existence of an apparatus of excrementitious secretion is one of the most prominent anatomical characters of the carabici; it consists of one or more clusters of *utriculæ secretoriæ,* the form of which varies according to the genera, of a long *efferential* canal, and of a vesicle or contractile reservoir, of an excretory conduit, the mode of whose excretion varies, and an excreted liquid which possesses ammoniacal qualities. The respiratory organ has *stigmata* or bivalve apertures, and *tracheæ* altogether *tubular.* The nervous system does not differ from that of the coleoptera in general.

The first, that of CICINDELETÆ, Lat., comprehends the genus

CICINDELA, L.

Which has at the end of the jaws an unguiculus, which is articulated at its base with them.

Their head is strong, with large eyes, mandibles very much advanced and extremely denticulated, and the ligula very short, and concealed behind the chin. Their labial palpi are composed distinctly of four articulations. They are generally hairy, as well as the maxillary. Most part of the species are exotic.

Some have a tooth in the middle of the emargination of the mentum. The labial palpi are separated at their origin, the first articulation being almost cylindrical, without any angular elongation at its extremity, and the external maxillary palpi are manifestly advanced beyond the labrum.

Here the tarsi are similar, and with cylindrical articulations in both sexes. The abdomen is broad, almost in the form of a heart, and entirely embraced by synostosized elytra, and the external edge of which forms a keel.

MANTICORA, Fab.

The two only known species inhabit Caffraria. (*Manticora maxillosa*, Fab. Oliv. Col. III. 37. 1. 2. Hist. Nat. des Coleops, d'Eur. and *Manticora pallida*, Fab.) These are the largest of the genus. One of them, *M. pallida*, is referred, though doubtfully, by Mr. M'Leay, to a new genus which he names PLATYCHILE, and which appears to us to differ little from the manticora but in not having the elytra synostosized. Annulosa, Jav. I. p. 9.

There the first three articulations of the two anterior tarsi are sensibly more dilated or broader in the males than in the females.

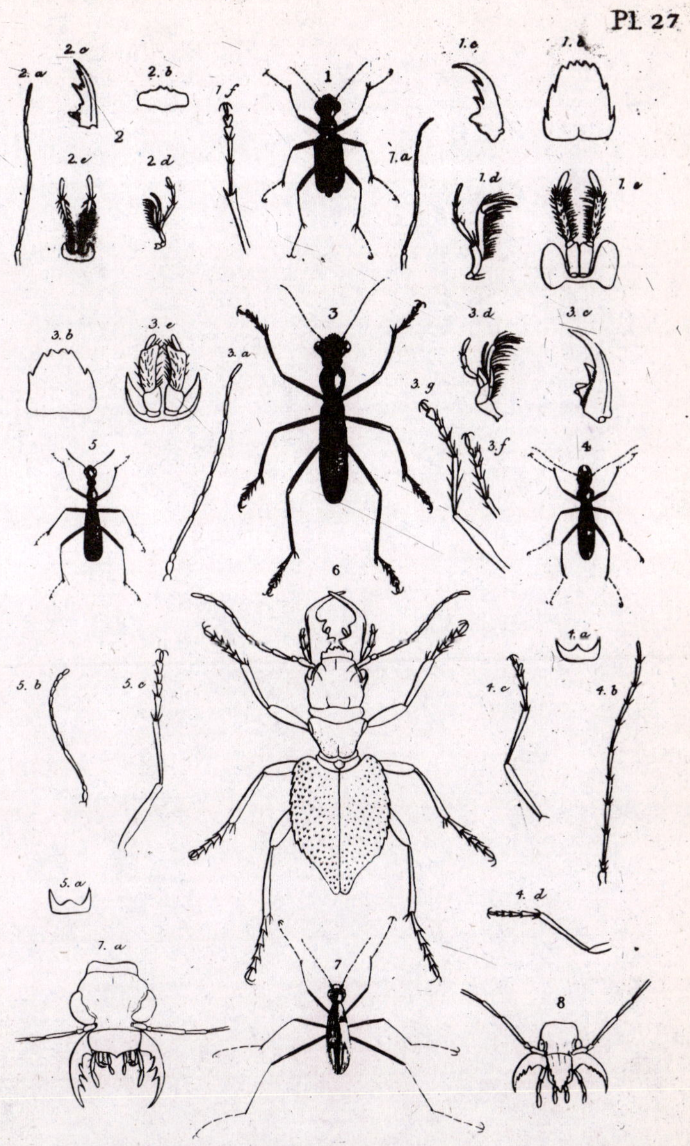

1. Therates *basalis. d'Urv.* 2. *Details of* Megacephala *carolina. Latr.* 3. Trycondyla *aptera. Latr.* 4. Clenostoma *Ichneumoneum. Dej.* 5. Colliuris *modesta. Dej.* 6. Manticora *maxillosa. Latr.* 7. Cicindela *tenuipes. Guer.* 8. *Head of* Oxycheila *tristis. Dej.*

Sometimes the body is simply oval or oblong, with the corselet almost squared, sub-isometrical, or more broad than long, and not globular, nor of a knotty form.

The third articulation of the anterior tarsi of the males, does not advance internally, and the following is inserted at its extremity.

Among these last, the species, whose labial palpi are sensibly longer than the external maxillary ones, with the penultimate articulation longer than the last, form two sub-genera.

MEGACEPHALA, Lat.

Which have the labrum very short, transverse, and the first articulation of the labial palpi much longer than the following, and projecting beyond the mentum.*

OXYCHEILA, Dej.

Of which the labrum is in the form of an elongated triangle, and the first articulation of the labial palpi is not much longer than the following, and projecting not beyond the emargination of the mentum.†

In the following species, the labial palpi are altogether of the length of the external maxillary ones, with the last articulation longer than the preceding. They also compose two sub-genera.

EUPROSOPUS, Lat. Dej.

In which the third articulation of the labial palpi is more

* *Cicindela Megalocepha'a*, Fab. Oliv. II. 33, 11, 12. *C. carolina*, Oliv. Ibid. xl. 22. *Megacephala Euphratica*, Hist. Nat. des Coleap. d'Eur, I. 1, 2, See, for other species, Spec. Gen. de M. Le Comte Dejean. Tom. I. p. 6, &c

† *Cicindela tristis*, Fab. Oliv. Coleop. II. 33. iii. 35. *Oxycheila tristis*. Dej. Spec. gen. des Coleop. I. p. 16. *Cicindela bipustulata*, Lat. Voy, &c. M. M. Humb. et Bonpl. Obser. d'Anat. et de Zool. No. 13, xvi. 1, 2,

thick than the last, and in which, the first three articulations of the anterior tarsi of the males, are a little elongated, flatted, careened below, and equally furnished with hairs on both sides. The eyes are very large, and these insects reside on trees.*

The Cicindelæ, proper (Cicindela), Lat.,

Are little different from the Euprosopi, but that the third articulation of the labial palpi is not sensibly more thick than the following. They also differ in their anterior tarsi, of which the first three articulations are in the males, very much elongated, more strongly ciliated on the internal side than the opposite, and without a keel underneath.

Their body is usually of a green, more or less deep, mixed with metallic and brilliant colours, with white spots on the cases. They frequent dry places exposed to the sun, run very fast, fly away the moment they are approached, and take ground at a little distance. If they continue to be disturbed, they have recourse again to the same means of escape.

The larvæ of the two indigenous species, the only ones which have been observed, dig themselves a cylindrical hole in the ground, tolerably deep, employing for this purpose their mandibles and feet. To fix it, they charge the top of their head with the particles of earth which they have detached, turn themselves round, climb by little and little, repose at intervals, fastening themselves to the internal sides of their habitation, with the assistance of the two mammellæ of their back, and when arrived at the orifice of the hole, cast out their burden. In the moment when they are in ambuscade, the plate of their head closes, exactly on a level with the soil, the entrance of their cell. They seize their

* *Cicindela, 4-notata,* Hist. Nat. des Coleop. d'Eur, I. 1. 6. *Euprosopus, 4-notatus* Dej. Spec. Jen. des. Coleop. I. p. 15.

prey with their mandibles, even shoot upon it, and precipitate it to the bottom of the hole, inclining their head downwards abruptly. They also descend there very quickly at the slightest appearance of danger. If they find themselves confined, or if the nature of the ground be unfavourable to their operations, they immediately proceed to form a new habitation. Their voracity extends even to other larvæ of their own species, which are established in their neighbourhood. They open the aperture of their dwelling, when they are about to change their skin, or be metamorphosed into the nymph-state. A part of these observations have been communicated to me by the late M. Miger, who has studied with much attention a great number of the larvæ of coleoptera, and has discovered many which have escaped the researches of former naturalists.

C. Campestris, Lin. Panz. Faun. Insect. Germ. LXXXV. iii. In length, about six lines, of a meadow-green above, with the labrum, white, feebly indentated in the middle. Five white points on each of the elytra. Very common in Europe in the spring.

C. Hybrida, Lin. Panz. ibid. iv. Which has on each elytrum two spots of a crescent form, and a white band; one of these spots situated at the external base, and the other at the end. The suture is coppery. Found in sand-pits, and never mixes with the preceding species.*

Another species of our own country, *C. Germanica*, Lin., and some others, have a form more narrow, and more elon-

* Add *C. Sylvatica*, Lin. Clairv. Ent. Helv. II. xxiv. A.; *C. Sinuata*, Fab. Clairv. *ibid.* B. b.; *C. Germanica*, Lin. Panz, Faun, insect. Germ. VI. v. *See* also, for these and other European species, The Hist. Nat. des Coleopt. d'Europe, par MM. Latreille et Le Comte Dejean, fasc. I. p. 37, et suiv. and as well for the same, as for a great number of exotics, the Spec. gen. of this last-mentioned naturalist.

gated, and seem to form a peculiar section. They do not fly away like the preceding, when about to be seized, but endeavour to escape by running very fast. M. Gotth. Fischer, in his entomology of Russia, has placed a species of them from Brazil, in his sub-genus Theratus (*T. Marginatus*).

All these species have wings, but some apterous ones are known, whose abdomen, moreover, is more narrow and oval, and in which the tooth of the emargination of the mentum is very small, and scarcely sensible. Such is that which we have represented in our Natural History of the Coleoptera of Europe (I. 1. 5.), under the name of *coarctata*, M. Le Comte Dejean (Spec. gen. des Coleop. II. p. 434,) has formed with them a new genus, that of *Dromica*.

Sometimes the body is long and narrow, and with the corslet elongated in a knotty form, but narrowed in front. The third articulation of the two anterior tarsi of the male, is in the form of a palette, and advanced interiorly. The following one is inserted externally near its basis—

CTENOSTOMA, Klug. *Caris*, Fisch.

This sub-genus has hitherto appeared to be peculiar to the intra-tropical countries of South America. The head is bulky, with the antennæ almost as long as the body, and almost setaceous. The external palpi are greatly projecting, and terminated by a thicker articulation, in the form of an elongated pear. The penultimate articulation of the external maxillary ones shorter than the following, the first two of the labial are very short, and the terminal lobe of the jaws is without any sensible unguiculus at the end. The abdomen is oval, contracted at the base, and pediculated. The feet are long and slender.

The Ctenostomata approach, in the relation of the size

of the palpi, the megacephala, and in other respects to the tricondyla and therates.*

The others have no tooth, in the middle of the emargination of the mentum. The labial palpi are contiguous at their origin, with their first articulation obconical, or in the form of a reversed pyramid, and dilated or prolonged internally at its extremity, in the manner of an angle or tooth. The external maxillary ones do not pass much beyond the labium. These species have been divided into three sub-genera.

THERATES. Latr. *Eurychile.* Bonnelli.

Similar in their general form to the circindelæ proper, but which are distinguished from them as well as from all the analogous sub-genera, by their internal maxillary palpi, very small and of an aciculary form. The tarsi are similar in the two sexes, with their penultimate articulation in the form of a heart, without emargination, and simply hollowed above for the insertion of the last.

These insects are exclusively peculiar to the most eastern islands of Asia, such as Java, the Sunda Isles, and those which are to the north of New Holland.

In the two following sub-genera, and in all which are peculiar to the East Indies, and the most remote islands to the east, the body is narrow and elongated, with the corslet almost cylindrical, and of a knotted form. The third or fourth articulation of the tarsi, is elongated internally in the manner of a lobe.

COLLIURIS, Lat. *Collyris*, Fab.

They are winged; the antennæ are more bulky towards

* See *Entomologiæ brasilianæ specimen* of H. Klug. Species gen. des. Coleop d'Eur, de M. Le Comte Dejean, tom. 1, p. 152, et Suiv.

the end. The final articulation of the labial palpi is almost hatchet-formed, and the preceding often curved. The corslet is almost cylindrical, narrowed, and strangulated in front, with the anterior edge widened. The abdomen, which is also almost cylindrical, is widened and enlarged posteriorly. The tarsi are similar in the two sexes, with the penultimate articulation prolonged obliquely at the internal side, equally large as the preceding, and this last in the form of a reversed triangle, with acute angles.

The species which I have described and figured, under the name of *longicollis*, is distinct from that so designated by Fabricius. It is the *colliuris emarginata* of M. Dejean. Spec. Gen. I. p. 165.

Tricondyla, Lat.

Here the wings are wanting, the antennæ are filiform, and the last articulation but one of the labial palpi, is the longest and the thickest of all. The corslet is in the form of a knot, sub-ovoid, strangulated, truncated, and double-edged at the two ends. The abdomen is ovalary, oblong, narrowed towards its base, and a little gibbous posteriorly. The first three articulations of the anterior tarsi are dilated in the males; the third is prolonged obliquely, in the internal side, in the manner of a lobe. The following is almost similar, but much smaller, and less prolonged—

The second tribe, that of the Carabici, Lat., comprehends the genus,

Carabus, L.,

Which has the jaws terminated simply in a point or crotchet, without articulation at its extremity. Their head is usually more narrow than the corslet, or at most, of the same breadth. Their mandibles, with the exception of those of a few, have none, or very few denticulations. The ligula is usually prominent, and the labial palpi have but three free articulations. In the cicindelæ, the radical articulation is disengaged, and it is on that account that the palpi have four articulations; but here it is entirely adherent, and forms only a support which we do not take into account. Many of them are deprived of wings, and have elytra only. They often shed a fœtid odour, and shoot from the anus an acrid and caustic fluid. Geoffroy presumes that the ancients designated them under the name of *Buprestis*, insects which they regarded as a very dangerous poison, particularly for oxen.

Carabi conceal themselves in the earth, under stones, the bark of trees, and are for the most part very agile. Their larvæ have similar habits. This tribe is very numerous, and the study of it extremely difficult.

We shall form a first general division, with those whose extreme palpi are not terminated like an awl; their last articulation is not united with the preceding to form a body, either oval and very much pointed at the end, or convex, with a narrow and acicular point at the end.

These carabi may be subdivided into those of which the two anterior legs have, at the internal side, a strong emargination, separating the two spines, which are usually placed one near the other, at the extremity of this side; and into those in which the legs have no emargination, or present only an oblique linear canal, not advancing on the internal side of these limbs.

We shall divide this subdivision into several sections.

First, the Truncatipennes, thus named in consequence of their elytra being almost always truncated at their posterior extremity. The head and corslet are more narrow than the abdomen. The ligula is most frequently oval or squatted, and rarely accompanied on the sides with projecting divisions.

Some have the crotchets of the tarsi simple, or without denticulations, disposed in the manner of a comb.

We shall commence with those whose head is not narrowed abruptly at its posterior extremity, and is not attached to the corslet, by a sort of neck suddenly formed, or a kind of rotula. The corslet is always in the form of a truncated heart. The external palpi are never terminated by an articulation much more thick, and hatchet-formed. The two anterior tarsi of the males are either not at all, or very little dilated. The penultimate articulation of these tarsi, and of the others, is never deeply bilobated.

The three following sub-genera have a common negative character, which is that of being deprived of wings.

Anthia,

Have a corneous ligula, oval, and advancing between the palpi, even as far as their extremity.

The labium is often large and denticulated or angular.

Their external palpi are filiform, with the last articulation almost cylindrical, or in a reversed and elongated cone. The emargination of the chin presents no tooth. The abdomen is oval, most frequently convex, and the elytra are almost entire or but little truncated.

These insects as well as those of the following sub-genus, have the body black, spotted with white, a colour formed by a sort of down, and inhabit the deserts or similar places, in Africa; though there have been found, in the southern parts of Spain and Italy, many insects of the north of Africa, there has never been discovered there, a single species of anthia or graphiptera.

The anthia, according to the observations of the late Leschenault de Latour, cast, by the anus, when they are dis-

Pl. 26

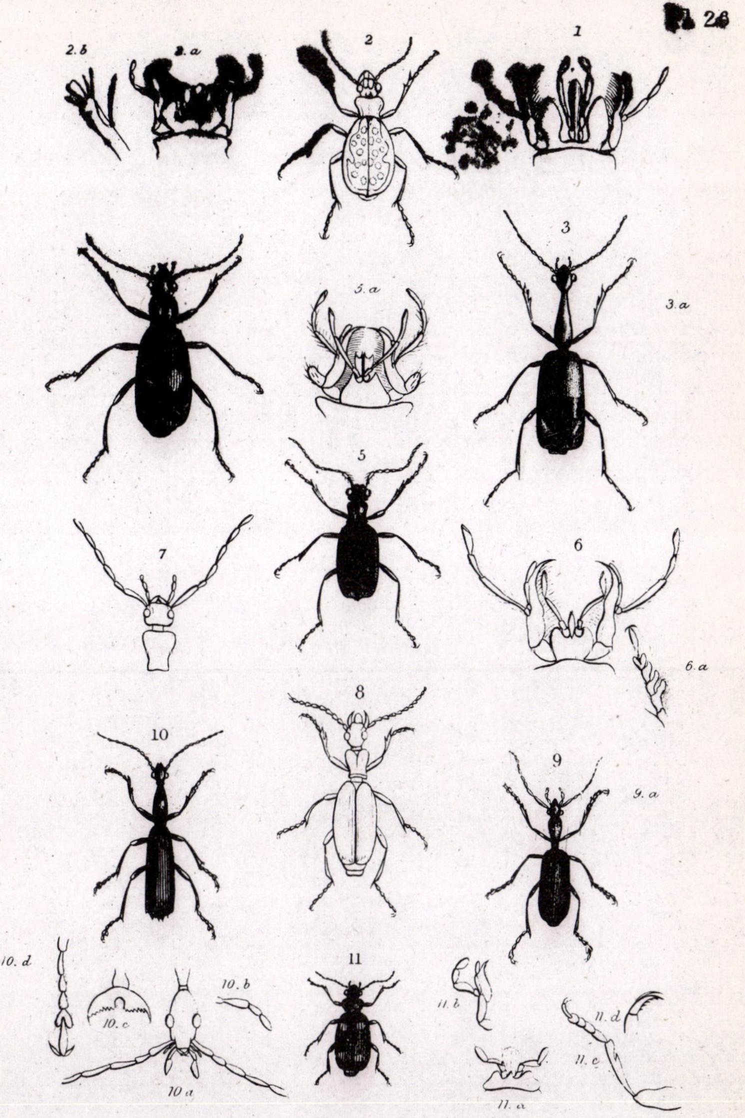

1. *Mouth of* Anthia *10 guttata*. *Latr.* 2. Graphipterus *multiguttatus*. *Latr.* 3. Casnonia, *Senegalensis.* *St Farg. et Serv.* 4. Brachinus *Jurinei*. *Dej.* 5. Trichognathus *marginatus*. *Latr.* 6. *Mouth and tarsi of* Galerita *americana*. *Latr.* *Head & corselet of* Zuphium *olens*. *Latr.* 8. Helluo *costatus*. *Latr.* 9. Dripta *ruficollis*. *Dej.* 10. Agra *splendida*. *Latr.* 11. Lebia *flavomaculata*. *Guer.*

turbed, a caustic fluid. The species are generally large, and in the males of some, the thorax is more or less dilated behind, and is terminated by two lobes.*

Graphipterus. Lat. *Anthia*. Fab.

Which had been confounded with the preceding, but which differ from them by having their ligula entirely membranous, with the exception of the central part, by having their antennæ compressed, and the third articulation of which is much longer; their abdomen, otherwise, is always flatted, orbicular, and one of the two spines terminating the posterior legs, is much larger than the other, and in the form of a plate.

The species of this sub-genus are exclusively peculiar to Africa, and smaller than the preceding.

Aptinus, Bon. *Brachinus*. Web. Fab.,

Have the final articulation of the external palpi a little more bulky, especially that of the labial, and a tooth in the middle of the emargination of the mentum. Their *ligula* otherwise resembles that of the graphipteri, but the lateral division, or paraglassi, form a small pointed projection; but what distinguishes them more particularly, as well as the following sub-genus, is that their abdomen, which is oval, and tolerably thick, encloses organs secreting a caustic liquor, issuing with an explosion from the anus, evaporating immediately, and with a penetrating odour. This liquor, when the animal is held between the fingers, produces a spot analogous to that produced by the nitric acid, and even, if the species

* *See* the second fasciculus of the natural history of the coleoptera of Europe, the first volume of the species of M. Le Comte Dejean, the excellent work of M. Schœnherr on the synonimy of the insects, and the Zoological part of the voyage of M. Caillaud, where I have described and figured the insects collected by him in Africa.

be tolerably large, a burning accompanied with pain. M. Leon Dufour has made us acquainted with the organs which secrete this fluid.

These insects are found, and often assembled in societies, at least in the spring, under stones. They employ the above-mentioned means of defence against their enemies, and are able to reiterate the explosion a good number of times. The largest species are found between the tropics, and in other warm climates, even as far as the limits of the temperate zone.

We shall cite 1st, *Brachinus displosor*, Duf. *Aptinus balista*, Dej. Hist. Nat. des Coleop. d'Eur, II. viii. 1. It is from five to eight lines in length, black, with the corslet fawn-colour, and the elytra furrowed. In Navarre, and different countries of Spain and Portugal.

Aptinus Pyrenæus, Dej. Hist. Nat. des Coleop. d'Eur, II. viii. 3. It is from three to four lines in length, of a deep black, with the antennæ, and palpi fawn-colour, and the feet of a reddish yellow. The elytra are furrowed. It has been discovered in the department of the eastern Pyrenees, by M. le Comte Dejean.

Brachinus, Web. Fab.,

Differ little from the aptini, but in being provided with wings, and that the emargination of the chin presents no tooth.

Some, and generally the largest, and for the most part exotic, have the elytra very distinctly furrowed, or with sides, and of this number is a species of the Antilles and Cayenne.

Brachinus complanatus, Fab. *Carabus planus*, Oliv. III. vi. 63. Its body is six or eight lines in length, of a reddish yellow, with black elytra, and presenting a humeral point, a sinuous band, traversing their middle, and a spot at their extremity of the colour of the body. It is also that of their external edge. The posterior angles of the corslet, are prolonged into a point.

The other Brachini have the elytra smooth, or slightly furrowed.

The following species are commonly found in the neighbourhood of Paris.

Brachinus crepitans, Fab. Hist. Nat. des Coleapt. d'Eur, II. Panz. Faun. Insect. Germ. xx. 5. Its middle length is four lines. It is fawn-coloured, with the elytra sometimes of a deep blue, sometimes of a bluish green, feebly furrowed, and the antennæ fawn, but having the third and fourth articulation blackish. The chest, with the exception of the middle, and the abdomen, are of this colour. With this species has been confounded that which M. Duftschmid has named *explodens*, Hist. Nat. des Coleop. d'Eur, II. viii. 7, and which is also very common. It is one half smaller, with the elytra blue, and almost smooth. That which M. Bonnelli has distinguished under the name of *glabratus* differs from it, only by the want of spots in the antennæ.

Brachinus Scolopeta, Fab. Hist. Nat. des Coleap. d'Eur. II. ix. 3, resembles the last altogether, but is distinguished from it by the suture of the elytra, which is of a fawn-coloured red, from the base, as far as the middle. The body is also proportionally wider, and of the same colour, as well above as below.

Another species, *Brachinus Bombarda*, Iliger. Hist. Nat. des Coleop. d'Eur. ix. 2, holds a middle place between the last and first. The elytra have around the shield a fawn-coloured spot, but which does not extend to the length of the suture. The department of the Herault, presents us two other pretty species, the one, *exhalans*, having the elytra of an obscure blue, with four yellowish points, and the other, *causticus*, altogether fawn, with a band along the suture, and a posterior blackish spot.

We had at first (Hist. Nat. des Coleop. d'Eur.) placed the genus *Catascopus* of Mr. Kirby, after the Brachini. We think, from a new examination, that it rather belongs to the section of the *simplicimana*. The posterior extremity of the elytra presents clearly a profound emargination, but it terminates in a point on the side of the suture, and is not truncated. Many species of this division present also the same sinus, although however less profound and less acute.

Between the Brachini and the Catascopi, M. Le Comte Dejean (Spec. I. p. 226,) places the genus *Corsyra* of Mr. Stevens, which has for its type the *Cymindus fusula*, of the Entomographie of Russia, by M. Fischer, I. xii. 3. It differs from this last in its tarsi, the crotchets of which are simple. The body otherwise is flatted, as in the preceding, and the other neighbouring sub-genera, short, tolerably broad, with the palpi filiform, the mentum unindented, the labrum transverse, the corslet broader than the head, and almost semi-circular.

But one species is known. The other *carabici* of the same division, and whose crotchets are likewise simple, are removed from the preceding by the form of their head, which is contracted abruptly from its origin, and presents the appearance of a neck or rotula.

First, come those whose tarsi are almost identical in the two sexes, sub-cylindrical, or linear, and of which the penultimate articulation for the most part, is profoundly emarginated, or bilobate.

Sometimes the external palpi are filiform, or but little swelled at the end, with the last articulation almost ovaliform. The head has the same form, and contracts gradually behind the eyes. The first articulation of the antennæ is always short, or but little elongated. The corslet is always narrow and elongated. The body is tolerably thick. The emargination of the mentum exhibits a tooth in its middle. The ligula is almost square, with the paraglossæ projecting and going to a point.

CASNONIA, Lat., *Ophionœa*, Klug.,

Whose corslet has almost the form of a truncated cone, or of a cylinder narrowed anteriorly.*

LEPTOTRACHELUS, Latr.,

In which this part of the body is pretty nearly cylindrical without any sensible contraction in front, in which the elytra are not truncated, and the tarsi have their penultimate articulation bilobate. *Odacantha Dorsalis*, Fab.

ODACANTHA, Payk., Fab.,

Similar as to the corslet, but with truncated elytra, and the articulations of the tarsi entire.

The species which serves as a type to the genus (*odacantha melanura*, Fab. Clavis Entom. Helv. II. v. Hist. Nat. des Coleop. d'Eur. II. x. 6), is three lines in length, of a greenish blue, with the elytra, their extremity excepted, of a reddish yellow. The base of the antennæ, the breast, and the principal part of the tarsi, are also of this colour. The end of the elytra is of a blackish blue. This species frequents aquatic places, and more particularly inhabits the north of France, Germany and Sweden. The *odacantha tripustulata* of Fabricius is a species of notoxus.

* Consult Entomol. Brasil of M. Klug; Spec. Gen. of M. le Comte Dejean, tom. I. p. 170; Hist. Nat. des Coleop. d'Eur. fasc. II. vii. 6. The species which is figured (*cynocephala*) forms, in consequence of the penultimate articulation of the tarsi, a peculiar division. It is found in Bengal. All the others, the principal of which is the *Attelabus Pennsylvanicus* of Linnæus, are American, and have all the articulations of the tarsi entire.

Sometimes the external palpi are terminated by a thicker articulation, in the form of a reversed or triangular cone. The head, immediately after the eyes, is abruptly narrowed, and of a triangular form, or that of a heart.

Some, whose body is flatted, and which Fabricius has placed with his galeritæ, have all the articulations of the tarsi entire, the corslet in the form of a heart truncated posteriorly, and the mandibles, as well as the jaws, of the ordinary length, or but moderately prominent.

The first articulation of the antennæ is in the form of a reversed, and elongated cone. The tongue is squared, and its lateral divisions are often as long as itself. A tooth is perceptible in the middle of the emargination of the mentum. These carabici, whose indigenous species are found under stones, the barks of trees, and most frequently in the neighbourhood of waters, form the following three sub-genera :—

ZUPHIUM, Latr.,

Which have the first articulation of the antennæ as long at least as the head, and the external maxillary palpi very much elongated. *Galerita olens*, Fab. Claire, Entom. Helv. II. xvii. A. a.; Hist. Nat. des Coleopt. d'Eur. Fasc. II. x. 3.

POLISTICHUS, Bon.,

In which, as in the following sub-genus, the first articulation of the antennæ is shorter than the head, and in which the maxillary palpi are of the usual length; but the second, third, and fourth articulations of the tarsi, those of the anterior two especially, are short, almost orbicular, and the tongue is terminated superiorly by a straight edge, has its lateral divisions projecting in the form of arched, narrow, and pointed earlets.*

HELLUO, Bon.,

Which are distinguished but little from the preceding sub-genus, except by their tongue, which is entirely corneous, rounded at the superior end, and without distinct divisions. The species are all exotic.†

The others, and which with those that immediately follow, appear to approximate much to the brachini, have the penultimate articulation of all the tarsi deeply bilobate; the mandibles and jaws long, narrow, and advanced; the body sufficiently thick, with the head in the form of a narrow and elongated triangle, and the corslet almost cylindrical, a little narrowed posteriorly.‡

The first articulation of the antennæ is very long, and narrowed at its base. The mentum is almost in the form of a crescent, without tooth in the centre of its emargination. The tongue is prominent, narrow, almost linear, and terminating with three spines, and accompanied by two paraglossæ. The under part of the tarsi is furnished with down. Such are the characters of

* *Galerita fasciolata*, Fab. Clairv. *ibid.* 4; *Polisticus discoideus*, ibid. 5. See the Spec. Gen. of M. le Comte Dejean, I. p. 194.

† *Helluo costatus*, Hist. Nat. des Coleop. de Eur. Fasc. II. vi. 5; *Galerita hirta*, Fab.—See the Spec. Gen. of M. le Comte Dejean, I. p. 283.

An unpublished helluo of Brazil ought, it appears to me, to form a new sub genus, in consequence of its filiform palpi, and of which the last articulation is cylindrical.

‡ The dryptæ have also relations with cychrus, and appear to connect the cicindeletæ with the section of grandipalpous carabici. Many sections of this family seem to be attached, like so many branches, to the cicindelæ. Most part of the other families of insects are in the same predicament, or form ramified trunks. In a word, continued series do not exist in nature.

Drypta, Latr. Fab.

All the known species belong to the old continent, or New Holland. Two are to be found in Europe, always on the ground. The most common is *Drypta emarginata*, Fab. Clairv. Entom. Helv. II. xvii. Hist. Nat. des Coleops. d'Eur. fasc. II. x. 1. It is about four lines in length, of a beautiful azure blue, with the mouth, the antennæ, and the feet, fawn-colour. The extremity of the first articulations of the antennæ, and the middle of the third, are blackish. The elytra have punctuated striæ. It is more common in the south of France than in the north; M. Blondel (the son) has, however, found it in abundance, in a locality, in the neighbourhood of Versailles.—N.B. For the other species *see* Hist. Nat. des Coleop. d'Eur. II. x. 2. and Spec. Gen. de M. le Comte Dejean, tom. I. pag. 182.

To those succeed carabici, very analogous to the preceding in their divisionary characters, but which are removed from them by the form of the tarsi. The first four articulations, or at least those of the anterior tarsi of the males, are very much dilated and bifid. The penultimate of all is, in the two sexes, constantly emarginated or dilated. The external palpi, and the first articulation of the antennæ, are always long.

Tricognatha, Latr.,

Have the last articulation of the external palpi in the form of a transverse and elongated cone, and a triangular and hairy projection at the external side of the jaws. The palpi are very long. The labrum presents two crenulations and three obtuse teeth. The summit of the tongue is armed with three spines. The four posterior tarsi are not dilated, at least in the females. The insect (*marginipennis*) serving as type, was brought from Brazil by the celebrated botanist, M. de Saint Hilaire.

GALERITA, Fab.

Which differ from the preceding sub-genera in their external palpi, the last articulation of which is triangular, or axe-formed, and in their jaws not being dilated at the external edge. The two anterior tarsi of the males are widened. The emarginations of the first four articulations are acute, and their internal divisions are larger and more elongated than the external. The lingua is tridentated at the summit, and its paraglossæ are very distinct. The emargination of the mentum is unidentated.

Some species, (*Galerita occidentalis*, Dej.; *G. Africana*, ejusd.,) from their ovaliform head, corslet more elongated and narrow, form a particular division. The majority are American.*

CORDISTES, Latr. *Colophæna*, Klug. *Odocantha*, Fabr.

Have the exterior palpi filiformed, and terminated by an orbiform and pointed articulation.

The first four articulations of all the tarsi are dilated; the first is in the form of a reversed and elongated cone, the lobes of the two following are equal, narrow and pointed; the fourth is in the form of a heart, or reversed triangle without emargination; the upper surface is excavated for the insertion of the following. The head is nearly ovaliform.

We shall terminate this section by two, in which the crooks of the tarsi are indented underneath like a comb, and we shall commence by those in which the head ovaliform or ovoide, is separated from the corslet by a sudden decided termination, forming a kind of knob or ball. The penultimate arti-

* See the second fasciculus de l'Hist. Nat. des Coleopt. d' Eur., and the first vol. of Spec. gener. de M. le Comte Dejean, and principally l'Entom Brasil. specimen de M. le Doct. Klug. All the described species are of South America.

culation of the tarsi is always divided up to its base into two lobes; the preceding are large, heart-shaped, or like a reversed triangle. The first articulation of the antennæ is but little elongated. All the known species are from the new continent.

CTENODACTYLA, Dej.

Their exterior palpi are filiformed, with the last articulation ovaliform. The body is but little elongated, flat, with the corslet nearly in form of an elongated heart truncated behind.*

AGRA, Fabr.

The exterior maxillary palpi are filiform, and the labial are terminated by an enlarged articulation securiform or triangular. The body is long and narrow, with the corslet formed like an elongated cone narrowed before.

The chin is suborbicular, with a tooth in the middle of the emargination. The ligula is nearly cylindrical, without distinct paraglossæ.†

Nevertheless, the head is not distinct from the corslet by a sudden termination, formed like a knot or ball;‡ the articulations of the tarsi are entire in many, and the first are but seldom dilated. The body is always flat. The paraglossæ are never prominent, and form merely a membranous margin, round or obtuse at the end.

* *Ctenodactyla Chevrolatii,* Dej. Spec. I. 227. of Cayenne.

† See the excellent monograph of this genus published by Dr. Klug, in the second fasciculus de l'Hist Nat. des Coleopt. d'Eur., and the first vol. of the Spec. Gener. of the Count Dejean. All the species are of the intertropical America.

‡ Slightly drawn back posteriorly in the Demetrias and the Dromias, but not fixed to the corselet by a ball.

Here the corslet is isometrical, or longer than wide, heart-shaped, truncated behind. The body is elongated. Such are

CYMINDIS, Latr. *Cymindis anomœus*, Fish. *Tarus*, Clairv. *Carabus*, Fabr.

Which have the exterior maxillary palpi filiform, or scarcely thicker at the extremity, with the last articulation almost cylindrical; the same of the labial, larger, almost in form of a hatchet, or reversed triangle in the males at least; whose head is not narrowed behind, and in which all the articulations of the tarsi are entire and nearly cylindrical.*

CALLEIDA, Dej.

Entirely similar to Cymindis, except as to the tarsi having the penultimate articulation bifid, and the preceding triangular. This sub-genus belongs to America.†

DEMETRIAS, Bon.

Analogous to the last by the tarsi, but having the head ovaliform, narrowed behind, and all the exterior palpi nearly filiform, with the last articulation nearly ovoide, or sub-cylindrical.

This sub-genus, like the following, is composed of very few species, inhabiting generally aquatic or humid and covered places, and almost all European.‡

DROMIAS, Bon.

Generally apterous, with the articulations of the tarsi entire, otherwise like Demetrias.§

* See the second and third fasciculus of the Hist. Nat. de Coleopt. d'Eur., and the first vol of Spec. Gener. of the Count Dej.

† Idem.

‡ Idem.

§ Idem.

In some, the corslet is sensibly wider than it is long, in the form of the segment of a circle, or a heart broadly and transversely truncated behind.

There are some in which the middle of the hind edge of the corslet is elongated backward. Such are

LEBIA, Lat. *Lebia*, *Lamprias*, Bon.

The exterior palpi are terminated by a somewhat larger articulation, nearly cylindrical, or ovaliform, and truncated behind. The first four articulations of the tarsi are nearly triangular, and the fourth is more or less bifid or bilobed.

These insects are prettily coloured. The commonest species in Europe is *Carabus cyanocephalus*, Lin., Fab. The *Bupreste blue, and corslet rouge*, Geoff.; Panz. Faun. Insect. Germ. 65. 5; Hist. Nat. de Coleopt. d'Eur., fasc. III. 12. 7. Its body is from two lines and a half to three lines and a half long. It is blue or green, very shining above, with the first articulation of the antennæ, the corslet, and the tarsi of a fulvous red, the extremity of the thighs black, and the elytra dotted, marked with light dotted striæ.

Another species of our environs is *Carabus hæmorrhoidalis*, Fab. Hist. Nat. de Coleopt. d'Eur., fasc. III. 13. 8., which is scarcely more than two lines long, whose body is fulvous, with the elytra black, and terminated by a yellowish fulvous spot; they have some light punctated striæ, and two more distinct points near a third at the commencement of the suture.*

In the ensuing the corslet terminates posteriorly by a straight line, without advancing at the middle.

PLOCHIONUS, Dej.

Which have the antennæ nearly grained, the last articulation

* The work above cited.

of the labial palpi large, nearly securiform, the first four tarsi short, in form of a reversed heart, the fourth being bilobed.*

Orthogonius, Dej.†

Have the tarsi formed like the last, but the antennæ are filiform, and the exterior palpi are terminated by an articulation nearly cylindrical.

Coptodera, Dej.

Have the same palpi as the last, the antennæ more or less granulated, the first three articulations of the anterior tarsi short, broad, and the same of the four posterior tarsi nearly filiform, the penultimate of all bifid, but not divided into two lobes. All the species mentioned by the Count Dejean (Spec. I. 273) are foreign, and for the most part American.

The second section, *Bipartiti*, *Scaritides*, Dej., which we may also call, from their habits, Diggers, is formed of Carabici, with elytra entire, and slightly sinuated at their posterior extremity; having the antennæ grained and bent, the head broad, the corslet broad, commonly cup-shaped, or nearly semicircular, separated from the abdomen by an interval which gives thereto a pediculated appearance; the feet are generally but little elongated, with the tarsi most frequently short, alike, or but little different in both sexes, without a brush underneath, and simply furnished with hairs, or covered lashes. The two fore-legs are indented on the outside, as if palmated or digitated in many, and the mandibles are often strong and indented. The emargination of the chin has a tooth. They all remain on the ground, hide themselves

* The same work.

† Dej. Spec. I. p. 279; the species are entirely exotic. Near this subgenus should be placed, apparently, that of *Hexagonia* of Mr. Kirby, Lin. Trans. XIV.

either in holes they dig, or under stones, and often do not quit their retreat, except at night. Their colour is generally uniform black. The larva of Ditomus Bucephalus, the only one which has been observed, has the form and habits of Cicindela. These insects inhabit especially the hot countries.

Three sub-genera, by which we will begin, form, by reason of their labial palpi, terminated by a larger articulation in form of a hatchet or triangle, a particular group. The last of these sub-genera conduct us to Scarites, while the first, which, with reference to the absence of the emargination on the internal side of both fore-legs, make an exception, appear to be connected with the first sub-genera of the family. They all have strong-toothed mandibles. The exterior maxillary palpi are terminated by an articulation a little larger ; the corslet is in form of a cup, or truncated heart ; the abdomen is pediculated.

Two of these sub-genera form a special subdivision in this group. Their anterior legs are not palmated. Their antennæ are composed of articulations nearly cylindrical, or in form of a reversed cone. The chin covers nearly all the under part of the head, as far as the labrum, and often has no transverse suture at its base. The body is very flat, and in many species is without wings. They are all of the old continent, or of New Holland.

Enceladus, Bon.

Their fore-legs have no emargination on the internal edge. The first articulation of their antennæ is a little elongated, and almost cylindrical ; the third is shorter than the second. The middle of the upper edge of the tongue is advanced like an angle or tooth. The corslet is nearly heart-shaped, broadly truncated, with the posterior angle a little dilated and pointed. The labrum is emarginated, or nearly bilobed.

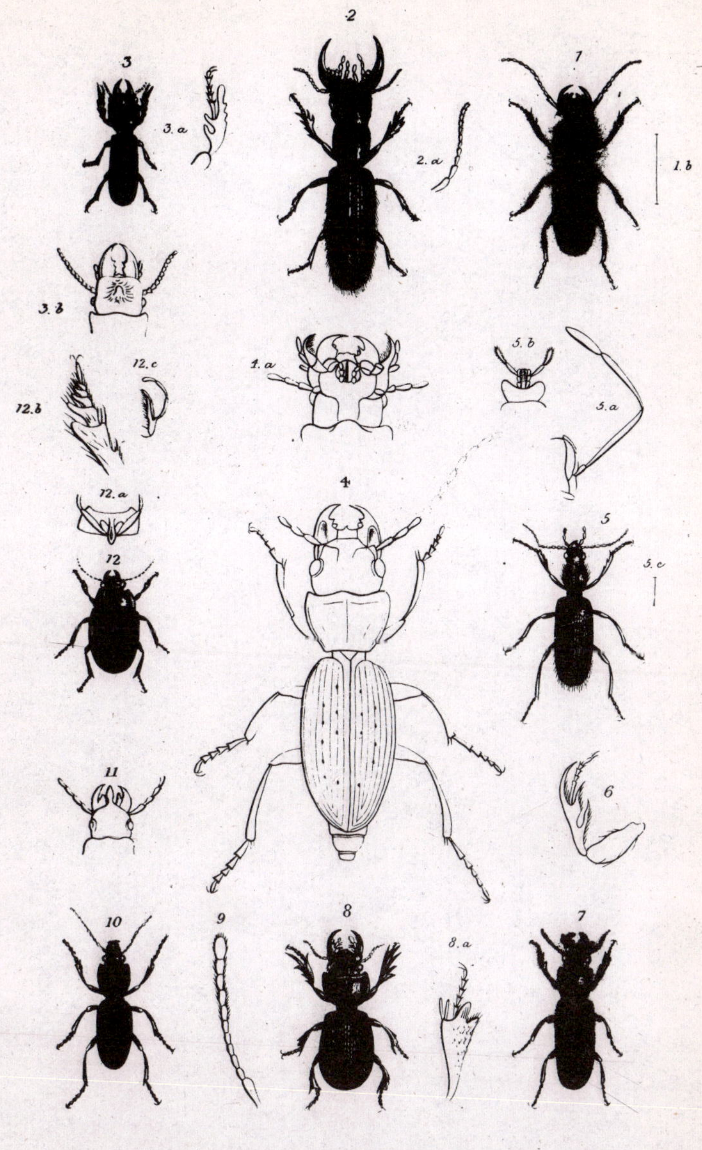

1. Siagona *Europœa*. *Dej.* 2. Oxystomus *Sancti-Hilarii*. *Lat.* 3. Scapterus *Guerini*. *Dej.*
4. Enceladus *gigas*. *Bon.* 5. Apotomus *rufus*. *Lat.* 6. *Fore tarsi of* Dischirius *thoracicus*. *Latr.*
7. Morio *simplex*. *Dej.* 8. Acanthocelis *ruficornis*. *Lat.* 9. *Antenna of* Ozæna *Rogerii*. *Dej.*
10. Ditomus *violaceus*. *Latr.* 11. *Head of* Ditomus *calydonius*. *Lat.* 12. Cyclosomus *flexuosus*. *Latr.*

The only described species, *Enceladus gigas*, Bon. Mem. de l' Acad. des Scien. de Turin, is from the coast of Angola.

Siagona, Latr.* *Cucujus, galerita*, Fab.

Have a decided emargination on the internal side of the two fore-legs; the first articulation of the antennæ is elongated into a reversed cone, and the second is shorter than the third; the top of the ligula is straight, without advancing; the corslet is nearly cup-shaped, nearly as long as it is wide, and without posterior projection, and the labrum is indented.

Some have the abdomen oval, and without wings; in others it is oval, truncated at the base, and these species have wings. M. Lefevre has discovered a new one in Sicily. All the rest, as well of this division as of the last, inhabit northern Africa, or the East Indies.†

The third sub-genus, by its moniliform antennæ, the teeth on the outer side of the two fore-legs, the ordinary proportions of the chin, the general form of the body, approach evidently to the Scarites.

Carenum, Bon.

The jaws are straight, without a terminal hook. The tongue is round at its summit. The last articulation of the exterior maxillary palpi is swelled out, and is twice the size of the preceding.

The only known species (*Scarites cyaneus*, Fab.) inhabits New Holland.

All other carabici of this section have not the labial palpi terminated by a larger and securiform articulation; the last is in the form of a reversed and elongated cone, or almost

* *Siagona rufipes*, Lat. Gener. crust. and insect, I. vii. 9; *Cucujus bifipes*, Fab.; *Siagona fuscipes*, Dej. Spec. I. 359.

† *Siagona atrata, depressa* (*Galerita depressa*, Fab.), *sejus*; *Gal. flejus*, Fab.; *Schupelii*, Dej.; *Scarites lævigatus*, Herbst. Col. 175, 6.

cylindrical, and thinner at the base; the corresponding articulation of the exterior maxillary is likewise nearly cylindrical; all these palpi are nearly of the same thickness throughout, or sometimes thinner at the extremity.

A first very natural subdivision, and which includes *scarites* of Fabricius, except the preceding species, will be composed of bipartite carabi, whose two anterior legs are palmated, or at least digitated at the end; that is to say, terminated exteriorly by a long spine-like point opposed to an internal strong spur. Their antennæ are grained with the second articulation, as long, and often longer than the following. The mandibles, those of a small number excepted, are robust, prominent, angular, or toothed on the inner side. The labrum is very short, transverse, and crustaceous. The ligula is, in general, entirely corneous, beset with hair or lashes, much emarginated, or widened at the top with the lateral angles advanced.

Some have the mandibles very strong, jutting out, and commonly toothed; the crustaceous labrum is much indented on the fore-edge; the tongue is short, not extending beyond the chin, entirely corneous, or crustaceous, beset with hairs, and widened at the upper edge. Their fore-legs are always palmated. The species are generally large.

One of these sub-genera,

Pasimachus, Bon.

Approximates to the last, with reference to the jaws, which are straight, and without a terminal hook.

The antennæ are of equal size. The body is very flat, oval, with the corslet heart-shaped, widely truncated behind, almost as wide at its posterior edge as before, and as the base of the elytra; this edge is nearly straight, and simply a little concave in the middle. This sub-genus belongs to America.*

According to Dejean (Spec. II. p. 471), after Pasimachus should follow the genus he has formed under the denomination *Scapterus*, an East Indian species, which has been communicated to him by one of our most zealous entomologists, M. Guérin, to whom it is dedicated. I am not aware whether the jaws resemble those of the preceding sub-genus, but the body has different proportions ; it is elongated and cylindrical. Their antennæ are in proportion shorter than common ; the second articulation is square, a little thicker than the others, which are short, nearly square, and increase progressively.

The following have the jaws arched, and crooked at the end ; the antennæ thickening insensibly toward the top ; the corslet always separated behind, from the base of the elytra, by a space, or a decided re-entering angle.

Here the exterior palpi are terminated by a sub-cylindrical articulation not pointed at the end.

ACANTHOSCELIS, Lat.

Are remarkable by their four hind-legs, which are in the form of an elongated battledore, arched, flat, and a little concave on the internal face, convex, and covered with little grains and spines on the opposite side, with the upper surface indented, and the posterior teeth large and compressed ; the trochanter of both posterior thighs is very large.

The body is short, broad, convex above, with the corslet transversal, rounded laterally, sinuous on the hinder edge. The spurs of the fore-legs are very long and the others nearly laminous.

* Refer to this sub-genus *Scarites Depressus* and *Marginatus* of Fabricius and of Olivier. See the first vol. of the Spec. of Dejean, p. 405 ; the entomological observations of M. Bonelli, and the work of Palisot de Beauvois on the insects collected by him in America and Africa.

The only species known (*Scarites ruficornis*, Fab.) inhabits the Cape of Good Hope.

SCARITES, Fab.

Have the four hind-legs straight, generally united, having small spines on their edges only; the intermediate have for certain on the outer side one or two teeth; the trochanter of the posterior thighs is much smaller than the thighs themselves. The mandibles are in form of an elongated triangle, strongly indented at their base. The second and third articulation of the antennæ are in form of a reversed cone, nearly uniform in size, and the following are grained.

Some have two teeth on the outer edge of the intermediate legs.

(*Scarites pyracmon*, Bonelli Dej. Spec. I. 367. *Scarites gigas*, Oliv. Col. III. No. 36, I. 1; Clairv. Entom. Helv. II. 9.) is about an inch long; wingless; flat; of a shining black, with the elytra a little enlarged behind; marked with very fine striæ, slightly dotted, the third stria having near its termination two deep distinct points. The head, according to Dejean, is much larger in the male than in the female; it has two impressions and little ridges in front. The corslet has posteriorly a tooth on each side, and three of them may be made out on the fore-legs. This is found on the borders of the Mediterranean, in the south of France, and east of Spain. M. Lefebvre de Cerisy, a distinguished marine officer and good entomologist, has published some observations on its habits.

Scarites terricola, Bonelli Dej. Spec. I. 398, has wings to the body eight or nine lines long, and black. The fore-legs have three strong teeth, followed by three other very small ones; the outer edge of the two following legs have only one each. The elytra are elongated, striated, and slightly rugose, and with three deep points near the third stripe. It is found with the last.

Scarites sabulosus, Oliv. Col. III. 36, 1, 8; Clairv. Entom. Helv. II. 9, 6. *Scarites lævigatus*, Fab. Dej., is very much like the last, but rather smaller; more depressed; wingless; with the elytra faintly striated. The fore-legs have only two indentations after the three common teeth. It inhabits the same locality as the first, and is found also in Sicily, whence it has been brought by M. Lefebvre.

Oxygnathus, Dej.

Essentially similar to Scarites as to the antennæ and palpi; but having in common with the two following subgenera, long, narrow, toothless mandibles, which enlarge considerably, like a pair of nippers; and the body narrow, elongated, and cylindrical. The antennæ are shorter than the head, and the mandibles united. The labrum is indistinct. The corslet is nearly square.

The typical species, *Scarites elongatus*, Wiedem; *Oxygnathus elongatus*, Dej. Spec. II. 474, is from the East Indies.

In others, the four exterior palpi, or, at least, the labii, are terminated by a spindle-shaped articulation, finishing in a point. The body is elongated and cylindrical, and the mandibles long narrow, without observable teeth, as well as those of Oxygnathus.

Oxystomus, Lat.

Have the labial palpi nearly as long as the outer maxillaries, bent, with the first articulation, and cylindrical; the following but little elongated, and the last spindle-shaped, long, and very pointed at the end. The antennæ are perfectly moniliform, from the middle of their length, with their first articulation as long as the three following together. (*Oxystomus cylindricus*, Dej. Spec. I. p. 410, of Brazil).

Camptodontus, Dej.

In which the labial palpi are evidently shorter than the outer maxillaries; not bent, and terminated like them by a spindle-formed articulation. The antennæ are for the most part composed of articulations, formed like reversed cones; the length of the first scarcely exceeds that of the two following taken together. (*Camptodontus Cayennensis*, ib. II. 477.)

The others, whose fore-legs are not indented on the outside, but simply didactylous at the end, have short mandibles, extending but little beyond the labrum, which is coriaceous and entire; the ligula extends beyond the emargination of the chin, bare, or but little velvety, with separate protruding membranaceous paraglossa. The outer palpi are terminated by an oval articulation, pointed at the end.

These Carabi are small, inhabit humid places, and are not strangers to the northern regions.

Clivina, Lat.

Have three strong teeth on the outer side of both fore-legs, and one on that of the two following.*

Dyschirius, Bon. *Clivina*, Dej.

Which have at most mere indentations, and small indistinct spines on the outer side of the two fore-legs; this outer side is in general elongated at its extremity into a long point, in form of a spine or finger, and opposed by another finger, formed by a strong spine on the inner side. The last articulation of the labial palpi is, in proportion, thicker than the

* *Tenebrio fossor*, Lin.; *Scarites arenarius*, Fab.; Clairv. Entom. Hebr. II. 8. The Clivina of M. Dejean, Spec. I. 411, 1—7.

corresponding articulation of Clivina, and is almost in a securiform knob. The corslet is in general globular.*

Our second and last sub-division of the Bipartiti will include those whose fore-legs are neither indented on the outside, or bidigitated at the end, and in which the second articulation of the antennæ is evidently shorter than the following. They approach very near by the manducatory organs to the last two sub-genera. They have, by some authors, indeed, been confounded with Scarites, whose appearance and habits they certainly possess.

Some have the body narrow, elongated, nearly paralleliped with the corslet nearly square; the antennæ entirely, or partially granulated; the last articulation of the outer palpi sub-cylindrical, and that of the labial palpi like a reversed cone. They are all exotic.

Morio, Lat.

Have the antennæ of equal thickness throughout, the labrum deeply emarginated, the outer palpi filiform, the thighs oval, and the legs triangular.†

Oxæna, Oliv.

The antennæ are thicker or swollen at the end, the labrum entire, the labial palpi terminated by a larger articulation, nearly hatchet-formed, or triangular; the thighs and legs narrow and elongated.‡

The others have the body oval or oblong, with the corslet

* Clivines, Nos. 8—21, of M. le Comte Dejean; but the eighth, or *arctica*, appears to have some of the character of Cephalotes.

† *Harpalus monilicornis*, Lat. Gener. Crust. and Insect. I. p. 206; *Morio monilicornis*, Dej. Spec. I. p. 430; *Scarit. Georgiæ*, Palis de Beau. vii. 15, 5; *Morio brasiliensis*, Dej. *ibid.*; *Morio orientalis*, ejusd. *ibid.*

‡ *Oxæna dentripes*, Oliv. Ency. Method.; *Oxæna rogeri*, Dej. Spec. p. 434; *Oxæna brunnea*, ejusd. *ibid*; *Oxæna gyllhenalii*, ejusd.

nearly heart-shaped or sub-orbicular; the antennæ are filiform, composed of articulations for the most part cylindrical, especially the last (the other being more slender at the base, and nearly in form of a reversed cone, and the last articulation of the outer palpi is nearly oval or spindle-shaped). The labrum is emarginated.

These belong to the warm and sandy countries of the eastern parts of the old continent.

DITOMUS, Bon. *Carabus*, *Calosoma*, *Scaurus*, Fab.

Whose palpi are shorter than the head, the corslet is heart-shaped, and the tarsi are short.

Some species, those to which M. Ziegler restores the generic denomination of *ditomus*, have the body more elongated, of the same width, with the head separated on each side from the corslet by a re-entering angle; and the males are commonly armed with one or two horns.*

The others, or those which compose the genus *Aristus*, of the same, have the body shorter, wider in front, with the head nearly contumous, with the corslet buried therein as far as the eyes; the anterior angles are pointed.†

APOTOMUS, Hoff. *Scarites*, Ross.

The anterior palpi, of which are very long, the corslet orbicular, and the tarsi filiform and elongated. The external maxillary palpi are much longer than the head, and terminated by an ovoid-cylindrical articulation. The same of the

* Dej. Spec. I. p. 439.; first division of Ditomus; *Carabus Calydonius*, from a note sent by him with an individual from the collection of M. Desfontaines, forms a very distinct species from *Ditomus calydanius* of M. le Comte Dejean The males have the mandibles forked, or as if divided into two horns; the horn of the middle terminates in a point, or rather in a lance-head; *Calosoma longicornis* of Fab. is probably the female of this or of an allied species.

† The second division of Ditomus of M. C. Comte Dejean, *ibid.* p. 444.

labials is in the form of an elongated spindle. I have perceived no tooth in the emargination of the chin.*

Our third section of Carabici, that of the Quadrimani, Dej., embraces those, which otherwise resembling the last in their elytra terminating in a point behind, have, in the males, the four anterior tarsi dilated. The first three or four articulations are in the form of a heart reversed, or triangular, and almost all terminated by sharp angles; their underpart is generally, with the exception of the *ophoni*, furnished with two ranges of papillæ or scales, with an intermediate linear vacancy.

The body is always winged, generally ovaliform, and arched above, or convex; with the corslet of greater breadth than length, or, almost, isometrical, squared or trapezoidal. The head is never abruptly narrowed behind. The antennæ are of the same thickness throughout, or but triflingly and insensibly thickened towards the end. The mandibles are never very strong. The external palpi are terminated by an articulation longer than the preceding, ovaliform or fusiform. The tooth of the emargination of the chin is always entire, and is sometimes wanting. The tongue, as well as in the two following sections, is always remarkably projecting, obtuse or truncated at the end, and accompanied with two distinct paraglossæ, membranous, and in the form of auricles. The feet are robust, with spinous legs, and the crotchets of the tarsi simple. The intermediate tarsi in the females themselves are short, and, with the exception of the dilation, conformed pretty nearly as in the preceding. These Carabici delight in sandy places, exposed to the sun.

This section is composed of the genus *harpalus*, such as

* *Scarites rufus*, Oliv. Col. III. 36, 11, 13, a. b. Rossi, Faun, *Druse*, I. iv. 3. *Apotomus rufus*, Dej. Spec. I. p. 450; ejus, *Apotomus testaceus*, ibid. p. 451.

M. Bonelli has confined it, in the table which presents the general distribution of the Carabici. The extent of that has been since diminished by new sections. They are subordinate to the three following subdivisions.

The first will have for characters, the emargination of the chin unindentated, the labrum emarginated, the head and the anterior extremity of the corslet as wide or wider than the abdomen.

N.B. If the Cyclosomi have the four anterior tarsi dilated, they will form a fourth division, in consequence of the two teeth in the emargination of the chin.

The first comprehends three sub-genera.

Acinopus, Zieg. Dej.

With filiform antennæ, composed of short articulations, but cylindrical, and with the corslet narrowed insensibly from front to back, with the posterior angles very obtuse and rounded. The labrum is strongly emarginated. The mandibles have no teeth; that of the middle of the emargination of the chin is broadly truncated.*

Daptus, Fisch., *Acinopus*, Dej.

With antennæ commencing with the fifth articulation, maniliform; with the corslet abruptly narrowed towards its posterior angles, which terminate in a point. One of the mandibles is advanced, and very much pointed. The four anterior legs, especially those of the males, are well furnished with small spines.†

* *Harpalus megacephalus*, Lat. Gen. Crust. et Insect. I. p. 206; *Carabus megacephalus*, Fab. Ross. Faun. Etrusc. Append, tab. III. ib.; *Acinopus megacephalus*, Dej. Catal.

† *Acinopus maculipennis*, Dej.; *Daptus pictus*, Fisch. Entom. de la Russ. II. xxvi. 2, xlvi. 2; *D. vittatus*, ejusd. ibid. 7, var.? *Ditoma vittiger*, Germ.; *D. chloroticus*, ejusd. ibid.

Near the dapti appears to come the genus *pangus* of M. Megerle, mentioned by M. le Comte Dejean, in the catalogue of his collection of coleoptera.

After studying one (*Pensylvanicus*) of the two species, which the latter refers to that genus, I have not been able to discover the characters which distinguish this section from the preceding.

The second division is composed of harpali, also having the emargination of the chin unindented, but whose body, more or less ovaliform or ovoid, is more narrow in front, and the labrum is entire, or merely a little concave. These are :—

HARPALUS, Dej.

A species the most common throughout all Europe, is *Carabus æneus* (Fab. Panz. Faun. Insect. Germ. LXXV. 3, 4.) Its body is almost four lines long, of a shining black, with the antennæ and the feet fawn-colour. The upper part of the corslet, and of the elytra, are most frequently green or coppery, and brilliant, and sometimes of a blueish black. The corslet is transverse, narrowed behind, finely edged on the sides, and at the posterior edge, with a punctuated sinking on each side, near the posterior angles. The elytra are striated, have an incision near their end, and small points sunk in the intervals of the external striæ. It has also received the name of *Proteus*, in consequence of the numerous changes in its colours.*

* *See*, for the species, the Catalogue of the Collection of M. le Comte Dejean, genus *Harpalus*, p. 14, and for their synonimy, Schœnherr, Synonimia Insectorum, and the Fauna of Austria, of M. Duftschmid. Fabricius has described but a small number, and among those we shall cite what he names: *caliginosus, ruficornis, binotatus, tardus, heros, analis, flavilabris, &c.* The *carabus signatus, hirtipes* of Panzer, also form a part of this sub-genus.

The absence of a sensible tooth in the emargination of the chin distinguishes the carabici of the third and last division of this section, and which, from the form of the body and the labrum, resemble otherwise those of the preceding division.

OPHONUS, Ziegl. Dej.

The males of which have the four anterior tarsi strongly dilated, or sensibly wider, and generally furnished underneath with numerous and crowded hairs, forming a continuous brush. The penultimate articulation is not bilobate. The last of the external palpi is truncated or very obtuse.

The under part of the body is very finely punctuated. The corslet is most frequently heart-formed, and truncated posteriorly.

STENOLOPHUS, Zieg. Dej.

Which differ from the Ophoni only in the form of the last articulation but one of the four anterior tarsi, at least in the males, and even of the posterior in some. It is divided as far as its base into two lobes.*

ACUPALPUS, Lat. *Stenolophus*, Dej.

In which the four anterior tarsi of the males differ little from the posterior, with the intermediate articulations rounded, almost grained, and hairy. The external palpi terminate with an articulation, pointed at the end.

These Carabici are very small, and appear to be connected with trechus. They are the stenolophi of the Comte Dejean, with the exception of the preceding. Among others we may cite the *Carabus Meridianus* of Linnæus and Fabricius, and the *C. Vespertinus* of Panzer, XXXVII. 21.

* *Stenolophus vaporariorum*, Dej. *ibid. Carabus vaporariorum*, Lin. Panz. Faun. Insect. Germ. XVI. 7; *Harpalus saponarius* Dufour. Senegal.

Pl. 25

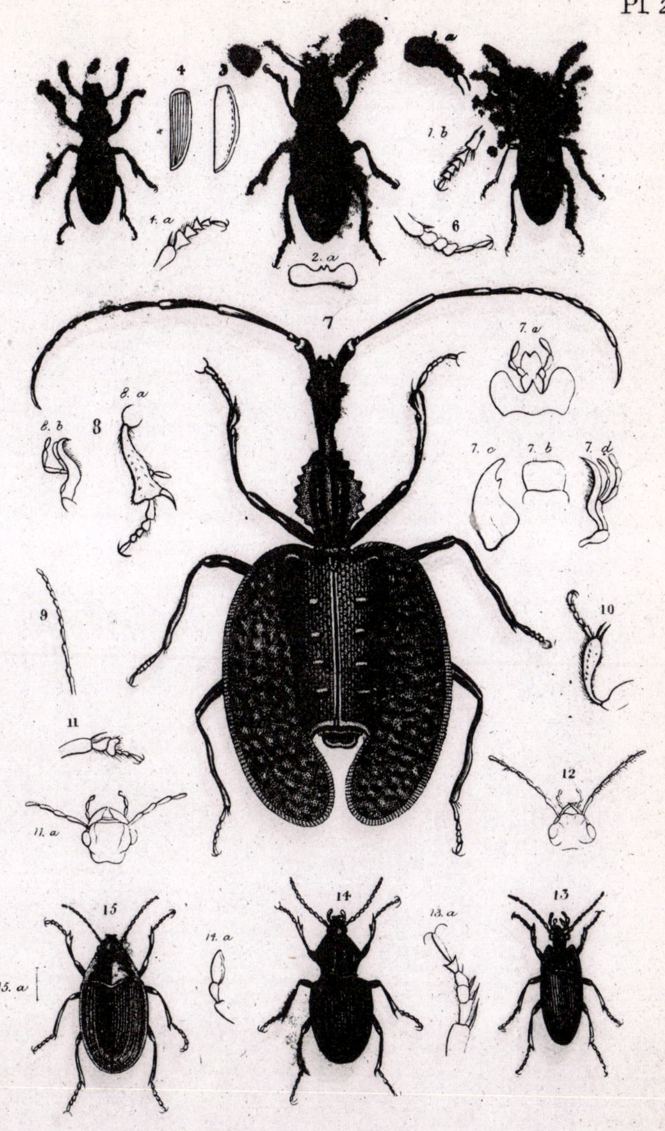

1. Harpalus *tricolor*. *Guer.* 2. Trigonotoma *viridicollis*. *Dej.* 3. *Elytra of* Feronia *navarica.*
Elytra and tarsus of Feronia *melanaria*. *Lat.* 5. Cephalotes *rufipes*. *Lat.* 6. *Ant. tarsus of* Patrob
rufipes. *Male.* 7. Mormolyce *phyllodes*. *Hag.* 8. *Ant. tarsu's & jaw of* Zabrus *gibbus*. *Lat.*
9. *Antenna of* Sphodrus *terricola* *Lat.* 10. *Ant. tarsus of* Feronia *hottentota*. *Lat.* 11. *Ant. tarsus*
& Head of Licinus *agricola*. *Male.* 12. *Head of* Loricera *pilicornis*. *Lat.* 13. Cynthia *abaxoi*
des. *Lat.* 14. Panagœus *fulgipennis*. *Lat.* 15. Omophron *suturalis*. *Guer.*

The fourth section, that of SIMPLICIMANI, approaches the preceding, as to the mode in which the elytra terminate; but the two anterior tarsi alone are dilated in the males; nevertheless, without forming a squared or orbicular pallet. Sometimes the first three articulations are by much the broadest, and the following is then always much smaller than the preceding. Sometimes this last and the preceding two are wider, almost equal, and in the form of a reversed heart or triangular. The first articulations of the four following tarsi are more narrow, and more elongated, almost cylindrical, or in the form of an elongated and reversed cone.

Some have the crotchets of the tarsi simple, or without denticulations.

Here the third articulation of the antennæ is, at the most, once longer than the preceding. The feet are generally robust, with the thighs thick, more or less ovaliform. The corslet, measured in its greatest transverse diameter, is as wide as the elytra.

Sometimes the mandibles are evidently shorter than the head, and do not pass the labrum but by one half, at the most, of their length.

We shall commence with those, all of whose external palpi are filiform.

ZABRUS, Clairv. Bon. *Pelor.* Bon.

Are distinguished from the following by the last articulation of their maxillary palpi, which is sensibly shorter than the preceding, and by the two spines which terminate their anterior legs.*

POGONUS, Zieg. Dej.

Which in the natural order appear to us very much approxi-

* *Carabus gibbus*, Fabr. Zabrus *gibbus*, Clairv. Entom. Helv. II. xi.

mating to the *amara* of H. Bonelli, but are removed from the other carabici of this division by the mode of dilatation proper to the two anterior tarsi of the males. The first two articulations only, the radical of which is the largest, are dilated. The following two are small and equal; their body is generally more oblong than those of the *amara*. These insects appear to inhabit almost exclusively the shores of the sea, or the banks of salt lakes.*

Tetragonoderus, Dej.

The anterior tarsi of the males are proportionally less dilated than in the following, their first articulations being more narrow, and more elongated, and rather in the form of a reversed cone than in that of a heart. These insects are peculiar to South America.†

Feronia, Lat.

In which the anterior tarsi of the male have their first three articulations strongly dilated, in the form of a reversed heart, and of which the second and third are rather transverse than longitudinal.

This sub-genus will comprehend a great number of generic sections, indicated in the catalogue of the collection of M. le Comte Dejean, such as the following: *Amara*, *Pœcilus*, *Argutor*, *Omaseus*, *Platysma*, *Pterostichus*, *Abax*, *Steropus*, *Percus*, *Molops*, *Cophosus*. This learned entomologist has since acknowledged, in the third volume of his *species*, the impossibility of determining them, and with the exception of the first, which he still preserves, he unites the

* *See* the Cat. of M. le Comte Dej. M. Germar has represented in his Fauna of the Insects of Europe, two species: *Pogonus halophilus*, X. 1; *Harpalus luridipennis*, VII 2, approximating *Pogonus pallidipennis*, of the first.

† *Harpalus circumfusus* of M. Germar, Insect Spec. Nov. I. 26.

others in a large general section which he terms, with me, *feronia*. But even as to *amara*, in vain have I sought in the antennæ, and the parts of the mouth, any characters which can precisely distinguish them from the other genera. That which is drawn from the tooth in the middle of the emargination of the chin, without speaking of its trifling importance, is very equivocal. This tooth, in all these carabici, has appeared to me to have at the end an emargination, but a little more distinct or deeper in some than in others. The antennæ of many are a little grained, or composed of articulations, more short, or more rounded towards the summit, but it is impossible to assign in a rigorous manner the limits of this distinction. The same may be said with regard to the concavity of the anterior edge of the labrum, and the form of the corslet.

The feroniæ may form three divisions; 1st. The species, generally winged, whose body, more or less oval, is a little convex, or arched above, with the antennæ more filiform, the head proportionally more narrow, and the mandibles a little less projecting. In their habits they approach the zabri and harpali. Such are the Amara, in which the corslet is transverse; some species, rather shorter, whose corslet is widened from front to rear, form the genus *Leirus* of certain authors. The *Scolytus flexuosus* of Fabricius would appear referable to this division, but, according to M. le Comte Dejean, the four anterior tarsi are dilated. It has appeared to me that they were more so internally than externally. This insect may form a proper sub-genus *(cyclosomus)*. *See*, for the preceding, the third volume of the *species* of this naturalist.

To the first division of Feronia also belongs Pœcilus, in which the corslet is almost as long as broad, and the antennæ rather short, have the third articulation compressed and angular. Also Argutor, similar to Pœcilus, but with antennæ proportionally longer, of which the third articulation is not angular.

2nd. The species generally winged, but whose body is straight, plane, or horizontal, above, with the head of nearly equal breadth with the body. They frequent green or humid places. Such is the genus PLATYSMA of M. Bonelli, to which we unite that of *Omaseus* of M. M. Ziegler and Dejean, and that of *Catadromus* of the younger Mr. Macleay.*

3rd. The third division of Feronia will be composed of species analogous to those of the preceding, in the assemblage of their characters, but which differ by the absence of wings.

Among these species, some, and the most numerous, and whose corslet is not always in the form of a truncated heart, have at the base of the elytra a fold, or transverse edging,

* Those whose body is very much flatted, with the corslet remarkably narrowed, posteriorly, in the form of a truncated heart, will form a first division; such is the *carabus picimanus* of M. Duftschmid, or the *C. Monticola* of some others. M. le Comte Dejean places it with *Pterostichus.* Some species of Brazil will also enter here. M. Germar (Insect. Nov. Spec. I. p. 21) has described one of these under the name of *Molops Corinthius.*

Those whose body is almost parallelipiped, with the corslet almost squared, little or not at all narrowed behind, will form a second division of this number; these are the *platysma nigra* of M. M. Bonelli and Dejean, the *Omaseus* of the latter (Catal. p. 12), and the *Carabus tenebrioides* of Olivier, the type of the sub-genus *catadromus*, of Mr. Macleay, the son, (Annal. Jav. I., p. 18. 1. v.) which do not differ from that of *Omaseus,* but in the tooth of the chin, which is much larger, and entire. Its elytra have at their extremity a large sinus, or rather an emargination. It is one of the largest species of this family.

The HARPALI, *nigrita, anthracinus,* and *atterrimus* of M. Gyllenhall, are *Omasei.* The last has the posterior angles of the corslet obtuse, which distinguishes them from all the others. We place in the same sub-genus the *carabus leucopthalmus* of Fabricius, or the *Melanarius of* Illiger, but it is apterous.

well marked, continuous, and extending as far as the suture.

Sometimes the corslet is almost square, or in the form of a truncated heart, with the posterior angles acute.

Those whose body is in a long or cylindrical square, with the corslet almost square, and but little narrower behind than before, form the genus Cophosus of M. M. Ziegler and Dejean. It has been established on one species *(Cylindricus)* of Austria.

We will join to this the *Omaseus melanarius* of M. le Comte Dejean, as well as another species of Germany, intermediate between the preceding and the *Cophosus Cylindricus*, and which is, I believe, the *Omaseus elongatus* of M. Ziegler.

Those whose body is generally oval, depressed, or but little convex above, with the corslet large, almost square and subisometric, always strongly edged laterally, as broad, or almost as broad at its posterior edge, as at the base of the elytra, compose the genus Abax of M. Bonelli.

Germany furnishes many species of them. That which has been named *Metallicus* and the *Molops striolatus* of M. le Comte Dejean, which have the antennæ composed of shorter articulations, or which are almost grained, seem proper to form a new genus, that of *Cheporus*.

N.B. The platysma, described and figured by M. Fischer, (Entom. de Russie, II. xx. 4, and 5,) are probably analogous to the above.

We find very often in the cold or humid parts of the forests of our environs. (*Carabus striola.* Fabri. *Carabus depressus.* Oliv. col. III. 35, IV. 46.)

Sometimes the corslet, always terminated posteriorly by two well marked angles, or acute ones, is sensibly narrowed behind. Its section approaches more or less to that of a truncated heart.

Among these species many have the body depressed or

plane above, and the antennæ composed of articulations tolerably elongated, rather obconic than turbinated. M. Bonelli distinguishes them generically under the name of Pterostichus.

They more particularly inhabit the high mountains of Europe and the Caucasus.

The environs of Paris furnish but a single species (*Carabus oblongo-punctatus*, Fab. Panz. Faun. Insect. Germ. LXX. iii. 2).

I think with M. Fischer, that the genus *myosodus* of M. Megerle does not differ essentially from that of *Pterostichus.*

Others, whose antennæ are almost grained, have the upper part of the body rather convex and proportionally broader, and the abdomen shorter. It is the genus *molops* of M. Bonelli, which evidently conducts to other very analogous feroniæ, but whose corslet is rounded at the posterior angles, and whose abdomen is ovaliform, the external angle of the base of the elytra being obtuse or not salient. The body and the antennæ are in general proportionally longer. These last species have been detached from Pterostichus to form a new genus, that of Steropus, Meg.*

We shall terminate, in fine, this sub-genus with some species generally rather large, whose corslet has almost always the form of a truncated heart, and of which the base of the elytra has no transverse fold, or presents at most but

* See as well for this as the preceding, the Catalogue of M. le Comte Dejean, and M. Germar (Insect. Spec. Nov. I. p. 26, et seq.). Some species, such as *Molops terricola* (*Scarites piceus*, Panz Faun. Insect. Germ. XI. 2). *Molops elatus* (*Scarites gagates*, ejusd. XI). *Steropus hottentota* (*Scarites hottentota*. Oliv. Col. III. 36, 11. 19,) have been ranged with *scarites*. *Carabus madidus*, of Fabricius, a species common enough in some southern departments, is a *steropus*. M. le Comte Dejean forms a new genus with *Sc. Hottentot*, in consequence of its anterior feet, the legs of which are arched, and of some other characters.

a smooth space, and without a posterior edge well terminated. Such is the character which appears to me the best to signalize the genus Percus of M. Bonelli. Neither the relative length of the two last articulations of the maxillary palpi, nor the inequality of the proportions of the mandibles, nor some slight sexual differences taken from the last rings of the abdomen, distinguish it clearly from the other sub-genera. These species exclusively inhabit Spain, Italy, and the larger islands of the Mediterranean. Some are flatted above.*

Myas,

Of M. Ziegler, resemble the Feroniæ, with which has been formed the genus *cheporus*; but their corslet is more dilated laterally, narrowed near the posterior angles, and presents immediately in front of them a small emargination. The labial palpi are terminated by an articulation evidently thicker and almost triangular. Two species are known, one of Hungary (*chalybæus*), and the other of North America, where it has been discovered by M. Leconte.†

Sometimes the mandibles are as long as the head, and advance strongly beyond the chaperon. The body is always oblong, with the corslet in the form of an elongated heart; some resemble the scarites, and others the Lebia.

* *Carabus Paykulii*, Ross, Faun. Etrusc. Mant. 1. Tab. V. f. C. *Percus Ebenus*, Charp. Hor. Entom. V. I. See also the Annals of Natural Sciences, and those of the Physical Sciences, by M. M. Bory St. Vincent, Drapier, and Van-Mons. I refer to the same sub-genus, the *Abax Corsicus* of M. le Comte Dejean.

† Some other species, analogous in the form of the labial palpi, but with stronger mandibles, of which the middle tooth of the mentum is much larger, and proper to the East Indies, form the genus *Trigonotoma*, of M. Dejean, whose characters are given in the third volume of his *species*. Here, again, it would seem proper to place the genus, *pseudomorpha*, of M. Kirby (Lin. Trans. XIV. 98.).

Cephalotes, Bon. *Broscus.* Panz.

Have antennæ whose length equals, for the most part, the half of that of the body, composed of short articulations, and the first of which is shorter than the two following taken together. The right mandible is strongly unidentated at the internal edge, and the labrum is entire. (*Carabus Cephalotes*, Fab. Panz. Faun. Insect, Germ. LXXXIII., i. Ind. Ent. p. 62.)

Stomis, Clairv.

In which the antennæ are longer than one half of the body, composed of elongated articulations, and of which the first is longer than the two following joined together. The right mandible presents towards the middle of its internal side a strong notch, and the labrum of which is emarginated. *Stomis pumicatus.* Clairv. Entom. Helv. II. vi.

The following sub-genus, that of

Catascopus, Kirb.

Is distinguished from the preceding two, to which it otherwise approximates by the relative length of the third articulation of the antennæ, in that the body is flatted, proportionally broader, with the corslet shorter, the elytra strongly emarginated laterally at their posterior extremity, and the labrum is elongated. The eyes are large and projecting. These insects have brilliant colours, and resemble at the first aspect, the cicindelæ and elaphri.*

* This sub-genus has been established by Mr. Kirby, on a species of *Carabus*, (*Catascopus Hardwickii*, Lin. Trans. XIV. iii. i. Hist. Nat. des Coleop. d'Europe, II. vii. 8.) of the East Indies, having the head and corslet green, the elytra of a greenish blue, with punctuated striæ, and the under part of the body almost blackish. Mr. Macleay (Annal. Jav. LI. p. 14.) places the catascopi in his family of Harpalidæ, after Chlenius, and refers to it

In others the length of the third articulation of the antennæ is triple, or nearly so, of that of the preceding. These organs, as well as the feet, are generally slender.

In these, the first four articulations of the anterior tarsi of the males are broad, and the penultimate is bilobate.

Colpodes, Macl.

This sub-genus, established by Mr. Macleay (the son), Ann. Jav. I. p. 17, t. i., fig. 3, appears to have great relations with the preceding and the following. According to him the labrum is in a transverse and entire square. The emargination of the mentum is simple, or without tooth. The head is almost of the length of the corslet. This last is almost in the form of a truncated cone, emarginated in front, with the sides rounded, and a litte bordered. The elytra are a little emarginated. The lobes of the penultimate articulation of the anterior tarsi of the male are larger. The

the *C. elegans* of Fabricius, ranged with the Elaphri, by M. Weber. He distinguishes them from another neighbouring sub-genus, which he establishes under the name of *Pericalus*, by its antennæ, the second and third articulations of which are almost of equal length, while here the third is the longest; by the mandibles, which are short, thick, and curved, instead of being advanced, and almost parallel; also, by reason of the palpi, which are short and thick, with the last articulation ovoid, almost truncated, while those of the pericali are slender and cylindrical. Finally, because here the head is broader than the corslet, which does not take place in *catascopus*. The eyes, besides, are very prominent and globular in Pericalus, which gives them some resemblance to Elaphrus and Cicindela. He has described but one species (*Pericalus cicindeloides*, 1. 2.); but we are yet ignorant what are the sexual differences, especially in relation to the tarsi. The form of the tongue of catascopus, and that of the legs, remove them from Elaphrus and Tachys. The insects approach much more to the Chlenia, Anchomenus, Sphodrus, &c. Many simplicimanous Carabici have the extremity of their elytra strongly sinuated at the end, and in this character are hardly distinguishable from the *Truncatipennes*.

body is a little convex. He cites but a single species—(*Brunneus*).

In those all the articulations of the tarsi of the two sexes are entire.

(Mormolyce, Hegemb.)

The body is very much flatted, foliaceous, and much more narrow in its anterior moiety. The head is very long, very narrow, and almost cylindrical. The corslet is ovaliform, and truncated at the two ends. The elytra are very much dilated, and arched externally, with a profound emargination at the internal side, near their extremity.

The only species known (*Phyllodes*) has been the object of a particular monograph, published by M. Hagembach, and is found in Java.

Sphodrus, Clairv. Bon. *Lœmosthenus*, Bon. *Carabus*, Lin.

Have the body depressed, but not foliaceous, with the head ovoid, the corslet in the form of a heart, and the elytra without external dilatation or internal emargination.

Many of these insects abide in cellars.*

The last of the simplicimani are distinguished from all the others, by the internal denticulations of the crotchets of the end of their tarsi.

Some have all their external palpi filiform, and the corslet either in the form of a heart, narrowed and truncated posteriorly, or in that of a trapezium, widening from front to rear.

* *Carabus leucopthalmus*, Lin. *Carabus planus*, Fab. Fanz. Paun. Insect. Germ. xi. 4. In the *Sphodrus terricola* (*Carabus terricola*. Payk. Oliv. Col. III. xxxv. 11. 124), the crotchets of the tarsi present some small denticulations, as in the following sub-genus.

CTENIPES, Lat. *Lœmosthenus*, Bon.

Whose body is straight and elongated, with the corslet in the form of a heart, narrowed and truncated posteriorly. The third articulation of the antennæ is elongated.*

CALATHUS, Bon.

In which the body is oval, arched above, with the corslet square, or trapezoid, and broader posteriorly.†

The others have the labial palpi terminated in a club-form, or in that of a reversed cone, and the corslet almost orbicular.

TAPHRIA, Bon. *Synuchus*, Gyllen.

The emargination of the mentum is bidentated, as well as in the preceding sub-genus (*Carabus vivalis*, Ilig. Panz., *ibid.* xxxvii. 19).

The fifth section, that of Patellimani, is distinguished from the preceding only by the manner in which the two anterior tarsi are dilated in the males. The first articulations (generally the first three), sometimes square, sometimes partly in this form, and the others in the form of a heart, or reversed triangle, but always rounded at their extre-

* The Sphodri, *janthinus*, *complanatus*, and many others of M. le Comte Dejean, which are distinguished from the true Sphodri, by the shortening of the third articulation of the antennæ, and the articulations of the crotchets of the tarsi. These two sub-genera are confounded together almost insensibly. M. Fischer has figured many species of one and the other under the generic denomination of Sphodrus, in the second volume of his Entomology of Russia.

† *Carabus Melanocephalus*, Fab. Panz. Faun. Insect. Germ. xxx. 19. *C. cisteloides*. ibid. xi. 12. *C. fuscus*, Fab. *C. frigidus*. ejusd.—See the Catal. de la Collect. de M. le Comte Dejean, and M. Germar, Insect. Spec. Nov. I. p. 13.

mity, and not terminated, as in the preceding sections, by acute angles, form an orbicular palette, or a long square, the under part of which is most frequently furnished with brushes, or crowded papillæ, without a vacancy in the middle.

The feet are usually narrow and elongated. The corslet is often more narrow in all its length than the abdomen. They frequent for the most part the banks of rivers, or aquatic places.

We shall divide the patellimani into those whose head narrows insensibly behind, or at its base, and into those in which the narrowing takes place abruptly behind the eyes, so that the head seems to be borne on a sort of neck or pedicle.

The first may also be subdivided into two.

Some, whose mandibles terminate always in a point, and the palette of the tarsi is always narrowed, elongated, and formed by the first three articulations, the second and third of which being square, have the labrum entire, or without remarkable emargination, and one or two teeth in the emargination of the chin. The anterior extremity of the head is not bordered.

Here the under part of the palette of the tarsi presents, as in the preceding, two longitudinal series of papillæ, or hairs, with one intermediate void, and not a crowded and continued branch. The external palpi are always filiform, and terminated by an articulation almost cylindrical, or cylindrico-ovaliform.

Sometimes the body is very much flatted.

Dolichus, Bon.,

Which approximate to the last sub-genera, and are removed from all the following by the crotchets of their tarsi being denticulated underneath. Their corslet is in the form of a truncated heart. *Carabus flavicornis,* Fab. Preysl. Bohem. Insect., I., iii. 6, and some other species of the Cape of Good Hope.

PLATYNUS, Bon.

Similar, as to the form of the corslet, but with the crotchets of the tarsi simple.

The wings are either wanting or imperfect in some of them.*

AGONUM, Bon.

In which the corslet is almost orbicular.†

Sometimes the body is of an ordinary thickness. The corslet is always in the form of a truncated heart.

ANCHOMENUS, Bon.

The under part of the palette of the tarsi is furnished with a thick and continuous brush. The external palpi, and above all, the labial are, in many, terminated by an articulation thicker, or wider, in the form of a reversed triangle.‡

We shall commence with those which are filiform.

CALLISTUS, Bon.

Have the tooth of the emargination of the chin entire, the external palpi terminated by an ovaliform articulation, and

* *Platynus Complanatus,* Bon. *Carabus Angusticollis,* Fab. Panz. Faun. Insect. Germ. lxxiii. 9. *Platynus blandus,* Germ. Insect. Spec. Nov. I. p. 12. *Carabus scrobiculatus,* Fab. *Harpalus livens,* Gyll.

† *Harpalus Viduus,* Gyll.; Panz, *ibid.* XXXVII, 18; *Carabus marginatus,* Fab. Panz. *ibid.* XXX. 14; *C. 6-punctatus,* Fab. Panz. *ibid.* XXX. 13; *C. parum-punctatus,* Fab. Panz. *ibid.* XCII. 4; *C. 4-punctatus,* Fab. Oliv. Col. III. 35. xiii. 158, per the Catalogue of M. Dejean. He forms the *A. rotundatum,* and some others into a new genus.

‡ *Carabus prasinus,* Fab. Panz. *ibid.* XVI. 6; *Carabus albipes,* Fab. Panz. *ibid.* LXXIII. 7; *C. oblongus,* Fab. Panz. *ibid.* XXXIV. 3.

pointed at the end, and the corslet in the form of a truncated heart.*

Oodes, Bon.

Resemble Callistus as to the tooth of the emargination of the chin, but have the last articulation of the external maxillary palpi cylindrical, and the same of the labial in a truncated oval. The corslet is trapezoidal, more narrow in front, and of the breadth of the base of the abdomen, at its posterior edge. *C. helopiodes.* Fab. Panz. *ibid.* xxx. 11.

Chlænius, Bon.

In which the tooth of the emargination of the mentum is bifid, which have the external maxillary palpi terminated by an articulation almost cylindrical, a little attenuated at the base, and the last of the labial in the form of a reversed and elongated cone.

Carabus Saponaceus of Olivier (Col. III. 36. iii. 26), which they employ in Senegal as a sort of soap, belongs to this sub-genus.†

In the following the external palpi are terminated by a broader and compressed articulation, in the form of a reversed triangle, or hatchet, and more dilated in the males. The tooth of the emargination of the mentum is always bifid.

* *Carabus lunatus*, Fab. Panz. Faun. Insect. Germ. XIV. 5; Dej. Spec. II. p. 296.

† *C Cinctus*, Fab. Herbst. Archiv. XXIX. 7; *C. festivus*, Fab. Panz. *ibid.* XXX. 15; *C. Spoliatus*, Fab. Panz. *ibid.* XXXI. 6; *Chlœnius velutinus*, Dej.; *Carabus cinctus*, Oliv. Col. III. 35. iii. 28; *C. holosericeus*, Fab. Panz *ibid.* XI. 9; *C. nigricornis*, Fab. Panz ibid. XI. 9. h. c. *C. agrorum*, Oliv. *ibid.* XII. 144; *C. 4-sulcatus*, Payk, and many other exotic species of Fabricius, such as the following: *tenuicollis, occulatus, posticus, micans, quadricolor, stigma, ammon, carnifex*, &c. *See* the second vol. of the Spec. of M. le Comte Dejean.

EPOMIS, Bonelli,

To which we join DINODES, in which the last articulation of the palpi is a little more dilated.*

The genus *Lissauchenus*, of Mr. Macleay (the son), Ann. Jav. I. i. 1., appears to me to differ but little from the preceding.

The others have, most frequently, the mandibles very obtuse, or, as it were, truncated and forked, or bidentated at their extremity. Their labrum is distinctly emarginated, or bilobate, and the anterior portion of the head from which it originates, is bordered, and often concave. The emargination of the chin presents no tooth. The palette of the tarsi of many of them is broad, and almost orbicular.

These have the mandibles terminated in a point, without emargination or tooth below it.

The palette of the tarsi of the males is formed by the first three articulations.

REMBUS, Lat.

The labrum is bilobate. The external maxillary palpi are filiform, and the last articulation of the labials is a little swelled, in the form of a reversed and elongated cone.

The head is narrow, relatively to the breadth of the body. The antennæ and palpi are slender.†

DICÆLUS, Bon.

The labrum is simply emarginated with an impressed and

* *Dinodes rufipes*, Bon. Dej. Spec. II. p. 372; *Carabus azureus*, Duft. *Chlænius azureus*, Sturm, V. cxxvii; *Epomis circumscriptus*, Dej. Spec. II. p. 369; *Carabus cinctus*, Ross. Faun. Etrusc. I. iv. 9; *Carabus crœsus*, Fab.

† *Rembus politus*, Dej.; *Carabus politus*, Fab.; Herbst. Archiv. XXIX. 2; *R. Impressus*, Dej.; *Carabus impressus*, Fab.

longitudinal line in the middle. The last articulation of the external palpi is larger, and almost hatchet-formed.

The body is almost parallelipiped, with the head almost as wide as the corslet, and the elytra strongly striated, and often careened laterally. The mandibles are arched below, at the internal edge, and, as it were, truncated afterwards, and terminating in a point. The species which are known are American.

Those have the mandibles very obtuse, emarginated at their extremity, or unindentated underneath.

Licinus, Lat.

Have the last articulations of the external palpi larger, and almost in the form of a hatchet. The palette of the tarsi of the males is broad, sub-orbicular, and formed by the two first articulations, of which the basilary is very large.*

Badister, Clairv. *Amblychus*, Gyll.,

In which the last articulation of the external palpi is ovaliform. That of the labial is simply a little thicker (often terminated in a sharp point). The palette of the tarsi is in a long square, and formed by the first three articulations.†

The last patellimani, or those which compose their second general division, have the head suddenly narrowed behind the eyes, and distinguished, as it were, from the corslet, by a sort of neck or pedicle. It is often small, with the

* *Carabus Agricola*, Oliv. Col. III. 35. V. 53; *C. Selphoides*, Fab.; Sturm, III. lxxiv. a; *C. Emarginatus*, Oliv. *ibid.* XIII. 150; *Carabus cassideus*, Fab.; *C. depressus*, Payk, Sturm, *ibid.* LXXIV. v. V.; *C. Hoffmansegii*, Panz. Faun. Insect. Germ. LXXXIX. 5. *See* the species of M. le Comte Dejean, II. p. 392, 401.

† *Carabus bipustu'atus*, Fab. Clairv. Entom. Helv. II. xiii.; *C. peltatus* Ilig. Panz. *ibid.* XXXVII. 20. See the second volume of the Species of Count Dejean, pp. 405, 411.

eyes projecting. In many the tongue is short, and advances but little beyond the emargination of the mentum.

Here, this emargination has no tooth. The mandibles are strong, and the labrum is strongly emarginated, and almost bilobate. Such are—

Pelecium, Kirby.

The last articulation of the external palpi is in the form of a hatchet. The tongue is short. The body is oblong, more narrow in front. The first four articulations of the anterior tarsi of the males are in the form of a reversed triangle, furnished with a brush underneath, and the fourth is bifid.

The species of this sub-genus, and the following, are proper to South America, *Pelecium cyanipes*, Kirb. Trans. Lin. Soc. XII. xxi. 1.

Here, the emargination of the chin presents a tooth. The mandibles are generally small and moderate in the others. The labrum is entire, or but very slightly emarginated.

Some approach to pelecium, in regard to the exterior palpi, terminated also by a larger articulation, in the form of a hatchet, or reversed triangle. Their head is always small, and the corslet is orbicular or trapezoid.

Cynthia, Lat. Aupar. *Microcephalus*,

In the males of which the first articulations of the anterior tarsi are in the form of a reversed triangle, and compose the palette; they are furnished with a brush underneath, and the fourth is bifid.

The head and mandibles are proportionally stronger than in the following sub-genus. The external palpi are less elongated, but more compressed at the end. The body is oval, with the corslet trapezoidal, wider posteriorly, plane, bordered, and furrowed longitudinally.

This sub-genus is established on certain Brazilian species, having, as well as the Diclus, the appearance of the *Abax* of Mr. Bonelli.

PANAGÆUS, Lat.,

In which the palette of the tarsi, proper to the males, is formed only by the first two articulations.

The head is very small, comparatively with the body, and the eyes globular. The mandibles, jaws, and tongue are also very small. The corslet is most frequently sub-orbicular.*

In the following sub-genera, and which terminate this section, the external palpi are filiform. The last articulation of the maxillary palpi is almost cylindrical, and the same of the labial is almost ovaliform, or almost in a reversed and elongated cone. The first sub-genus, that of

LORICERA, Lat.,

Is very remarkable. Its antennæ are setaceous, curved, with the second articulation, and the following four shorter than the last, and furnished with bundles of hairs. The mandibles are small. The jaws are barbed externally. The labrum is rounded in front. The labial palpi are longer than the maxillary. The eyes are very much projecting. The corslet is almost orbicular, or heart-formed, broadly truncated, and rounded at the posterior angles. The first

* *Carabus crux-major*, Fab. Clairv. Entom. Helv. II. xv.; *Carabus notulatus*, Fab.; *Cychrus reflexus*, Fab. Oliv. Col. III. 35, vii.; *Carabus angulatus*, Fab. Oliv. *ibid.* vii. 76; *Panagée a quatre taches*, Cuv. Regne An. IV. xiv. 1. See the article *panagée*, of the Encyc. Method., and the second vol. of the Spec. of Count Dejean, p. 283, and following.

three articulations of the anterior tarsi are dilated in the males.*

PATROBUS, Meg.

Have filiform antennæ, straight, and without bundles of hairs, with the fourth and the following articulations equal, and almost cylindrical. The mandibles are of the usual size. The labrum is in a transversal square, with the interior edge straight. The length of the labial palpi does not exceed that of the maxillary. The corslet is in the form of a truncated heart, with the posterior angles acute. The first two articulations of the anterior tarsi alone, are dilated in the males. The eyes are less projecting, and the neck is less narrow than in the preceding sub-genus.—*Carabus rufipes*, Fab. *C. excavatus*, Payk, Panz. *ibid.* xxxiv, 2. M. le Comte Dejean, in the catalogue of his collection, mentions two other species, one of Portugal, and the other of North America.

We shall now pass to the carabici, whose anterior legs have no emargination in the interior side, or which have but one, but commencing very near the extremity of these legs, or not advancing on their anterior face, and forming only an oblique and linear canal. The tongue is often very short, terminating in a point in the middle of its summit, and accompanied with paraglossæ, also advancing in a point. The mandibles are robust. The last articulation of the external palpi is usually more large, compressed in the form of a reversed triangle or hatchet in some, almost spoon-formed in others. In the males it is often more dilated, which is peculiarly sensible in the *procerus*. The eyes are projecting. The elytra are entire, or simply sinuated at their posterior extremity. The abdomen is usually voluminous compara-

* *Loricera ænea*, Lat.; *Carabus pilicornis*, Fab. Panz. Faun. Insect. Germ. XL. 10; Oliv. Col. III. 35, xi. 119. Dej. Spec. II. p. 293.

tively to the other parts of the body. These carabici are for the most part of large size, ornamented with brilliant metallic colours, run very fast, and are extremely carnivorous. They will compose a peculiar section, the sixth of the genus, and which we shall name Grandipalpi, a denomination more characteristic than that of *abdominales*, which we formerly applied to it.

A first division will have for characters, the body always thick, without wings; labrum always bilobate; last articulation of the external palpi always very large; emargination of the chin without tooth; internal side of the mandibles entirely, or almost entirely denticulated in its length.

In some the mandibles are arched, strongly denticulated in all their length, and the lateral external extremity of the first two legs is prolonged into a point. The last articulation of the external palpi is a semi-oval longitudinal, with the internal side arched. The internal maxillary palpi are straight, with the last articulation much larger than the first, and almost ovoid. The emargination of the chin is of no great depth. Such are the characters of

Pamborus, Lat.

But a single species is known (*Pamborus alternans*), Cuv. Reg. An. V. xiv. 2, Dej. Spec. II. 18, 19, and which was brought from New Holland by M. M. Peron and Lesueur.

In others the mandibles are straight, simply arched, or crooked, and dilated at their extremity. The two anterior legs are not prolonged in the manner of a spine at their lateral extremity. The last articulation of the external palpi is much larger than the preceding, concave above, almost in the form of a spoon. The chin is profoundly emarginated, proportionally more elongated than in the following subgenera, thick on the sides in most of them, and as it were divided longitudinally into three spaces. The elytra are

Pl. 1

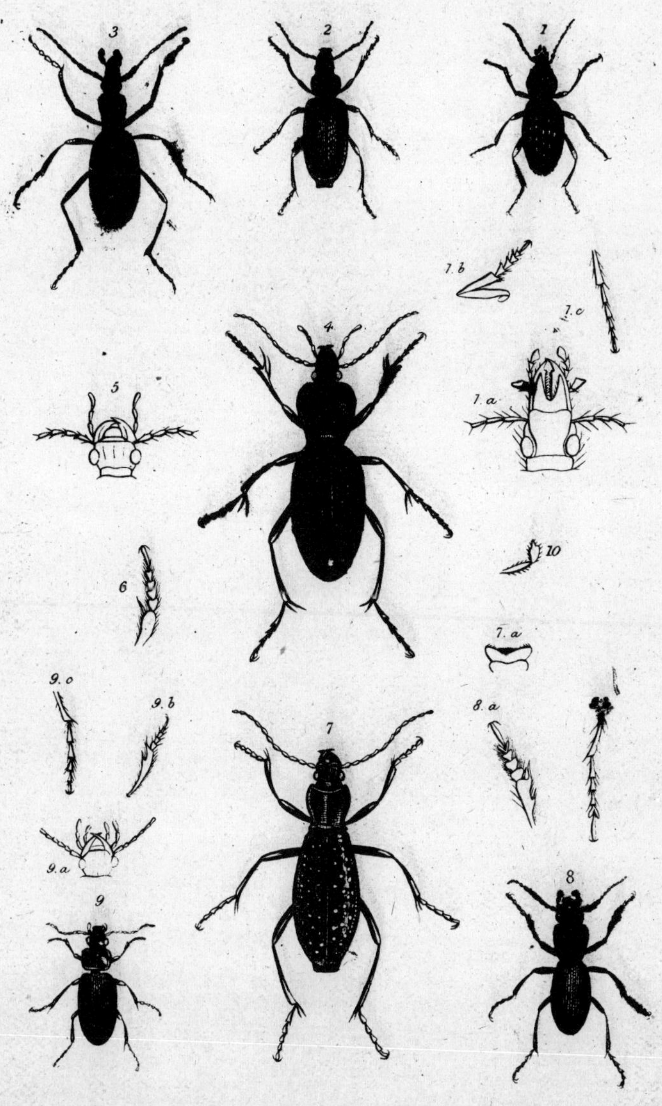

1. Sphœroderus *nitidicollis. Chevrol.* 2. Calosoma *rufipennis. Dej.* 3. Cychrus *italicus. Bonelli.* 4. Pamborus *alternans. Lat.* 5. *Head of* Pogonophorus. 6. *Anterior tarsus of* Pelophila. 7. Carabus *rutilans. Lat.* 8. Pelicium *cyanipes. Kirby.* 9. Masoreus *luxatus. Dej.* 10. *Maxillary palpi of* Bembidion.

synostosized, careened laterally, and embrace a part of the sides of the abdomen. These carabici compose the genus *Cychrus* of Paykull and Fabricius, but which has been since modified in the following manner.

Those whose tarsi are similar in both sexes, whose corslet is in the form of a truncated heart, more narrow behind, or almost orbicular, and not raised on the sides, with the posterior angles either obliterated or rounded, have alone preserved the generic denomination of

CYCHRUS.*

Those in which the males have the first three articulations of the anterior tarsi dilated, but feebly so, and palette-formed. The corslet is in the form of a trapezium, broad, emarginated at the two ends, raised on the sides, with the posterior angles acute and recurved, compose another generic section, that of

SCAPHINOTUS, Lat. Dej.

Cychrus elevatus, Fab. Knock. Beytr. I. viii. 12., Dej. Spec. II., p. 17.

Other species in fine having the appearance of Cychrus, but whose anterior tarsi have, in the males, the first two articulations very much dilated, and forming with the following, which is less so, and the figure of which is that of a heart, a palette, are, according to M. Dejean,

SPHÆRODERUS.

The species of these last two sub-genera are peculiar to America.

* *Cychrus Rostratus*, Fab. Panz. Faun. Insect. Germ. LXXIV. 5; Clairv. Entom. Helv. II. xix. A; *C. attenuatus*, Fab. Panz. *ibid.* II. 3; Clairv, *ibid.* xix. B.; *C. Italicus*, Bon. Observ. Entom. (Mem. de Acad. de Turin.) See for the other species, Spec. de M. le Comte Dejean, II. p. 4, et seq.

The second division of this section contains carabici, having, like the preceding, the body thick, most frequently without wings, but the chin is furnished, in the middle of its emargination, with an entire or a bifid tooth, and the mandibles are, at the most, armed with one or two teeth, situated at their base.

The corslet is always in the form of a truncated heart. The abdomen is most frequently ovaliform.

Some, the labrum of which is occasionally entire, have all the tarsi identical in the two sexes.

TEFFLUS, Leach,

Are the only ones of this division of which the labrum is entire, or without emargination.

Tefflus of Megerle (*Carabus Megerlei*, Fab. Voet. col. II. xxxix. 49) is about two inches long, and inhabits the coast of Guinea, and the eastern extremity of the Senegal. It is altogether black, with the corslet wrinkled, and the elytra divided by longitudinal ribs, and having raised points in their furrows. The last articulation of the external palpi is very large, in the form of an elongated hatchet, with the internal side curvilinear. The tooth of the emargination is very small. The third articulation of the antennæ is at least three times longer than the second.

PROCERUS, Meg.

Have the labrum bilobate. All the known species are equally of a large size, either entirely black, or of that colour underneath, and blue or greenish above, with the elytra very much chagreened. They generally inhabit the mountains of the eastern and southern countries of Europe, and those of the Caucasus and Mount Libanus.*

* *Carabus scabrosus*, Fab.; *C. Gigas*, Creutz Entom. I. ii. 13; *C. scabro-*

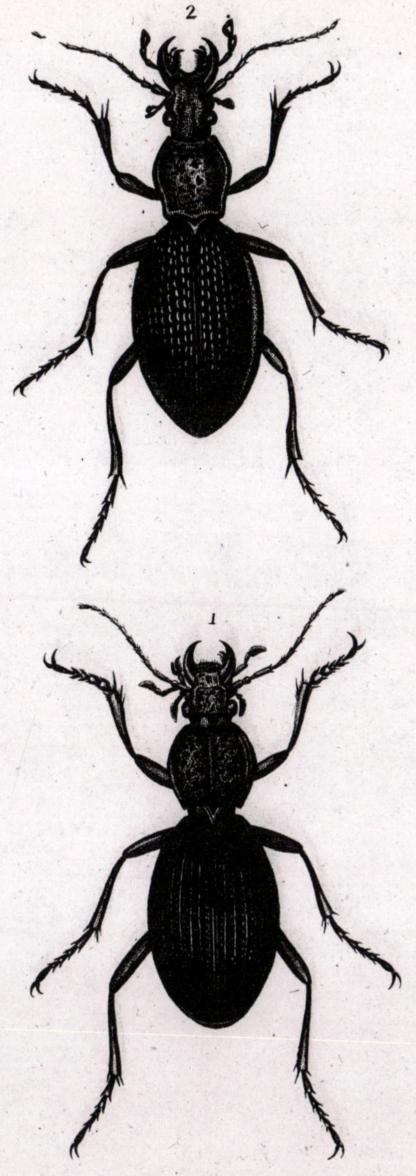

1. Tefflus *Megerlei. Leach.* 2. Procerus *Olivieri. Dej.*

The others, and the labrum of which is always divided into two or three lobes, have the anterior tarsi very sensibly dilated in the males.

These are never winged; their mandibles are smooth, and one or two teeth have been remarked at their base, or at least at the base of one of them. The corslet is in the form of a truncated heart, subisometrical, or more long than broad. The abdomen is ovaliform.

PROCRUSTES, Bonn.

The labrum of which is tribolate, and the tooth of the emargination of the chin is bifid.—*Carabus coriaceus*, Fab. Panz. Faun. Insect. Germ. LXXXI. 1. *See* second vol. of the Spec. of Count Dejean, n. 26, and following.

CARABUS, Lin. Fab. *Tachypus*, Web.

In which the labrum is simply emarginated or bilobate, and of which the tooth of the emargination of the chin is entire.

M. le Comte Dejean has described one hundred and twenty-four, which he has distributed into sixteen divisions. The first thirteen comprehend those whose elytra are convex or gibbous, and the last three those in which they are plain, and of which M. Fischer forms two genera, *plectes* and *cechenus*, founded on the relative proportions of the head and corslet.*

sus, Oliv. Col. III. 35, vii. 83, described and figured long ago by Mouffet, Ins. *theath.* 159; *P. tauricus*, Dej. Spec. II. 24; *Carabus scabrosus*, Fisch. Entom. de la Russ. I. ii. i. b. d. f.; *Procerus caucasicus*, Dej. *ibid.* 25; *Carabus scabrosus*, Fisch. *ibid.* c. e. M. Labillardiere has found, on Mount Libanus, another, but an unpublished species.

* *Carabus Hispanus*, Fab. Germ. Faun. Insect. Eur. VIII. 2; *C. cyaneus*, Fab. Panz. Faun. Insect. Germ. LXXX. I. 2; *C. Creutzeri*, Fab. Panz. ib. CIX. 1.; *C. depressus*, Bonell.; *C. osseticus*, Dej.; *plectes osseticus*, Fisch. Entom. de la Russie, II. xxxiiii. 3; *C. Fabricii*, Panz. *ibid.* CIX. 6; *C.*

The consideration of the surface of the elytra furnishes the other secondary characters of these divisions, and such has been the method of M. M. Clairville and Bonelli. The major part of these species inhabit Europe, the Caucasus, Siberia, Asia Minor, Syria, and the North of Africa, as far as about the thirtieth degree of north latitude. Some are also found at the two extremities of America, and it is probable that the mountains of the intermediate countries also possess some others.

Among the species with convex and oblong bodies, one of the most common is (*C. Auratus Lin.*) Panz. Faun. Insect. Germ. LXXXI. 4, vulgarly called the *Gardener*. Nearly an inch long, of a golden green above, black underneath, with the first articulations of the antennæ, and the feet fawn-colour. The elytra are furrowed, unindentated at the exterior edge near their extremity, especially in the female, with three smooth ribs on each.

This Carabus disappears in the South of Europe, or is no longer found but in the mountains.*

irregularis, Fab. Panz. *ibid.* V. 4; *C. pyræneus*, Dufuur. The last two enter into the genus *cechenus*, of M. Fischer. Their head is proportionally broader than those of the preceding species, or of the plates of M. Fischer.

* Add, *C. auro-nitens*, Fab. Panz. *ibid.* IV. 7; *C. nitens*, Fab. Panz. LXXXV. 2; *C. cœlatus*, F. Panz. *ibid.* LXXXVII. 3; *C. purpurascens*, F. Panz. *ibid.* IV. 5; *C. catenatus*, F. Panz. *ibid* LXXXVII. 4; *C. Catenulatus*, F. Panz. *ibid.* IV. 6; *C. affinis*, Panz. *ibid.* CIX. 3; *C. Scheidlerii*, F. Panz. *ibid.* LXVI. 2; *C. Monilis*, F. Panz. *ibid.* CVIII. 1; *C. consitus*, Panz. *ibid.* 3; *C. cancellatus*, F. Panz. *ibid.* LXXXV. 1; *C. arvensis*, F. Panz. *ibid.* LXXIV. 3, LXXXI. 3; *C. morbillosus*, F. Panz. *ibid.* LXXXI. 5; *C. granulatus*, F. Panz. *ibid.* 6; *C. violaceus*, F. Panz. *ibid.* XXXIX. 7; *C. glabratus*, F. Panz. *ibid.* LXXIV. 4; *C. convexus*, F. Panz. *ibid.* 5; *C. hortensis*, F. Panz. *ibid.* V. 2; *C. nodulosus*, F. Panz. *ibid.* LXXXIV. 4; *C. Sylvestris*, F. Panz. *ibid.* V. 3; *C. gemmatus*, F. Panz *ibid.* LXXIV. 2; *C. cæruleus*, Panz. *ibid.* CIX. 2; *C. concolor*, F. Panz. *ibid.* CVIII. 2; *C. Lin-*

Those are for the most part winged. Their mandibles are striated transversely, without perceptible teeth on the internal side. The corslet is transverse, equally dilated and rounded laterally, without elongations at the posterior angles. The abdomen is almost squared. Their external palpi are less dilated at their extremity. The second articulation of the antennæ is short, and the third elongated. The four hinder legs are arched in many of the males.

CALOSOMA, Web. Fab. *Calosoma, Callisthenes*, Fischer.

This sub-genus is much less numerous than the preceding, but its species extend from the north to the equator.

Carabus Sycophantha, Lin. Clairv. Entom. Helv. II. xxi. A.

From eight to ten lines in length, of a violet black, with the elytra of a golden green, or very brilliant copper colour, very finely striated, and having each three lines of little deep and distant points.

Its larva lives in the nest of *processionary caterpillars*, on which it feeds. It will eat a great number of them in one day. Some other larvæ of its own species, as yet young and small, attack and devour it, when by dint of having gormandized it has lost its activity. They are black, and are sometimes found running on the ground, or on trees, and more especially on the oak.*

næi, Panz. *ibid.* CIX. 5; *C. angustatus*, Panz. *ibid* 4. *See* for the synonimy of these species, and others of the same sub-genus, the second vol. of the species of M. Le Comte Dejean, pp. 30, 189.

* Add, *C. inquisitor*, Fab. Panz. Faun. Insect. Germ. LXXXI. 7; *C. reticulatum*, F. Panz. *ibid.* 9; *C. indagator*, F. Clairv. Ent. Helv. II. xxi. B. *C. scrutator*, F. Leach, Zool. Misc. xciii; *C. calidum*, F. Oliv. Col. III. 35, iv. 45, and ii. 21. The *C. porculatum* of Fabricius is an *helops*. *See* the second volume of the Species of M. Le Comte Dejean, p. 190, and following.

The third and last division of the grandipalpi presents an assemblage of characters which mark it distinctly from the preceding. The most part of them have wings. The anterior tarsi of the males are always dilated. The labrum is entire. The external palpi are merely a little dilated, or a little thicker towards their extremity, with the last articulation in the form of a reversed and elongated cone. The internal side of the mandibles presents no notable teeth. That of the middle of the emargination of the chin is bifid. The middle of the upper edge of the tongue rises into a point. The anterior limbs of many have at the interior side a short emargination, or one of the two spurs inserted above the other, so that in this point of view these carabici are ambiguous, and may come, as well as those of the following section, immediately after the patellimani. They generally frequent humid and watery places. Some of them, such as Omaphron, seem even to unite this tribe with the following, or the aquatic carnivora.

Some, whose body is flatted, or gibbous and orbicular, have eyes of the usual size, the antennæ linear, and composed of articulations generally elongated, and almost cylindrical, the external side of the jaws barbed, and the two internal spurs of the two anterior limbs on a level from their origin. These legs have but a simple longitudinal canal.

Sometimes the body is of an oblong oval, flatted, with the corslet in the form of a truncated heart, narrowed behind. The sentellum is distinct. The first three articulations of the anterior tarsi of the males are dilated.

Pogonophorus, Lat. Gyll. *Leistuss*, Froel. Clairv. *Carabus*, Fab. *Manticora*, Panz.

Remarkable for the elongation of their external palpi, and the labial being longer than the head ; for their mandibles, the external side of which forms a salient and flatted angle ;

and finally, for their tongue, which is advanced and terminated by three spines. Their head is suddenly narrowed behind the eyes, and the articulations of their antennæ are long and slender. All the known species are European.*

NEBRIA, Lat.

Which do not differ from pogonophorus, but in negative characters, or in that the palpi are much shorter, that the external side of the mandibles is little or not at all dilated, and forms only a very small auricle, not advancing beyond the basis of the jaws; that the tongue is short, and the head presents no vestiges of a neck. The antennæ are also proportionally more thick, and composed of shorter articulations.†

ALPÆUS, of M. Bonelli.

Are only apterous nebriæ, a little more oblong, and which more especially inhabit the high mountains. (*C. Helwigii*, of Panzer, Faun. Insect. Germ. LXXXIX. 4, is an Alpæus. See the Spec. of M. Dejean, p. 22, et seq.)

Sometimes the body convex, or gibbous above, is almost orbicular, with the corslet very short, transverse, very much emarginated in front, wider, and lobate posteriorly. The sentellum is not apparent. The first articulation of the two anterior tarsi of the males, and sometimes the same of the

* *Carabus spinibarbis*, Fab.; *Leistus cœruleus*, Clairv. Entom. Helv. II. xxiii. A. a.; *C. Spinilabris*, Fab.; *Leistus rufescens*, ibid. B. b.; *C. rufescens*, Fab.; *Carabus terminatus*, Panz. Faun. Insect. Germ. VII. ii. *See* for the other species the second volume of the Spec. of Count Dejean, p. 112, et seq.

† *Nebria arenaria*, Lat. Gen. Crust. et Insect. I 2. vii. 6; *Carabus brevicollis*, Fab. Panz. *ibid.* XI. 8; Clairv. *ibid.* 1. XXII. B; *C. Sabulosus*, Fab. Clairv. *ibid.* A. Panz. *ibid.* XXXI. 4; *C. picicornis*, Fab. Panz. *ibid.* XCII. 1; *C. psammodes*, Ross. Faun. Etrusc. Mant. I. v. M.

intermediate tarsi, as in the *O. Melangé*, is alone sensibly dilated.

OMOPHRON, Lat. *Scolytus*, Fab.

This sub-genus is composed of a small number of species which are found on the banks of waters in Europe, in North America, in Egypt, and at the Cape of Good Hope. M. Desmarest has described the larva of the most common species. Its form approaches that of the larva of dytiscus. The anatomical observations of M. Dufour appear to confirm these relations.

The others, whose body is tolerably thick, have very large and projecting eyes; antennæ thickening a little towards their extremity, and composed of short articulations, for the most part in the form of a reversed cone. One of the two spurs of the internal extremity of the two anterior legs is inserted more high than the other, with a notch between. The first three or four articulations of the anterior tarsi of the males are but little dilated in the majority. The palpi are never elongated. These insects inhabit the neighbourhood of rivers, and all belong to Europe or Siberia.

Sometimes the labrum is very short, transverse, and terminated by a right line. The last articulation of the external palpi is almost in the form of a reversed cone, thicker, and more truncated at the end. The mandibles advance remarkably beyond the labrum. The anterior tarsi of the males are sensibly dilated.

ELAPHRUS, Fab. *Elaphrus*, *Blethisa*, *Pelophila*, Dej.

Some, and the largest (*Blethisa*, Bonelli), have the corslet more broad than long, plain, bordered laterally, almost square, and a little contracted towards the posterior angles.

In one set the first three articulations of the anterior tarsi are strongly dilated, and cordiform in the males. These are the *pelophila* of M. Dejean. (*Carabus Borealis*, Fab. *Nebria borealis*, Gyll. Panz. Faun. Insect. Germ. LXXV. 8.)

In another the first four articulations of the anterior tarsi of the males are slightly dilated, these are the *Blethisa* of the same. (*Carabus multipunctatus*, Fab. Panz. *ibid.* XI. 5.)

The others have the corslet at least as long as broad, convex, and in the form of a truncated heart. The body is proportionally more convex than in the preceding. The first four articulations of the anterior tarsi are slightly dilated in the males. These composed exclusively his genus ELAPHRUS.

C. uliginosus, Fab.? *elaphrus riparius*, Oliv. Col. II. 34, I. i. A. E. is about four lines in length, of a blackish bronze, very much punctuated, with impressions or small fassets on the forehead and corslet, and others with a violet ground, raised in their contour, and joined one to another on the elytra. The tarsi are of a blueish black, but the legs are sometimes of this colour, and sometimes reddish. These last individuals have been considered as belonging to a peculiar species (*Cupreus*), by M. M. Megerle and the Count Dejean. It is very rare in the environs of Paris, but common in the other parts of France, in Germany, Sweden, &c.

Cicindela riparia, Lin. *Elaphrus riparius*, Fab. Clairv. Entom. Helvet. II. xxv. A. a. *Elaphrus paludosus*, Oliv. Col. II. 34, 1, 4, a. b. Panz. Faun. Insect. Germ. xx. 1. About one-third smaller than the preceding, very finely punctuated, and of a dull copper, mingled with green above, with circular impressions, nipple formed at the centre, green, disposed in four lines, and a coppery spot polished and shining near the section on each elytrum. Common in the environs of Paris. (*See* for the other species the second volume of the species of Count Dejean, p. 268, et seq.)

Sometimes the labrum is almost semicircular and rounded in front, the external palpi are terminated by a suboval articulation, narrowed into a point at the end. The mandibles are advanced a little beyond the labrum. The tarsi are identical in the two sexes.

The anterior extremity of the head forms a small muzzle. The upper part of the body is plane, with the corslet trapezoid, almost as broad as the head, and a little narrowed posteriorly.

NOTIOPHILUS, Dumer. *Elaphrus.* Fab. Oliv.*

Our second general division of this tribe (SUBULIPALPI) is distinguished from the preceding by the form of the external palpi, of which the last articulation but one, in the form of a reversed cone, is united with the following, and comprises with it a common ovaliform, or fusiform body, terminated, either insensibly or suddenly, in a point, or oval-like form. The two anterior legs are always emarginated at the internal side. These insects resemble the last very much, both in forms and manner of life.

BEMBIDION, Lat. *Bembidium*, Gyll. Dej.

Have the last articulation but one of the external palpi large, swelled, and somewhat cone-formed, and the last much more slender, very short, conical or acicular. The first articulation of the two anterior tarsi is dilated in the males.

M. M. Ziegler and Megerle, have divided this sub-genus into many others, but without giving the characters, and resting solely, as it appears, on the changes of form in the corslet.†

* *Cicindela aquatica*, Lin.; *Elaphrus aquaticus*, Fab. Panz. Faun. Insect. Germ. XX. 3; *Elaphrus biguttatus*, Fab., and to which M. le Comte Dejean refers his *C. semipunctatus*. Consult the species of this last, II. p. 276, et seq.

† This sub-genus may be thus divided: some have the corslet less de-

The following species is ranged by M. le Comte Dejean, with his tachypus—

Cicindela, flavipes. Lin. Panz. Faun. Insect. Germ.

pressed, at least as long as broad, much more narrow behind than in front, a truncated heart, with the posterior angles very short, or but little elongated.

Those in which this part of the body presents at the posterior angles no very marked impression, and whose eyes are very large, and cause the head to appear broader than the corslet, form the genus *Tachypus*, of M. Megerle.

Those whose eyes, as well as in all the following, have less projection, so that the corslet is not broader than the head, but otherwise exhibits the same characters, are the *Bembidium* proper, of M. le Comte Dejean.

With M. Megerle he ranges, in the genus *Lopha*, those whose corslet, having the same form and the same proportions, presents at each posterior angle a very well marked impression, so that these angles are well bordered.

The other Bembidia have the body more flatted, the corslet more broad than long, and proportionally less narrowed posteriorly. Its posterior angles have always a strong impression, and a small oblique keel.

Some species whose corslet, though narrowed near the posterior angles, is nevertheless less so than in the others, so that the posterior edge is but little more narrow than the anterior, compose the genus *Notaphus* of the same, and of M. Megerle.

Among those whose corslet is notably narrowed behind—sometimes its length is only a little greater than its breadth, and it is in the form of a truncated heart—such are the *Peryphus* of these naturalists. Sometimes, much shorter proportionally, its form approximates to that of a cup, or a heart, much widened; in some it is even rounded at the posterior angles. These species constitute, with them, the genus *Leja*. The tachypus, in consequence of the extraordinary projection of their eyes, of their other relations with the Elaphri, are tolerably distinct. But it is not so with the other genera. It is impossible to determine them by rigorous characters. Those which may be drawn from the respective and comparative lengths of the second and third articulations of the antennæ still appear to me uncertain. See the Catalogue of the Collection of Coleoptera of M. Dejean.

XX. 2, very similar to the *Elaphrus riparius;* two lines in length; the corslet a little more narrow than the head, in the form of a truncated heart, as long as broad. The eyes large; under part of the body of a blackish green, upper part bronzed, and marbled with coppery red; two large points sunk near the suture, on each case; base of the antennæ, palpi and feet, yellowish. Very common in the environs of Paris.*

TRECHUS. Clairv.

Which have the last articulation of their external palpi as long or longer than the preceding, of the same thickness at its origin, so that these two articulations form, when united, a fusi-form body.†

* Add *Carabus tricolor*, Fab. ejusd. *C. modestus—cursor—biguttatus—quatuor-guttatus—guttula; C. minutus*, Panz. Faun. Insect. Germ. XXXVIII. 10; *C. pygmæus*, Panz. *ibid.* 11; *C. articulatus*, Panz. *ibid.* XXX. 21; *Cicindela qudrimaculata*, Lin.; *Carabus pulchellus*, Panz. ibid. XXXVIII. 8; XL. 5; *C. doris*, Panz. *ibid.* 9; *Elaphrus rupestris*, Fab. Panz. *ibid.* XL. 6; *C. decorus*, Panz. *ibid.* LXXIII. 4; *C. ustulatus*, Lin. Panz. *ibid.* XL. 7. 9; *C. bipunctatus*, Lin. Oliv. Col. III. 35. xiv. 163; *Elaphrus ruficollis*, Panz. *ibid.* XXXVIII. 21; *Elaphrus impressus*, F. Panz. *ibid.* XL. 8; *Elaphrus paludosus*, *ibid.* XX. 4.

† *Trechus rubens*, Clairv. Entom. Helv. II. ii. B. b. The *carabus merulianus*, which he represents in the same plate, A. a. is a stenolophus, *Carabus micros*, Panz. Faun. Insect. Germ. XI. 4. The *G. Masoreus* of MM. Ziegler and Dejean appears to me to border on that of Trechus. The species on which it is founded is very near the *harpalus collaris* of M. Gyllenhall. The maxillary palpi terminate, as well as those of Trechus, in the manner of a spindle, only the last articulation but one is much shorter than the following. The antèrior tarsi are slightly dilated in the males. This insect appears to connect the trechus with various smaller species of Stenolophi of M. Dejean.

BLEMUS of the same naturalists, are some of the species of Trechus, more narrow and more elongated, with subisometric corslet, in the form

The pentamerous, aquatic, carnivorous Coleoptera, form a third tribe, that of HYDROCANTHARI, or *Swimmers*.

Their feet are adapted to natation; the last four are compressed, ciliated, or plate-formed, and the last two are remote from the others. The mandibles are almost entirely covered. The body is always oval, with the eyes not much projecting, and the corslet more wide than long. The crotchet which terminates the jaws is arched from its base. Those of the tarsi are often unequal.

These insects compose the genus *Dytiscus* and *Gyrinus* of Geoffroy. They pass the first and last state of their life in the fresh and tranquil waters of lakes, of marshes, ponds, &c. They swim very well, and repair from time to time to the surface to respire. They easily remount thither by holding their feet in a state of repose, and suffering themselves to float. Their body being reversed, they elevate the hinder part a little out of the water, raise the extremity of their elytra, or incline the end of their abdomen; so that the air may insinuate itself into the stigmata which they cover, and from thence into the tracheæ. They are very voracious, and live on small animals, which, like them, make their constant residence in the same element. They do not remove from it, but during the night, or at its approach. When they are taken out of the water, they shed a most nauseous odour. The light sometimes attracts them into the interior of houses.

Their larvæ have the body long and narrow, composed of twelve rings, of which the first is the larger, with the head strong, and presenting two powerful mandibles, curved into an arch, and pierced near their point with small antennæ

of a reversed and truncated triangle, and with mandibles considerably larger, and prolonged beyond the labrum. They are found under stones, on our maritime coasts, and even in the sea itself.

with palpi, and six simple eyes on each side, approximating closely. They have six feet, tolerably long, often fringed with hair, and terminated by two small claws. They are agile, carnivorous, and respire either by the anus or by species of fins which imitate gills. They come out of the water to be metamorphosed into nymphs.

This tribe is composed of two principal genera.

DYTISCUS,

Which have the antennæ like threads, longer than the head, two eyes, the anterior feet shorter than the following, and the last most frequently terminated by a compressed tarsus, going in a point. They swim with much swiftness by the assistance of their feet, furnished with long hairs, and partilarly with the last two. They dart on other insects, aquatic worms, &c. In the majority of the males the four anterior tarsi have their first three articulations widened, and spungy underneath. Those of the first pair are especially remarkable in the larger species. These three articulations form there a large palette, the inferior surface of which is covered with small bodies, some like papillæ, and others larger, and in the form of suckers, &c. Some females are distinguished from their males by furrowed cases. The larvæ have the body composed of from eleven to twelve rings, and covered with a scaly plate. They are long, big-bellied towards the middle, more narrow at the two extremities, particularly behind, where the last rings form an elongated cone, furnished on the sides with a fringe of floating hairs, with which the animal pushes the water and causes his body to advance, which is usually terminated by two conical threads, barbed, and mobile. Between them are two small cylindrical bodies, pierced with a hole at their extremity, and which are aeriform conduits, to which the two trachea lead. Stigmata, however, are distinguishable on the side of the abdomen.

The head is large, oval, attached to the corslet by a neck, with mandibles very much arched, and under the extremity of which De Geer has perceived a longitudinal cleft, so that in this respect these organs resemble the mandibles of the larvæ of antlions, and serve as suckers. The mouth, nevertheless, presents jaws and a labrum, with palpi. The first three rings have each a pair of feet, tolerably long, the leg and tarsus of which are bordered with hairs, which are also useful in swimming. The first ring is larger or longer, and defended underneath, as well as above, by a scaly plate.

These larvæ suspend themselves to the surface of the water by means of two lateral appendages at the end of their tail, and which they keep dry. When they desire suddenly to change place, they give to their body a prompt and vermicular motion, and strike the water with their tail. They are nourished more particularly with the larvæ of gnats, stipulæ, libellulæ, asselli, &c. When the time of their transformation is come they quit the water, gain the shore, and bury themselves in the ground; but it must be always moistened, or very humid. They there form an oval cavity, in which they shut themselves up.

According to Rœsel, the eggs of the *dytiscus marginalis* exclude the young in ten or twelve days after the laying. At the end of from four to five days the larva is already four or five lines in length, and moults for the first time. The second change of skin takes place at the end of an interval of the same duration, and the animal is as large again. The length of two inches is the limit of its growth. In summer it has been observed to change into the nymph-state at the end of fifteen days, and into the perfect insect, fifteen or twenty days afterwards. Besides the cloaca of the insects of this family, the dytisci have a cœcum tolerably long, which is perceptible in the larva state.

This great genus is subdivided as follows:

Some have the antennæ composed of eleven distinct articulations, the external palpi filiform, or a little thicker at their extremity, and the base of their posterior feet, as well as that of the others, uncovered.

Sometimes the thickness of the antennæ gradually diminishes from their origin as far as their extremity; the last articulation of the labial palpi is simply obtuse at its extremity, without emargination. Such are

Dytiscus (Proper),

In which all the tarsi have five very distinct articulations, and of which the anterior two have, in the males, the first three articulations very broad, and forming altogether a pallet, either oval or transverse, or orbicular.

D. Latissimus, Lin. Panz. Faun. Insect. Germ. LXXXVI. i. Nearly an inch and a half long, and very distinct by the compressed and trenchant dilation of the external margin of the wing-cases whose edge is yellowish. The corslet is edged all around with the same colour. The cases are furrowed and ribbed in the female. Found in the department of the Vosges, in the North of Europe and in Germany.

D. Marginalis, Lin. Panz. *ibid.* 3. About one-fourth smaller, having also a yellowish border all around the corslet, and a line of the same colour on the external and not dilated edge of the wing-cases. Those of the female are furrowed from their base, as far as about two-thirds of their length.

Fabricius tells us, that when turned upon its back, it replaces itself by leaping, in its ordinary position.

Esper preserved, for three years and a half, in a glass vessel, a dytiscus of this species, and always in good health. He used to give it every week, and sometimes oftener, a bit of raw beef, about the size of a nut, which it would attack

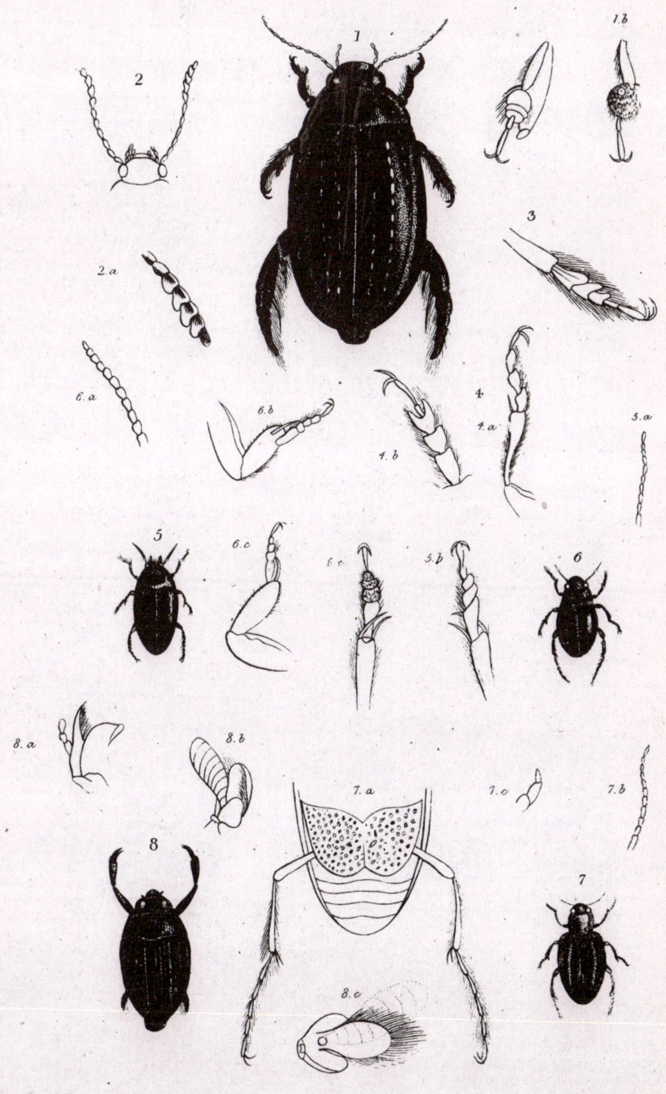

1. Dytiscus *Lherminieri. Chevr.* 2. *Id. serricornis Payk.* 3. *Ant tarsus of* Colymbetes. 4. *Id. of* Hydroporus *planus. Latr.* 5. Hygrobia *Hermanni. Latr.* 6. Noterus *crassicornis. Clairo.* 7. Haliplus *elevatus.* 8. Gyrinus *sulcatus. Dej.*

with avidity, and suck the blood out in the most complete manner. It is capable of fasting for four weeks at least. It kills the brown hydrophilus, though as big again as itself, by piercing it between the head and the corslet, the only part of the body which is without defence. According to Esper it is sensible to the changes of the atmosphere, and indicates them by the height to which it rises in the bottle.

D. Rœselli, Fab. Rœs. In. II. Aquat. Class. I. ii. More narrow, or more oval, and more depressed than the preceding. The external edge of the corslet and wing-cases is yellowish. These cases are very finely striated in the female. Found in the neighbourhood of Paris, and in Germany.

D. Serricornis, Payk. Nov. Sect. Acad. Scient. Stockh. XX., 1. 3. Very singular from the irregular form of the antennæ in the male, of which the four last articulations form a compressed mass, and denticulated like a saw.*

COLYMBETES, Clairv.

In which all the tarsi have also five articulations very distinct, but the anterior four have, in the males, their first three articulations equally dilated, and form together only a

* Dr. Leach has founded on this character his genus, AGABUS—(Zool. Misc. III. pag. 69, and 72.) Some slight differences in the form and relative proportions of the articulations of the external maxillary palpi have also determined him to establish some others, such as those of HYDATICUS (*D. Hybneri, transversalis, stagnalis 4-vittatus*) of ACILIUS (*D. Sulcatus*), and of TROGUS (*D. lateralis*). The last alone may be preserved in consequence of some other characters. The hinder feet have the legs short, very broad, and their tarsi are terminated only by a single crotchet. To the species described above, add, *D. Sulcatus*, Fab. Clairv. Entom. Helv. II. xx; *D. Castalis*, Oliv. Col. III. 40. 1. 7; *D. punctatus*, ibid. 1. 6. 1. and 1. e.; *D. aciculatus*, ibid. III. 30; *D. lævigatus*, ibid. 23; *D. tripunctatus*, ibid. 24; *Ruficollis*, ibid. ii. 20; *D. vittatis*, ibid. 1. 5; *D. griseus*, ibid. ii. 12; *D. sticticus*, ibid. ii. 2; *D. circumflexus*, F.

small palette in the form of a long square. Their antennæ are at least of the length of the head and corslet. The body is perfectly oval, and more broad than high. The eyes are either not at all, or very little prominent.*

HYGROBIA, Lat. *Hydrachna*, Fab. Clairv. *Pœlobius*, Schœnh.

Which again have the tarsi with five distinct articulations, and of which the four anterior ones are dilated almost equally at their base, in the males, into a small palette formed like a long square, but whose antennæ are shorter than the head and corslet; which have the body oval, very thick in the middle, and the eyes prominent.†

* *D. fuscus*, Panz. Faun. Insect. Germ. LXXXVI. 5.; *D. Cinereus*, F. Panz. *ibid.* XXXI. ii.; *D. zonatus*, F. Panz. *ibid.* XXXVIII. 3; *D. bipunctatus*, F. Panz. *ibid.* XCI. 6; *D. fenestratus*, F. Panz. *ibid.* XXXVIII. 16; *D. Chalconatus*, F. Panz. *ibid.* 17; *D. ater*, F. Panz. *ibid.* 15; *D. guttatus*, Payk. Panz. *ibid.* XC. i.; *D. fuligenosus*, F. Panz. ibid. XXXVIII. 14; *D. bi-pustulatus*, F. Panz. *ibid.* CI. 2: *D. Stagnalis*, F. Panz. *ibid.* XCI. 7; *D. transversalis*, F. Panz. *ibid.* LXXXVI. 6; *D. abbreviatus*, F. Panz. *ibid.* XIV. i.; *D. maculatus*, F. Panz. ibid. 7.; *D. agilis*, F. Panz. *ibid.* XC. 2; *D. adspersus*, F. Panz. *ibid.* XXXVIII. 18; *D. minutus*, F. Panz. XXVI. 3. 5.; *D. leander*, Oliv. *ibid.* iii. 23; *D. varius*, Oliv. *ibid.* ii. 17; *D. bimaculatus*, Oliv. *ibid.* 18. *See* Clarville, Entom. Helv., tom. II., genus *Colymbetes.*

Some small species having no distinct scutellum, and whose anterior tarsi are but little dilated in the males, compose the genus

LACOPHILUS of Dr. Leach.—He quotes the following: *D. hyalinus*, Marsh; *D. interruptus*, Panz.; *D. minutus*, Lin.; *D. marmoreus*, Oliv. *See* his Zool. Miscel. III. p. 72.

† *Hydrachna Hermanni*, Fab. Latr. Gen. Crust. et Insect. I. vi. 5; Clairv. Entom. Helv. II. xxvii. A. a.; *H. uliginosa*, Clairv. *ibid.* B. b.

These insects, and the Halipli form in the method of Dr. Leach, Zool. Misc. III. p. 68, a particular group, having for characters, a scutellum, all the feet adapted for walking, five articulations in all the tarsi, and two crotchets at the end of the last.

Hydroporus, Clairv. *Hyphydrus*, Lat. Schœnh.,

In which the four anterior tarsi almost similar, and spungy underneath, in both sexes, have but four distinct articulations, the fourth being nullified, or very small and concealed, as well as a part of the last in a deep fissure of the third.

They have no apparent scutellum, but the preceding, with the exception of some small species, have a very sensible one.

We might detach from them some species (The Hydrachnæ, *gibba*, *ovalis*, and *scripta* of Fabricus; *Hyphydrus Lyratus*, Schœnh. Synon. Insect. II. iv. i.), whose body is very gibbous, or almost globular, and of which the least articulation of the four anterior tarsi is very small, and projects but little beyond the preceding. (*Hyphydrus*, Lat.) The others have the body oval and less thick.*

Sometimes the antennæ are a little dilated, and wider towards the middle of their length; the last articulation of the labial palpi has an emargination, and appears forked.

Noterus, Clairv.

The scutellum is wanting; the tarsi have five distinct articulations, the first two of the four anterior ones are dilated in the males, and form an elongated palette. The first articu-

The Hygrobiæ have the external palpi a little swelled at their extremity, two strong spurs, and approximating to each other, at the end of the legs; and their anterior tarsi capable of being folded under the legs, on which they depend.

* The Dytiscus: *inæqualis*, *reticulatus*, *confluens*, *picipes*, *pictus*, *geminus*, *lineatus*, *halensis*, *duodecim—pustulatus*, *dorsalis*, *sex—pustulatus*, *palustris*, *depressus*, *lituratus*, *planus*, *erythrocephalus*, *nigrita*, *granularis*, of Fabricius. *See* Schœnherr. Synon. Insect. tom. II. genus *Hyphydrus*; Panzer, Index Entom. genus *Hydroporus*, and Clairv. Entom. Helv., tom. II. same genus.

lation of the two anterior tarsi is covered in the same individuals by a broad spur, plate-formed. The pectoral piece which bears the hinder feet, has on each side a deep groove. *Dytiscus crassicornis*, Fab. Clairv. Entom. Helv. II. xxxii.

The others have but six distinct articulations on the antennæ. Their external palpi terminate oval-like, or by a more slender articulation, and going into a point. The base of their posterior feet is covered with a large lamina, of a buckler-form.

The body is gibbous underneath, and ovoid, as in hygrobia. But they have no scutellum, and all their tarsi are filiform, with five distinct articulations, and almost cylindrical, and have pretty nearly the same form in both sexes. These are

HALIPLUS, Lat. *Hoplitus*, Clairv. *Cnemidotus*, Ilig.*

The second genus, or that of

GYRINUS,

Comprehends those whose antennæ are club-formed, and shorter than the head. The first two feet are long, advanced in the form of arms, and the other four very much compressed, broad, and fin-formed. The eyes are four in number.

The body is oval, and usually very shining. The antennæ, inserted in a cavity in front of the eyes, have the second articulation prolonged externally in the form of an auricle, and the following articulations (of which but seven are visible, the first and last of which are the longer), very short, very

* The Dytisci: *fulvus, impressus, obliquus*, of Fabricius. *See* Latreille, Gener. Crust. et Insect. I. p. 234; Clairv. Entom. Helv. tom. II. genus *Hoplitus*, XXXI. Panz. Ind. Entom. genus id., and Schœnherr. Synom. Insect. II. genus *Cnemidotus*.

crowded, and unite into a mass, almost in the form of a spindle, and a little curved. The head is sunk in the corslet, as far as the eyes, which are large, and divided by a border, so that there appear two above and two below. The labrum is rounded, and very much ciliated in front. The palpi are very small, and the interior of the maxillary is wanting, or a mere abortion in many species, more especially in the largest. The corslet is short and transverse. The elytra are obtuse or truncated at the posterior end, and leave the anus uncovered, which terminates in a point. The two anterior feet are slender, long, folded in double, and almost at a right angle with the body, in a state of contraction, and terminated by a very short and very compressed tarsus, the under part of which is furnished with a fine and thick brush, in the males. The other four are broad, very slender, as it were membranous, and the articulations of the tarsi form small leaves, disposed like furbelows.

The gyrini are in general of small or middle size. They are seen from the first days of spring to the end of autumn, at the surface of dormant waters, and even on those of the sea, often assembled in troops. They appear there, from the effect of the light, like brilliant points, swimming or running with an extreme agility, making turns, circular, oblique, and in all directions, from which comes the name of *aquatic flea*, and *tourniquet*, given them by some writers. Sometimes they repose and appear perfectly motionless, but the moment they are approached they escape by swimming, and immediately sink under water with the greatest celerity. The last four feet serve them as oars, and the first two for seizing their prey. Placed on the surface of the water the upper part of their body always remains dry, and when they dive, an air-bubble, like a silvery globe, remains attached to their hinder part. If they are seized, a milky liquid oozes from their body, and is spread over it, which produces the disagreeable and penetrating odour which they then exhale, and which remains a long time on the fingers. They copulate on the surface of the water. Sometimes they remain at the bottom, fixed to the plants. It is there in all probability that they conceal themselves to pass the winter.

M. Leon Dufour has published, in the Annals of Natural Science, some anatomical observations on these insects. The narrow intestine is remarkable for its length. The cœcum is not lateral like that of the dytisci. The genital male organs differ from those of the other carnivora.

G. Natator, Lin. Panz. Faun. Insect. Germ. III. 5; De Geer. Insect. IV. xiii, 4. 19. Three lines in length, oval, very smooth, very shining, of a black, bronzed above, black underneath, and with fawn-coloured feet. The scutellum is triangular, very pointed, a little longer than broad. The elytra are rounded at the end, with small sunken points, forming regular and longitudinal lines.

The female lays her eggs on the aquatic plants. They are very small, in the form of little cylinders, and of a white a little yellowish. The larva has the body long, slender, linear, and composed of thirteen rings, of which the first three have each a pair of feet. The head large, in an elongated oval, and very flatted, presents the same parts as those of the larvæ of the dytisci; but here the fourth ring and the following seven, have on each side a conical, membranous, and flexible thread, and barbed at the edges. The twelfth ring has four similar threads, but much larger, and more directed backwards. Two very fine tracheæ traverse the whole length of the body, and receive from each thread an arterial vessel. The last ring of the body is very small, and terminated by four long and parallel crotchets. This larva lives in the water, and proceeds from it in the commencement of August, to pass into the nymph state. It forms, with a matter which it derives from its body, and similar to grey paper, a small oval shell, pointed at the two ends, which it affixes to the leaves of a reed, and shuts itself up there.

This species is very common in Europe.*

* *See* for the other species, Olivier, Col. III. No. 41, and Schœnherr. Synom. Insect. II. No. 55. We find also in the Environs of Paris, the Gyrini, *minutus* and *bicolor*, of Fabricius. The largest species, and all exotic, have no sensible scutellum, and their palpi are but four in number.

Mr. Macleay (the younger), Annal. Jav. I. p. 30, forms a peculiar genus, under the name of Dineutes, with the species whose labrum is not ciliated, whose palpi are club-formed, which have all the anterior feet of the length of the body, and the antennæ terminated in a point. He names but one species (*Politus*).

SUPPLEMENT

ON THE CARNIVOROUS FAMILY

OF THE

ORDER COLEOPTERA.

On this immensely numerous order of the insect class, we can afford but a moderate portion of supplementary matter. Not that the study of coleopterous insects is devoid of interest; on the contrary, it is replete with all that may gratify curiosity, and communicate instruction. But with not a great many exceptions, the animals now under our consideration arrest our attention, more from peculiarities of structure and appearance, than those of habitudes and manners. On the former it is not our intention to dilate, further than may be necessary to supply any omission in the very ample details of the text; on the latter we shall endeavour to glean as much information as possible.

This order is not one of a new creation—Aristotle may, with propriety, be considered as its founder, having been the first to name and define it. The name is derived from πτερον, a wing, and κολεος, a sheath or case, and Aristotle embraces within his definition all insects with the character of increased wings, "οσα το πτερον εχει εν κολεω." *As many as have the wing in a sheath.* To this have been added, by subsequent authors, such characters as the folding of the wings, and the straight suture of the elytra. These characters, however, are not universal in the order; some beetles are destitute of both wings and sheath, and many, though possessing the latter, have not the former. Such is the case with *Meloë* and many Carabi. Nor does the transverse folding of the wings always exist. It is not found for instance in *Buprestis*, *Molorchus*, and some others. The straight suture of the elytra, again, is wanting in *Meloë*.

Notwithstanding these exceptions, it must be owned, on the whole, that the coleoptera constitute one of the most natural groups in the animal kingdom. It unites an immense number of insects that have the greatest possible relations to each other, and are separated from all the rest by numerous and strongly distinguishing characters. A reference to the text will be quite sufficient to satisfy any inquirer on this subject.

This order corresponds to the *Eleutherata* of Fabricius, a name devised from the disposition of the jaws, which are free, or do not support that appendage called *galea*, which characterizes the mouth of the Orthoptera, which the same author termed ulonata.

The coleoptera, for the most part, remain a very long time under the larva form, sometimes even three or four years, while they scarcely live for some weeks in the perfect state. It is only under the first form that their growth takes place, during which they change their skin several times. All such differences are referable to the degree of nourishment. Each family of coleopterous insects experiences modifications according to the nature of the climate, the quality of the food, and other particulars which belong to the admirable order which nature has observed in the respective relations of all her productions. Thus, the larvæ of the herbivorous coleoptera, as those of *chrysomela*, *crioceris* and *galeruca*, acquire their full growth in a few months, and it is under the form of eggs, that the species is continued and exists during the winter. Others, such as *prioceris*, the *lamellicornes*, &c., pass many winters under ground, where they feed on roots, or in the interior of the trunks of trees, sheltered from the vicissitudes of the seasons. This is observable in the horn-beetle, the May-bug, the capricornes. In fine, there are coleoptera, such as rhinocerus, whose larvæ feed and undergo transformation in fruits and the seeds of vegetables. It is in this dwelling, in the very centre of their aliments, that these insects, in the nymph state, pass all the cold season; and they do not acquire wings and begin to propagate their race, until the time when fecundation takes place in the plants, in whose germs their eggs are to be deposited.

All the larvæ of the coleoptera change their skin. Moulting takes place many times, pretty nearly as in the caterpillars of the lepidoptera. Four or five such changes of skin have been counted in the larvæ of *Tenebrio*.

In the nymph state the coleoptera take no nourishment. They are perfectly inactive and motionless, though all their parts are clearly distinct. Immediately after their transformation, all these nymphs are of a white, more or less transparent, or yellowish, and exceedingly soft. The majority of them cover themselves in cavities, the walls of which they have consolidated, so as to form of them a kind of shell. Under a sort of very thin epidermis, the corneous sheaths, which are to form all the articulations of their body, by lodging the muscles and viscera, are consolidated, and variously coloured, until the insect has acquired sufficient strength to break its shell, and appear in the open light; that is, if it be intended that it should seek its food at that period of the day when the heat and light of the sun exercise all their influence; or, in the shades of night, if, like the lampyrides, the photophygi, and the ligophili, darkness and obscurity are necessary to the preservation of their race.

Without any particular regard to method, we shall add a few observations to the structural details of the text.

The elytra, or sheaths of the membranaceous wings, can only be removed from the body at a right angle. They do not strike the air in flight. Once extended they remain fixed, and their separation constantly precedes the developement of the wings, properly so called. Their form, consistence, and colour, vary much. They sometimes embrace the abdomen, and are as it were soldered together completely by the suture, which always involves the total absence of the wings. This is observable in anthia and tachypus, among the creophagi, in many brachycera, and weevils; in lamia, blaps, eurychorus, pimelia, some alurni, and chrysomelæ. At other times, as in many galerucæ, and in meloë, the elytra, very well distinct and separate, do not protect the entire abdomen, and nevertheless do not cover any membranaceous wings, which are entirely wanting. While in the rhipiphora, molorchus, ædemera, and sitarides, the elytra narrowed, shortened, and not united in their entire length by a suture, do not suffice to cover the extent of the membranaceous wings.

The shape of these organs is very various. Their general form is more or less oblong, or inclining to the elliptical. This is the case when they are taken together in their state of repose. Considering them separately, the shape inclines to that of an isoceles triangle, the external side being curvilinear. When truncated, they are of a quadrangular form in general, sometimes trapezoid, sometimes in an oblong, and sometimes in a perfect square. Their proportions are very different; but for the most part they are twice as long as broad. This, however, is not always the case. In the *Buprestis gigas*, for instance, the length is three times the breadth. In many Staphylinidæ they are as wide as they are long, and sometimes wider. Though the general rule is, that they are narrower at the apex than the base, yet, nevertheless, in some instances, as in certain species of Lycus, the reverse is a fact. In other instances, their width is every where the same. Sometimes their form is convex, and sometimes extremely flat.

The mode in which these peculiar organs, of which we are treating, are often sculptured, is very curious, and often very beautiful. Some are marked with striæ or furrows, others with slightly raised ridges. That these elevations and depressions, not only in the elytra, but on other parts, are not intended for the mere purposes of ornament, Mr. Kirby maintains, with much appearance of reason. "By means of these," remarks that gentleman, "many important purposes that do not at first sight strike the observer, may be served, such as giving firmness to the crust in those places where it is most wanted; diminishing its power of resistance in others, so that it may yield somewhat to the action of the muscles; increasing or deducting from the weight of the body, so as to produce a proper equipoise during its motions, whether on the earth, in the air, or in the water. The depressions of the outer surface of the crust, in many instances, produce an elevation of it in the interior, and so afford a useful point of attachment to certain muscles. This observation seems more especially applicable to those excavations which are common to particular tribes or genera; thus the dorsal longitudinal channel to be met with on the prothorax of most of the ground beetles, (*Eutrechina*,) on the inside of the crust, has a corresponding ridge. In *Locusta Dux*, also, a Brazil locust, the same part has four transverse channels, corresponding with which, on the inside, are as many *septa* or ridges to which muscles are attached; and those larger impressed puncta denominated *puncta ordinaria*, which distinguished the same part in *Geotrupes* and many of the *scarabœidæ* within are elevated so as to

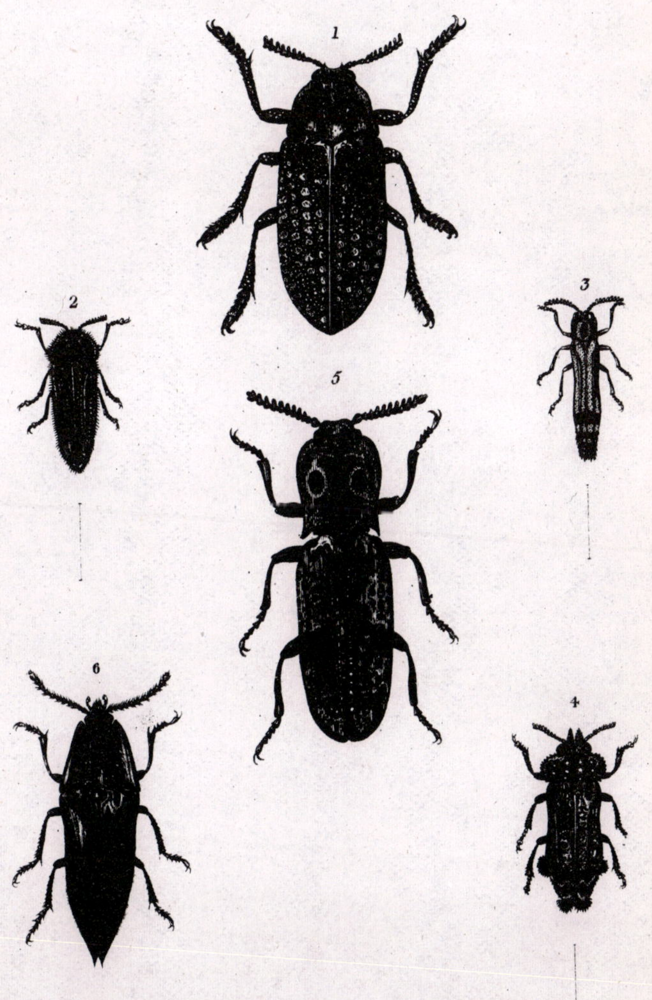

1. Buprestis *Whitehillii.* *Hope.* 2. Buprestis *flavomarginata.* *G. R. Gray.* 3. Buprestis *bifasciata.* *G. R. Gray.* 4. Buprestis *penicillata.* *Kuhl.* 5. Elater *lusciosus.* *Hope.* 6. Elater *aureolus.* *Hope.*

form a sort of ginglymous articulation with the base of the anterior coxæ. The other impressed puncta so often to be seen on the different parts of various insects, which, sometimes, so entirely cover the surface, that scarcely any interval is discoverable between them, though in many cases they appear to be mere impressions which attenuate, but do not perforate the crust, yet in others, perhaps, equally or more numerous, they are real *pores* which pass through the integument. If, for instance, you take the thoracic shield of the cock-chafer (*Melolontha vulgaris*), and after removing the muscle, &c., hold it against the light with the inner side towards the eye, you will see the light through every puncture; or take the elytra of *Geotrupes stercorarius*, or any common beetle in which these organs have punctate striæ, and examine them under a lens on the inside, and you will distinctly see that the punctures pass through the elytrum, and the membrane that lines it. It is not improbable that in the case last mentioned these pores may be of use, as the spiracles are usually closely covered by the elytra for the better transmission of air to those respiratory organs."

The colour of elytra varies considerably. A curious observation is made on this subject by Mr. Kirby that the majority of the coleoptera which feed upon putrescent matter are darker and more dismal in their appearance than those which feed on living prey, or fresh substances. Many families or tribes are distinguished by the peculiar colours of their elytra, which, as a secondary character, is useful in classification.

Before dismissing this part of our subject it may be requisite to say a word or two on the uses to which the elytra are destined to serve. Their most evident use is to protect the wings from injuries, to which they might otherwise be liable, in the haunts which many of those animals frequent, either in search of food, rest, or for the purpose of depositing their eggs.

The wings are, therefore, in general, very carefully folded up under the elytra, and when the latter are short, this folding is more complicated. There are cases, however, in which the wings are only partially protected by the elytra, and in such instances it is probable that, from the peculiar habits of the animals, the wings are in less danger of being torn or otherwise injured. They also protect the upper part of the body and the abdomen. The coleoptera not having the spiracles covered in the latter part, by the inosculations of the segments, as most other insects have, may require some peculiar covering for this part, when they are not on the wing. The elytra, too, are not wholly without their use in flight. Many flying beetles of considerable bulk, require more support in the air than the extent of their wings would appear to afford, and though elytra do not move much in flight, they may, by presenting a broad surface to the air, act in some measure as sails, and assist the progress of the insect, especially in flying before the wind.

The membranaceous wings are inserted equally on the upper part of the breast, within the elytrum, which covers them. They are veined with anastomoses, pretty nearly like those of the hymenoptera, a peculiarity by which they are distinguished from those of the neuroptera. With one or two exceptions, which have been remarked in the rhipiphora and the molorchi, these membranaceous wings are bent on their external edge. They form there an angular articulation, which permits the wing, generally double the length of the elytrum, to conceal itself underneath, and to fold itself crosswise by a sort of hinge movement, which distinguishes these wings from those of the orthoptera. The nervures of these wings present very great varieties in the different genera. Thus in the cicindelæ there is constantly seen in the bend of their articulation a more transparent space, bounded by a sort of fibrous ring, representing an eye, or circular hole. An elastic ligament brings back the wing to the state of extension or flexion, pretty nearly like the articulation of the leg in the birds called grallæ.

The tarsi of the coleoptera have been studied with very great attention by entomologists, being of so great value in determining the principal subdivisions of this very numerous order. M. Dumeril makes a curious remark, that the number of articulations to the tarsi is constantly the same in the middle and the anterior feet, so that it is sufficient to count the number of the articulations of the anterior feet to know that of the middle feet, and reciprocally.

The name of CICINDELÆ, a Latin word employed by the Romans to designate a brilliant insect, has been employed by the majority of authors to indicate insects very different from each other. Sometimes, and most commonly, it has been applied to the glow-worm, and sometimes to the cetoniæ, the cantharides, the nitidulæ, and all the insects with metallic reflections. Geoffroy, remarking that the cantharides of the shops had been erroneously termed cicindelæ, and that they had been ranged in the same genus with the telephori which have not the same number of articulations on the tarsi, thought he would do well to separate the cantharides from the telephori, to which he gave this name of cicindela, "which," says he, "was formerly that of a genus approaching to the glow-worm, and, perhaps, of this same genus to which we now restore it." Our author, however, was not ignorant that Linnæus had assigned this name of cicindela to the insects now under our immediate consideration.

The cicindelæ are very carnivorous coleoptera, ornamented most frequently with brilliant golden colours. They are met with in sandy places, where they run with the utmost swiftness on their prey, which they devour alive. Most of them, when seized, emit an agreeable odour, slightly partaking of that of musk. This, however, is not peculiar to those insects, for many others, inhabiting sandy places, are found to do the same. Although the cicindelæ fly remarkably fast they stop at but a small distance from the place which they have quitted. In flying, some of them, such as *sylvatica*, emit a loud humming noise. They all appear to be endowed with excellent sight.

The architecture of the larvæ of this genus is very curious and interesting. Their operations were first observed by M. Geoffroy, and the larvæ themselves subsequently well described and figured by M. Desmarest, in the "Bulletin des Sciences." This larva or grub is to be found during spring, and also in summer and autumn, in sandy situations. It is of a cylindrical form, is long, of a soft contexture, and a whitish colour; the head is squared, presents six or eight eyes, and is large. These larvæ have powerful jaws, and on the eighth ring of the body there are two tubercles thickly covered with red hairs, and armed with a recurved horny spine.

They hollow in the earth or sand vertical holes of about eighteen inches deep, and resting on their tubercles, and bending themselves into a form resembling the letter z, they mount and descend something in the manner of chimney-sweepers. The cavity which they form is cylindrical, of greater width than their body, and the entrance is perpendicular. Their large head serves to transport the sand or earth from the hole which they are digging. When come to the opening of the hole they fling out the earth and sand. According as this larva gets deeper down, it returns to the orifice by climbing up by means of the tubercles on its back. This is an operation neither quick nor easy, and the animal in performing it is forced to rest several times. When it finds the particular ground it has chosen not adapted to its labours, it removes, and chooses another place to work in. When it has completely succeeded in forming a habitation, it sticks at the entrance, by means of the hooks of its tubercles, which answer the purpose remarkably well by forming a support, and the insect exactly closes the hole with its broad and flat head, which answers all the purpose of a trap-door. Here the grub remains immoveable, ready to seize the first insect which passes, which is very speedily devoured. Its voracity is enormous, and even its own species do not escape from its devouring jaws. It carries its prey to the bottom of the hole, where it always remains itself, in case of danger. It is easy to observe its manœuvres by placing it in narrow tubes of glass, but it is rather difficult to be caught. The French naturalists succeeded by introducing a straw or twig into the hole, which acts as a kind of sound, and prevents the sand from falling in and choking up the hole, while the earth or sand around it is very carefully cleared away. The grub was usually found at the bottom of the hole, in a zigzag posture.

Relative to the systematic arrangement of the Cicindelidæ, it may be stated, in addition to the text, that after *Manticora*, M. Eschscholtz has added a new genus *Omus*; the characters of which are, that the labrum is transverse, narrow, bisinuated; the labial palpi are of an equal length with the maxillary, with the two basal joints short, and the last securiform. The anterior tarsi of the male have the three basal joints transversely dilated, eyes round, and the elytra angulated. The only species known is the *Omus Californicus.*

The *Megacephalæ* of the old Continent are apterous, while those of the New World have wings. This distinction has induced MM. Saint Fargeau and Serville to form them into distinct sub-genera; we will add one from Mexico, which is nearly allied to the *Megacephala Carolina*, *Fabr.* This new species we name *M. Mexicana.* It is of a brilliant green colour, with the back copperish; with a lunar mark on each elytron, which is narrower at the tip than those of the other described species; the antennæ, legs, and mandibles, yellow; the tip of the last black; length six lines. In the splendid collection of J. G. Children, Esq.

To the species of *Oxycheila*, already described, we add a new one under the name of *Oxych. binotata*. This species is purplish black with a small yellow spot on each elytron, the three first joints of the antennæ black, the others yellow, also the legs, except the apex of the femora alone being black; the palpi yellow; this may probably be the *O. distigma* of M. Gory.

The Baron Dejean, in his Iconographie, has established a genus under the name of *Tresia*. The characters are, that the labrum is very large, in the form of a semioval plate, concealing the mandibles; the labial palpi very long, much longer than the maxillary, the first joint long, second short, third very long, cylindrical and slightly curved, the last very long and securiform; the only species is the *Tresia Lacordairei, Dej.*, which is with the elytra transversely wrinkled, green, tinged with blue, the labrum pale, testaceous; with the breast, abdomen, and femora rufous; this insect is found in the woods of Brazil, on the trees, and flies like the *Muscidæ*, from leaf to leaf, with great rapidity.

We figure for the first time the splendid *Cicindela octonotata* of *Weidemann*, which is of a brilliant indigo blue, with four spots on each elytron, with the suture green. The first very small at the shoulder, the others are larger, with the third bilobed; the antennæ blue; the labrum, mandibles and palpi, yellow. This insect inhabits all parts of India.

We have likewise figured a new species belonging to the genus *Dromica*, from the fine collection of the Rev. F. W.

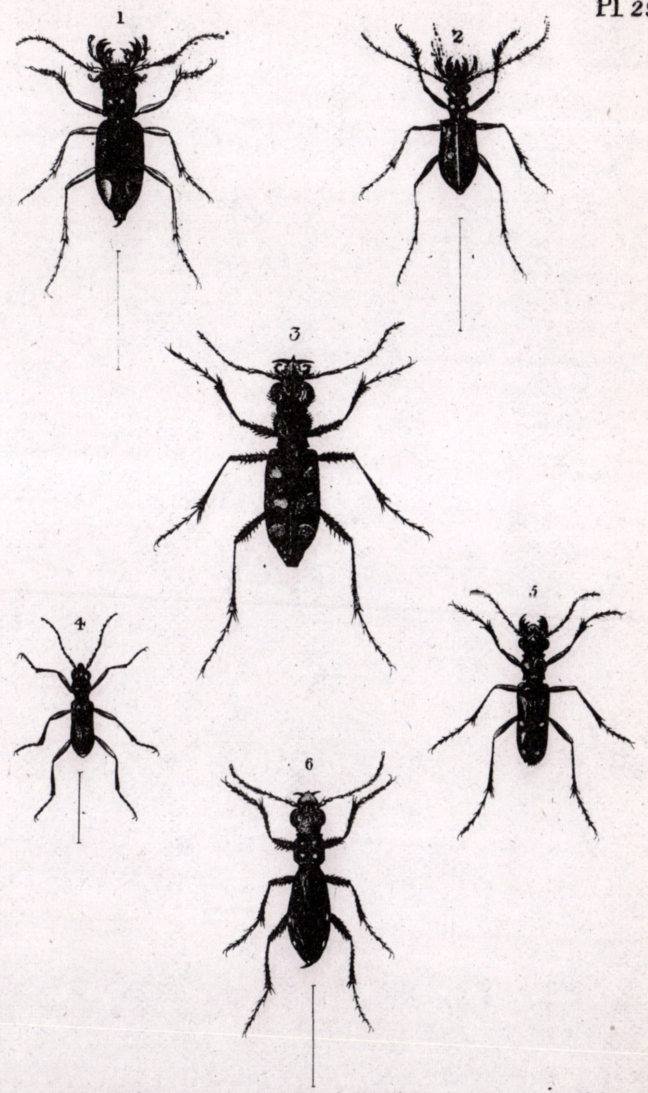

1. Megacephala *Mexicana.* *G.R. Gray.* 2. Oxycheila *binotata.* *G.R. Gray.* 3. Cicindela *octonotata.* *Weidm.* 4. Iresia *Lacordairei.* *Dej.* 5. Euprosopus *quadrinotatus.* *Lat.* 6. Dromica *tuberculata.* *Hope.*

Hope, which that gentleman names *D. tuberculata.* It is above, greenish black; the thorax and elytra tuberculate, with two small yellow spots on each side of the elytra; beneath black, with the upper lip and palpi yellowish white, and is an inhabitant of Africa.

The name of Carabus is very ancient in the language of naturalists, having been employed by Aristotle to designate certain species of crabs, and by Oppian, who appears to denote by the term Κάραϐος that species of *octopus* which walks upon its head, τῳ κάρα βαινῶ. Be this as it may, Linnæus is the first author who has applied this term to insects, and though Geoffroy set his face against the usage of it, it has nevertheless maintained its place in the science, and it is most probable that it will continue to do so, as it is now restricted.

Geoffroy gave to most of those species, now comprehended under this name, that of *Buprestis*, an expression which signifies the power of causing oxen to swell, and there is room to believe that if any insects can produce such an effect, it is those with which we are engaged. Nevertheless, this term buprestis has been otherwise employed, and in particular applied to the species to which Geoffroy gave the Latin name of *Cucujus.* Finally, to fill up the measure of confusion and difficulty, the latter name has been applied by Fabricius to insects equally remote from both the genera now mentioned.

All the tribe of carabici, in their first and final state, feed on living prey, and particularly on insects, which they catch by running. The species of this tribe are all terrestrial, their feet being solely formed for running. Their larvæ are likewise wholly terrestrial, and respire in the ordinary way. In this they are distinguished from the Hydrocanthari, a third tribe of the same family of carnivora, whose larvæ are aquatic, and receive the air by the anus, or by lateral fins resembling gills. The carabici differ from the cicindeletæ, in that the point or crotchet of the superior extremity of their jaws is not articulated with them. They emit, as we have already seen, a fœtid and caustic liquid from the anus, which is often volatile, as in brachinus. According to Baron Cuvier their intestine is terminated by a widened cloaca, provided with two small sacs, from which this humour is secreted. They are for the most part very agile. Many have no wings under their elytra, and those which do possess them make much less use of them than the cicindeletæ. They remain, in general, under stones, or in the earth; some conceal themselves under the barks of trees. The habits of the larvæ are exactly the same.

It is necessary to apprize our readers that the sub-genera of this tribe of carabici supply but very scanty materials indeed for this part of our book. We shall take them, however, in succession, omitting all mention of those which present nothing of popular interest, and confining ourselves to the few of whose habits any thing is known. Our performance here will, therefore, be necessarily somewhat meagre, but the difficulty of it will be best appreciated by such as have attempted a similar task.

And first, of the Brachini. Solander had observed that the *carabus crepitans* of Linnæus, *B. crepitans*, Fab., an insect very common in Europe under stones, caused to issue from the anus with an exploding noise when it was disturbed, or desirous of terrifying its enemy, a liquor of a penetrating odour, and exhaling itself in vapour. It has been since remarked that this property was common to many other species of an analogous form, as well indigenous as exotic. It is with these insects that M. Weber has formed the genus Brachinus.

In M. Latreille's general history of crustacea and insects, the brachini, with agra and lebia, comprise a family to which, from the phenomenon above alluded to, he gives the French name "*Bombardiers*." M. de Clairville has criticized this denomination, because it does not apply to all the insects comprehended in the division, and even says that the *harpalus prasinus* combats its enemies by the same means as the *brachinus crepitans*. On this M. Latreille makes a singular remark, that this harpalus possesses this property in Switzerland, but not in France, as appeared from this eminent naturalist's researches on a very great number of individuals. M. Bonnelli was not more fortunate in discovering this property among the harpali of Italy.

The following is Mr. Kirby's account of this very curious phenomenon :—" The most famous for their exploits in this way are those which, on this account, are distinguished by the name of Bombardiers. The most common species (*B. crepitans*), which is found occasionally in many parts of Britain, when pursued by its great enemy, *Calosoma inquisitor*, seems at first to have no mode of escape, when suddenly a loud explosion is heard, and a blue smoke, attended by a very disagreeable scent, is seen to proceed from its anus, which immediately stops the progress of its assailant; when it has recovered the effect of it, and the pursuit is renewed, a second discharge again arrests its course. The bombardier can fire its artillery twenty times in succession, if necessary, and so gain time to effect its escape. Another species *B. Displosor*, makes explosions similar to those of *B. crepitans*, when irritated it can give ten or twelve good discharges; but afterwards, instead of smoke, it emits a yellow or brown fluid. By bending the joints of its abdomen, it can direct its smoke to any particular point. M. Leon Dufour observes that this smoke has a strong and pungent odour, which has a striking analogy with that exhaled by the nitric acid. It is caustic, reddening white paper, and producing on the skin the sensation of burning, and forming red spots, which pass into brown, and, though washed, remain several days."

The *brachinus bimaculatus* also produces a very strong

explosion, and the action of the vapour, which it then sheds, occasions a very sensible degree of pain, and one of long duration.

The *brachinus displosor* is found, like the other species in general, under stones, and under heaps of rotten plants in the dry and elevated grounds of Navarre, Arragon, and Catalonia. Mr. Kirby says, above, that its caustic smoke reddens white paper, but M. Leon Dufour says that it is blue, in which this change is wrought. Probably both are right. Be that as it may, the yellow or brown fluid, which becomes fixed immediately, and in the form of a crust, lets escape, immediately after its emission, some air-bubbles, and presents an appearance of fermentation. If the animal be disturbed about the corslet, the surface of the elytra is soon powdered over with a sort of acid dust, resulting from the explosions. These properties are common to the two sexes.

There are some anatomical observations given by M. Dufour, relating to these insects, that are well worthy of the attention of the reader, and of which we shall make a brief abstract.

The apparatus of the organ which produces the smoke, is double, that is to say, there is one on each side in the abdominal cavity. It consists of two very distinct bodies, one of which is the *preparatory organ*, the other the *conservatory organ*. The first is more interior, and presents itself under two different aspects, according as it is contracted or dilated. In the first case it is a whitish body, irregularly rounded, soft, appearing glandular, placed under the last rings of the abdomen, opening by one end into the reservoir, and constantly terminating at the other in a very long and narrow thread. In the second case, or when it is dilated, it resembles an oblong membranous diaphanous sac, filled with air, occupying then the entire extent of the abdomen, and appearing free, with the exception of the extremity which opens into the reservoir. The second organ, or the *conservatory*, which is also the reservoir, presents a spherical body of the bulk of a turnip-seed, brown or reddish, of a papyraceous consistence, constant in its form, hollow interiorly, and placed under the last dorsal ring, exactly above the rectum. It opens by a pore on one side of the anus. It is contiguous to that of the opposite side, but both one and the other are very distinct. Their interior is clothed with the same crust, which is fixed on the back of the animal when it can no longer produce explosions. A membranous tube, very short, moved unquestionably by a sphincter muscle, serves to expel the smoke. M. Dufour has observed in the carabi and blaps, an organ similar to that which he names preparatory, but which is never swollen with air.

M. Dufour observes, on the digestive organ of this insect, that the alimentary tube is about as long again as the body. It commences with a straight, cylindrical œsophagus, occupying the length of the corslet. The stomach, which comes after, is lodged in the breast. When dilated, and filled with air, it has the appearance of a small ovoid balloon, having longitudinal stripes, and the intervals of which, slightly convex, are divided transversely by other stripes, short and whitish. When it is very much dilated all these lines disappear. If contracted, its parietes are thick, its surface is wrinkled, verrucose, and granular, and this organ then resembles an ear of maize, furnished with its grains. One line below the stomach is a small swelling, almost globular, and formed of a slender, smooth membrane, without any muscular appearance. The intestine succeeds, and presents a cylindrical tube, bristling with small papillæ. It makes a circumvolution on itself, and before it terminates by the rectum, it presents a swelling, almost similar in all respects to the stomach. The rectum is a line in length.

Since the publication of the Memoir from which these

remarks are extracted, M. Dufour has discovered the hepatic vessels, which are four in number, and the epiploon, consisting, as in many other insects, of fatty whitish shreds, and of very various forms.

We may observe here that the *B. Displosor* belongs to the sub-genus *Aptinus* of the text.

In Odacantha. The *O. melanura* lives in shady and humid places, in the cold or temperate regions of Europe. Dr. Leach has found great numbers of them in England, in places planted with reeds and rushes. Its other habits are unknown to us. It is the only species which is indigenous.

In the sub-genus Drypta the form of the masticating organs, the jaws, being elongated, and armed with powerful crotchets, proves the very carnivorous character of these insects, and their capability of pursuing and seizing their prey, even in its deepest retreats. They are found on the ground, and especially in places somewhat humid. They inhabit only the warm regions of the ancient continent.

The Lebiæ differ from the Brachini in their tongue, their flatted body, and particularly in the absence of those organs of crepitation, for which those last-mentioned carabici are so remarkable. Notwithstanding the assertions of M. Clairville, often too staunch a disciple of Fabricius, this last character of the Brachini is of more importance in the natural order than that furnished by the relative proportions of the antennæ, in their second and third articulations. But both Lebia and Cymindis have another character exclusively peculiar to themselves, in this tribe. The crotchets of their tarsi are denticulated underneath, in the manner of a comb. Accordingly, these insects often remain under the barks of trees, and have thus the facility of crooking or suspending themselves in a vertical position. All the individuals are winged.

We have figured a new species of *Anthia*, under the name

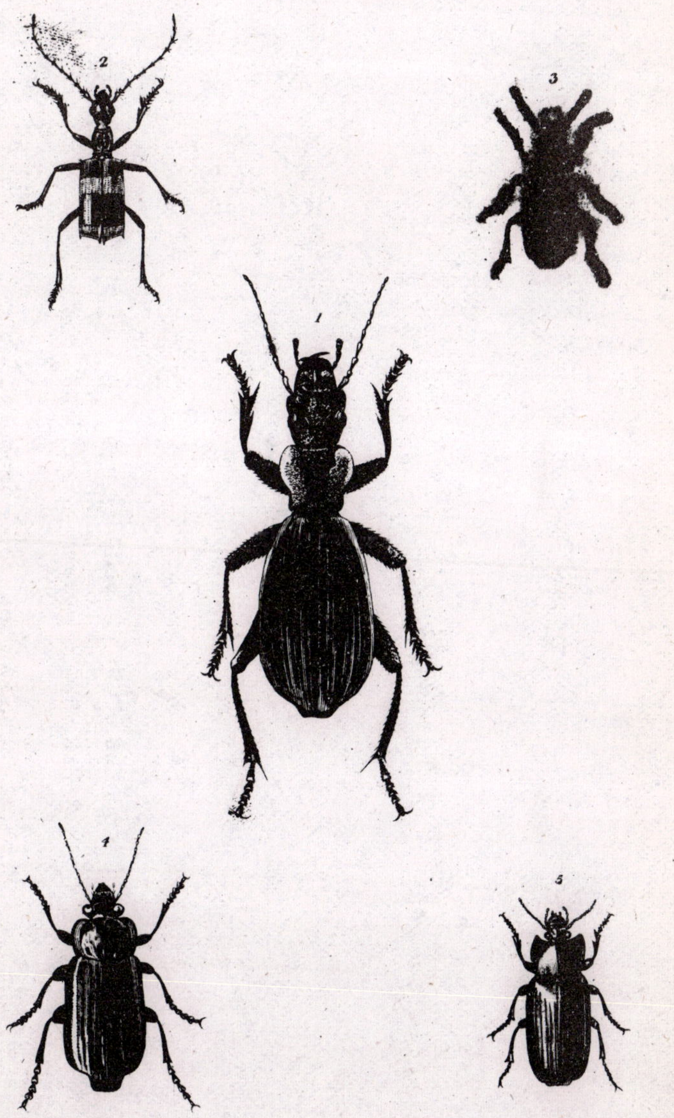

1. Anthia *Burchellii*. *Hope*. 2. Cordistes *cinctus*. *G.R.Gray*. 3. Pleuracanthus *sulcipennis*. *G.R.Gray*. 4. Orthogonius *Hopei*. *G.R.Gray*. 5. Axinophorus *Brasiliensis*. *G.R.Gray*.

Pl 34

1. *Details of* Pleuracanthus *sulcipenni.*
2. ______ Axinophorus *Brasiliensis.*
3. ______ Cordistes *cinctus.*
4. ______ Scaphinotus *elevatus.*
5. ______ Cnemacanthus *gibbosus.*

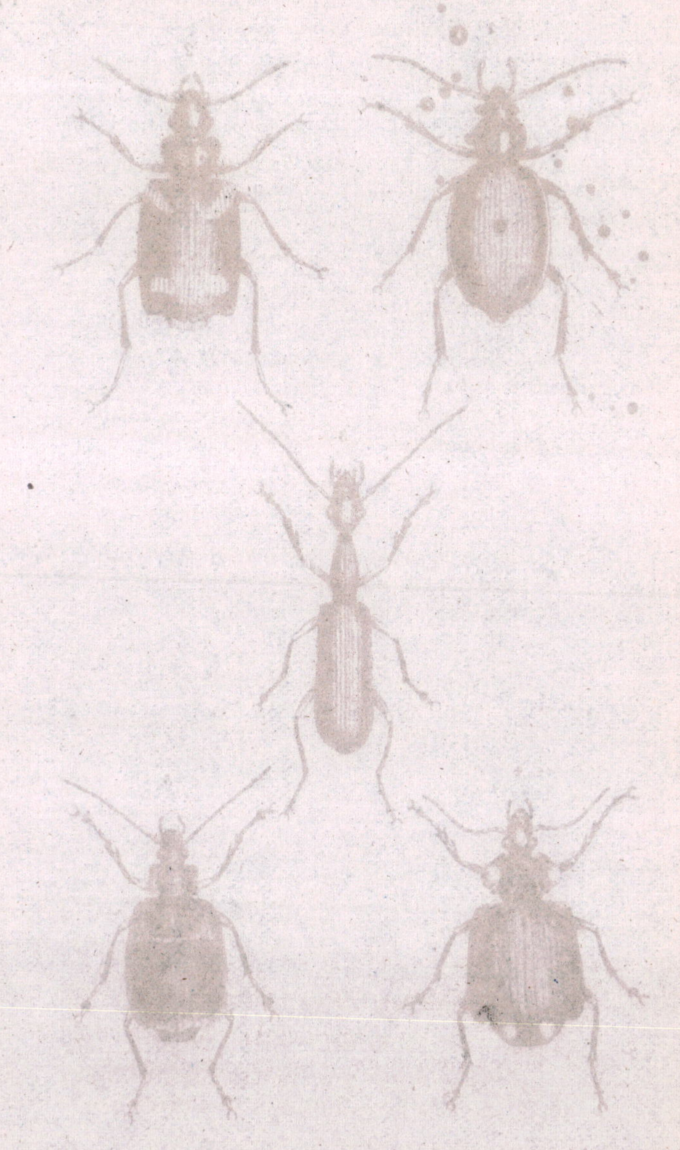

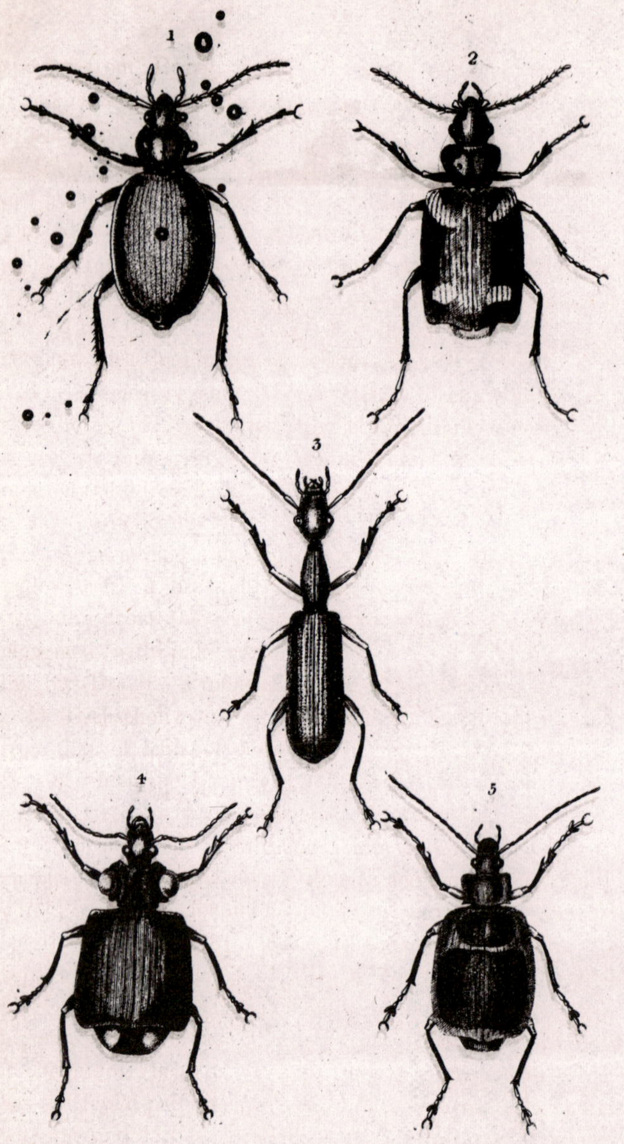

1. Loxocrepis *ruficeps*. 2. Plochionus *4notatus*. 3. Rhagocrepis *Riedelii*.
4. Physodera *Dejeanii*. 5. Lia *dorsalis*.

of *Burchellii*, which is black; with the three first joints of the antennæ, the margins of the thorax, and the sulcations on the elytra covered with short yellowish hair; length two inches. This insect was discovered by the celebrated traveller Mr. Burchell, after whom it has been named, by the Rev. F. W. Hope, in whose cabinet it is contained.

The Baron Dejean has formed a new genus under the name of *Axinophorus*, with the following characters. The maxillary palpi with the last joint rather cylindrical, and those of the labial very strong, securiform; the antennæ short, and filiform; the labrum short, and leaving the mandibles exposed, a strong simple tooth in the middle of the margination of the mentum; tarsi with the joints rather cylindrical; the body very flat, rather in the form of a long square; thorax rather transverse, rather straight anteriorly, elytra oblong, feet very short. The Baron mentions two species, one from North America, the other from Brazil; a third species we will describe under the name of *Ax. Brasiliensis.* It is obscurely piceous, shining, with the thorax marginated, and with six longitudinal simple striæ on each elytra, the central ones not reaching the base; palpi, legs, antennæ, with the margins of the thorax and elytra pitchy red; length nine lines. Inhabits Brazil. From the collection of J. G. Children, Esq.

The genus *Leptotrachelus*, of M. Latreille, appears to us to be synonymous with that of Rhagocrepis of M. Eschscholtz, which the latter has described as follows. The claws simple, palpi acute, penultimous joint of the tarsi bilobate, the elytra with the apex rounded. He mentions one species under the name of *R. Riedelii,* which is ferrugineous, the head brown, the base of the antennæ and femora yellow, the antennæ and tarsi black, the elytra with crenated striæ. This species inbabits Brazil. Length four lines.

After *Helluo* of Bonelli we will add a genus formed on a

species from Brazil, under the name of *Pleuracanthus*. The characters are, the labrum crescent-shape, a small tooth in the front and slightly dentated on each side; the mandibles moderate, not notched; maxillary palpi long, the last joint sub-securiform; the mentum large, deeply emarginated, with a rather large central simple tooth; the labium oblong, rather produced in the centre in front; labial palpi short, filiform; the antennæ short, hirsute, slightly enlarging toward the tip; the tarsi rather dilated; thorax transverse cordate, with the hinder angles acute.

The species we name *sulcipennis*, is black, with the head and thorax red, punctured; the elytra deeply sulcate, with a greyish pubescent; length nine lines. In the collection of J. G. Children, Esq.

To the delicately formed genus of *Cordistes*, we have pleasure in adding a new species, *C. cinctus*. Pale testacous yellow, with a broad black fascia crossing the elytra, and extending from the centre nearly to the tip; this fascia is margined with pale cream-colour; length six lines; and inhabits Brazil. From the cabinet of Mr. Children.

M. Eschscholtz has separated the three following genera from that of lebia of Latreille, namely:—*Lexocrepis*; the type of which the Lampryas ruficeps of Mr. Macleay, and characterized by its simple claws; filiform palpi, with the tip sub-truncated; the penultimous joints of the tarsi, produced on one side, and the elytra sinuate acuminated at the tip. The type is an inhabitant of Java, and is four lines in length. It is red, elytra cyaneous, the eyes black. *Lia*, the type which is the lebia dorsalis of *Dejean*, claws serrated, palpi filiform, subacute, tarsi, with the penultimous joint bipartite; elytra broad, obliquely truncated; thorax produced in the middle, behind. The next genus is *Physodera*, which has the labial palpi compressed, dilate, truncate; the thorax with the middle base produced; the elytra broad, truncate, the last joint of the tarsi bipartite, serrate. The type he has named *Phys. Desiani*, the length of which is four lines, and is found at *Mariella*.

In the same plate is figured M. Eschscholtz's species of *plochionus*, which he names *quadrimaculatus*, and is of an æneous green colour, the elytra olivaceous, with four pale spots, antennæ and legs yellow. From Brazil; length, three lines.

To *Orthogonius* we will add a new species, which we name *Hopei*, with the following description:—the elytra reddish brown, with punctated striæ, the head black, the thorax black, with the exterior margins and the femora reddish brown. The antennæ, tibia, and tarsi piceous, length eight and a half lines; it is from India. In the cabinet of Mr. Hope.

We pass on to Scarites. This genus of Fabricius, forms, in the method of M. Bonelli, a small sub-family, composed of the genera, *scarites*, *pasimachus*, *carenum*, *clivina*, *dyschirius*, and *apotoma*. The first three are distinguished from the others in consequence of the very marked denticulations of the mandibles, and the form of the tongue, which is short and broad. In carenum the external palpi are thicker at their extremity. In scarites the jaws are terminated in an arched or crooked point, while in pasimachus they are straight. The body of these last insects is also shorter, and proportionally broader, and almost of an oval figure.

The scarites are peculiar to the warm climates of both continents, and do not appear to possess the carnivorous habits characteristic of the other insects of the same family. They generally remain on the ground, and often concealed in the holes or crevices presented to them by the soil. Some are winged, and others apterous. All the known species are of a shining and uniform black. Some of them are to be found in the southern provinces of France.

We will add a new species of pasimachus, which we name Mexicanus. It is dark green, with the margins lighter, the thorax rather cordiform, with the antennæ, legs and mandibles, black. The length is eleven lines and a half; it is an inhabitant of Mexico. In the collection of Mr. Hope.

After *Ozæna* of *Olivier*, we have established a new genus, to which it is nearly allied, under the name of *Goniotropis;* it has the upper lip short, transverse, with a row of hairs near the interior margins; mandibles strong, dentated; the maxillary palpi with the last joint cylindrical, truncate; the last joint of the labial palpi is subsecuriform, the mentum with the margination in the middle slightly produced, and the antennæ the length of the head and thorax, hirsute, moniliform, with the last joint acute, with the anterior femora and tebia strongly toothed; the thorax subcordiform; the elytra with a hook at each angle. The species is from Brazil, therefore we name it *Braziliensis.* It is piceous, black, shining, delicately punctated, the tarsi and suture of the elytra piceous. The length is nine lines. In the collection of J. G. Children, Esq.

In Ditomus, to which M. Ziegler, a very distinguished German entomologist, gives the name *Aristus,* the body is elongated and depressed, and separated into two parts, by a remarkable strangulation, which divides the corslet from the abdomen. Their antennæ are filiform, tolerably long, with the second articulation shorter than the third. The head is thick, and the corslet large, and crescent-formed. It is attached to the abdomen by a short pedicle or stalk. The abdomen is almost square, rounded at the end, and entirely covered by the elytra. The palpi are filiform, and terminated by an oval articulation.

These insects inhabit the dry and sandy lands in the southern countries of Europe and Africa. They remain either in cylindrical, and tolerably deep holes which they have

Pl.12

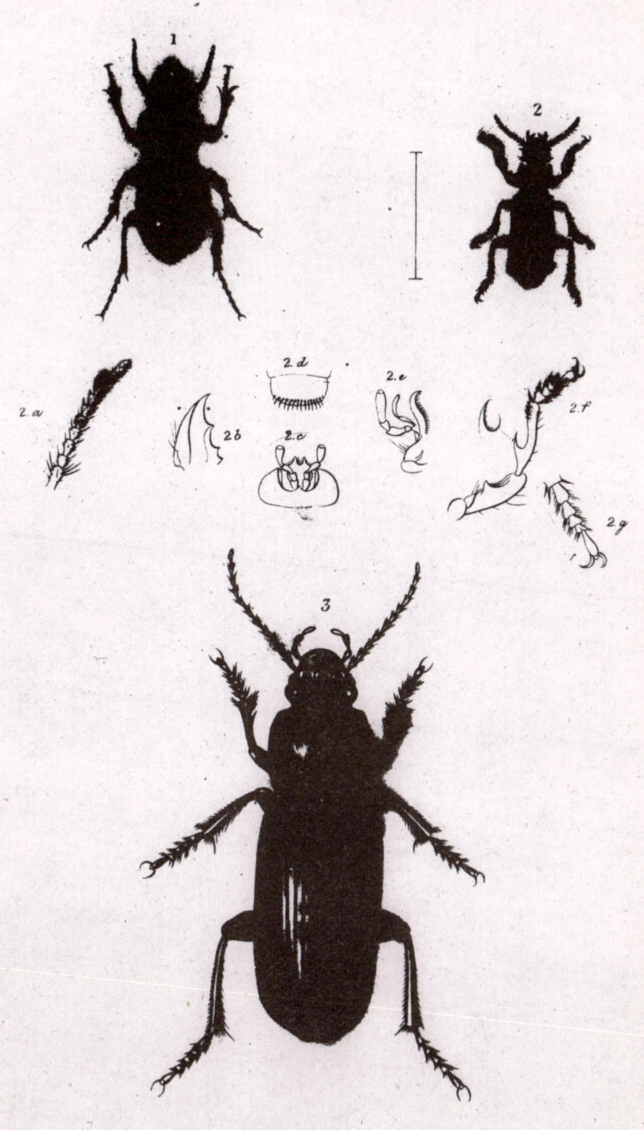

1. Pasimachus *Mexicanus*. *G.R.Gray.* 2. Goniotropis *Brasiliensis*. *G.R.Gray.* 3. Catadromus *tenebrioides* *Mac Leay*

dug, or in the crevices of the earth, and also under stones. In the fine weather they issue forth from their retreats, and always at the hottest moments of the day, but without removing far from their habitation. Their walk is rather slow. The species called *aristus bucephalus*, has been seen to climb on the blades of meadow grass, and carry off the little balls from them. All of them are winged. Some males differ from their females, by prominences of a conical figure on the anterior extremity of the head. Their larvæ have a singular resemblance to those of the cicindelæ, and live in the same manner.

Two species of this sub-genus are to be found in the neighbourhood of Paris; one is the *scarites bucephalus* of Olivier, of which Fabricius has made a Scaurus. The second species (*Aristus fulcatus*) is almost one-half smaller and of a more flatted form.

The Harpali are, in general, found upon the ground, under stones, or in holes, and most frequently in dry, or but moderately humid places. The numerous spines with which their anterior limbs are provided, give them the facility of digging retreats for themselves in light and sandy soils. They run tolerably fast, and without fearing, like many species of Feronia, the light of the sun. Their larvæ inhabit the same places. They have a conico-cylindrical form, and present at the posterior extremity of their body a membranous tube, terminating in an elongation of the anal region, and two fleshy appendages articulated and tolerably long. The head is thick, and armed with two tolerably strong mandibles, and almost similar to those of the perfect insect.

The insects of the sub-genus loricera, live after the manner of the carabi (which shall be described below), and are found under stones in humid places, and on the edge of rivers.

After *Cephalates* of *Bonelli*, we have figured a new genus which we call *Cnemacanthus*. The characters are, that the antennæ is moderate, moniliform; the mandible slightly dentated, the labrum short, transverse; the maxillary palpi with the last joint fusiform, truncate; and the last joint of the labial truncated, oval. The mentum, with the middle of the margination slightly produced. The anterior tibia dilated, with two strong spines. The thorax globlous, and the body globlous oval.

The species we name *gibbosus*. It is black, tinged with bronze green; the elytra with a simple stria; the legs black; the antennæ and tarsi black. The length is seven lines and a half: it is from Africa. In the cabinet of Mr. Children.

We pass on to the *Epomis*, of which we have figured a new species, under the name of *Gorgi*. The head and thorax metallic green, punctured; the elytra with broad hairy striæ; the margins, the first joint of the antennæ, palpi and legs, yellow; with the three last joints of the tarsi black. The length is sixteen lines: it is an inhabitant of Senegal.

The CARABI are much more restrained as to species, than the genus to which that name was given by Linnæus, Degeer, and Olivier. It comprehends but a portion of the species, distinguished by the qualification of *majores*. These insects are found only in the cold and temperate climates of Europe, Asia, and America. Their body is usually elongated and adorned with brilliant and metallic colours, such as bronze, golden-green, violet, or coppery.

Of all the carnivorous coleoptera, terrestrial and indigenous, the carabi and procrustes are the largest. They are very voracious and agile insects. They are often to be met with running on the ground, in fields, gardens, and woods.

Many species inhabit exclusively mountains and elevated countries. They conceal themselves under stones or moss, feed upon larvæ, caterpillars, or perfect insects, which they

Pl. 15

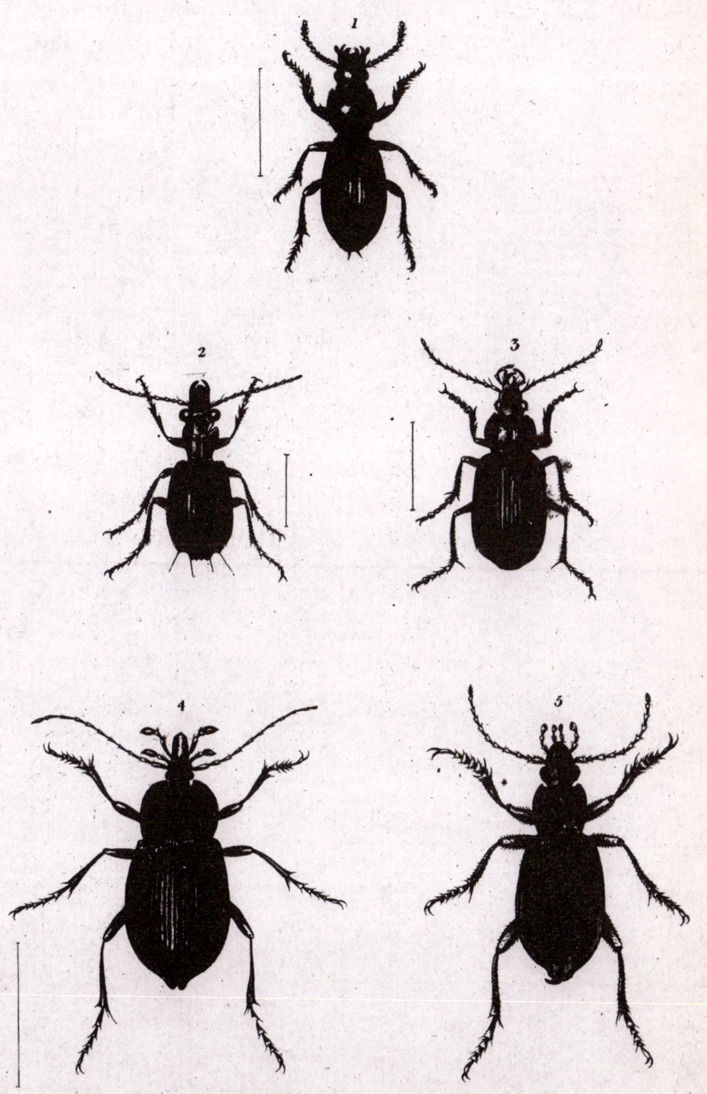

1. Cnemacanthus *gibbosus*. *G. R. Gray.* 2. Pericalus *cicindeloides*. *Mac Leay.* 3. Colpodes *brunneus*. *Mac Leay.* 4. Scaphinotus *elevatus*. *Latr.* 5. Epomis *Goryi*. *G. R. Gray.*

seize with their powerful mandibles, and often devour in common. They shed a very strong and disagreeable odour, which approaches that of tobacco, and some poisonous, or suspected plants. When they are taken they cause to issue from the anus, a blackish liquor, very irritating and acrid, the odour of which is still more disagreeable than that shed from their body. Their larvæ have not yet been observed or described, but, in all probability, they resemble those of procrustes, which are known.

The coleoptera, designated by the ancients under the name of carabus, appear to be very different from those now so called; since, according to those authors, they proceeded from larvæ living in dry wood. They are, probably, capricorns, or other analogous insects. It is presumable that the earlier naturalists comprehended our carabi, properly so called, among the cantharides, as they are, by persons ignorant of natural history, at the present day. Olivier, who in this respect follows the sentiments of Geoffroy, says that the ancients regarded those insects as a poison to such cattle as swallowed any of them mixed up with the grass which they fed on in the fields and meadows. They believed them capable of inflaming the intestines of these animals, by their caustic character. It was on account of this maleficent quality that they bestowed on them the name of *buprestis*.

Hippocrates, Pliny, and the ancient physicians, attribute to them a virtue but little inferior to that of the cantharides. They made use of them in various maladies, in hydrophy, in *tympanitis*, and especially in some maladies to which women are more especially subjected. They used to administer them internally, in very small doses, and sometimes employed them in pessories, mingled with aromatic substances.

The Calosoma have some general resemblance to the last sub-genus. They are tolerably large insects, and often adorned with metallic colours of the most brilliant dye.

Reaumur has given us the history of the larvæ of one species, the *Carabus Sycophanta* of Linnæus, which lives in the nest of the processionary caterpillars, and is their most redoubtable enemy. The caterpillar, which it attacks and pierces through the belly, writhes, agitates, torments itself, and uses every endeavour to escape to no purpose. The pertinacious *sycophant* never abandons it until it is entirely devoured. The largest caterpillar is insufficient to support it for a day. It kills and eats several of them in the course of the same day, and when gluttony has rendered it incapable of moving, it is attacked by other larvæ of its own species, which are yet young and rather small, which pierce its belly and eat it, although there is no deficiency of caterpillars for them to prey upon.

This species, and the *Calosoma Inquisitor*, are to be found in the neighbourhood of Paris. They usually remain on trees, and principally upon oaks, where they give chase to the different insects on which they feed. It appears that both, even in the perfect state, principally attack caterpillars.

In his grand work on the Coleoptera, Olivier has not endeavoured to define the limits of the genus Carabus of Linnæus and Fabricius; but he has remarked that many species present, in regard to the organs of manducation, some important differences; and he cites, as an example, the *Carabus Spinibarbis* on which M. Latreille has founded his genus POGONOPHORUS.

These insects, though of the same family as the Manticoræ, should not be confounded with these last carnivorous coleoptera, as they have been by M. M. Jurine and Panzer. They are even widely removed from them in a natural series, and properly come in the place assigned them in the text.

They are particularly found in spring, under the barks of old trees, about the trunk. But a small number of species are known, all belonging to Europe, and of small size.

The Nebriæ approximate to the carabi, properly so called, in many points, though they are distinguished from them by many others. They are carabici of a middling size. These insects do not present the brilliant and metallic colours which distinguish most of the carabi. They are for the most part black or brown. Others have the ground yellowish, and varied more or less with black. The greater number of species inhabit cold and elevated situations, and in general humid. Their metamorphoses are unknown.

The genus Omophron, which had been confounded with that of *Carabus*, was by Fabricius designated *Scolytus*, a name previously bestowed by Geoffroy upon a very different genus of coleoptera, and adopted subsequently by Olivier and M. Latreille. The confusion which was the result of all this, determined M. Latreille to designate the *Scolytus* of Fabricius by the name of *Omophron*. M. Clairville has put this genus at the head of his *adephagous* aquatic coleoptera, as forming the passage from them to the terrestrial adephagi,—but, says M. Latreille, he should have pointed out the characters of these two sections, and not have confined himself to distinctions purely nominal. Though the insects of this genus inhabit the borders of aquatic places, and though they are even found sometimes in the water, it is not less true that they appertain, by the assemblage of their characters, notwithstanding certain relations, to the terrestrial carnivorous coleoptera. The mandibles of the larvæ of *Omophron limbatus*, discovered and described by M. Desmarest, resemble those of the carabici, and do not present near their extremity that aperture which is observable in the mandibles of the hydrocanthari, or aquatic carnivorous coleoptera—a character which seems to indicate that those organs perform the office of a sucker.

These insects live on the banks of waters. They remain in the sand, between the roots of plants which grow there, and do not appear to issue forth until towards evening. The larva of that species which is best known (*Omophron limbatus*) inhabit the same places where certain aquatic plants are found. Its body is elongated, depressed, and of a conical form, having its greatest breadth on the side of the head. It is composed of a dozen rings, which are of a dirty white. The head is of a trapezoid form, rusty brown colour, and presents two small black eyes; two antennæ, small, setaceous, with five articulations, and placed in front of these organs. The mouth is formed of two strong mandibles, arched and denticulated. The two jaws have each two palpi, and an under lip also furnished with two palpi. The first three rings have each a pair of feet. These feet are scaly, directed backwards, and terminated by two sharp claws. At the end of the last ring is a filiform appendage, raised, with four articulations, the last of which finishes with two hairs. This larva is very agile, and raises the posterior extremity of its body like the *Staphylini*, *forficulæ*, &c.

The *Omophron limbatus* is found in humid sand, under stones, in the south of France, and sometimes even in the neighbourhood of Paris.

The Elaphri have a good deal of the physiognomy of the *cicindelæ*, both in the general form of their body, their colours, the projection of their eyes, and the celerity of their movements. But other characters more essential appear to remove them from that genus, and approximate them to those carabici of the sixth section, which conduct them by shades to the hydrocanthari. They are found only in humid places, and on the borders of marshes and rivers. They run with the most extreme rapidity, and feed on small insects, and probably on aquatic larvæ. Their metamorphoses have not yet been observed.

The insects of the genus Bembidion are small coleoptera, which have great relations with the elaphri, and, like them, frequent the sandy banks of waters, where they run remarkably fast; but they differ from them by the manner in which their palpi are terminated. This genus is very numerous, and almost all the species described to the present day are European. Fabricius and Olivier have ranged them among the elaphri and carabi.

We now arrive at the aquatic coleoptera in which the genus DYTISCUS stands foremost. Linnæus formed two sections of this genus, according to the clavate, and perfoliated or setaceous form of the antennæ. Geoffroy, under the name of hydrophilus, separated those in which these organs exhibit the first of these characters. The other species, or those in which the antennæ are setaceous, have preserved for a long time the general name of Dytiscus, but this division is now much more restrained than it was originally.

The dytisci seem to be amphibious insects. Though the water appears to be their principal element, and although they live there almost continually, they have also the faculty of living on the land and flying in the air. The size of these insects varies very considerably; some are more than an inch and a half in length, while others are scarcely larger than fleas, and some are found of all the mean sizes between those two extremes. The dytisci are carnivorous, and exceedingly voracious. They live only on other aquatic and terrestrial insects which they catch, and to which they give chase continually. They seize them with their anterior feet, as with hands, and then lift them to their mouth to devour them. Although they can live a long time under water, they have need, however, to respire the air, and they usually do this from time to time. They bring themselves to the surface, and to do so they have only occasion to keep their feet in a state of repose, and to suffer themselves to float; being lighter than the water they swim above it at once. It is their hinder part which is then always applied to the surface, and even almost above the water. They afterwards raise the elytra a little, or lower the end of the abdomen. When the insect wishes to return to the bottom of the water it brings the belly and the elytra promptly together, and stops the vacuum which was between them, so that the water can no longer penetrate there.

The dytisci live in all kinds of fresh waters, in rivers, lakes, but more especially in marshes and ponds. They swim with very great celerity. It is usually at the approach of night that they issue from the water to fly and transport themselves from one pond or marsh to another. Accordingly these insects, and many others, which, like them, are amphibious, are found in the smallest pieces of water, even in those which are formed by the descent of rain. They make a humming noise in flying, like the scarabei. In coupling, the male employs the two remarkable palettes which are presented by the anterior tarsi, to hold himself fixed to the female. The inferior surface of these palettes presents, in large spaces, a great number of small concave pieces, in the form of the calix of flowers, or little cups, and some of which, particularly the largest, have, at their bottom, a little elevation. We may again consider these calices as species of cupping-glasses or suckers. The mode in which the females deposit their eggs is unknown.

The skin which covers the larvæ of the dytisci is generally scaly, and something like the plates of a tortoise. The two last rings of the body are furnished towards the sides with a sort of fringe, which being placed on a ridge, or a line somewhat elevated, seem to be formed for swimming. When the larva suddenly wishes to change place in the water, or fly the approach of some large insects which might devour it, it gives a prompt and vermicular movement to its body, striking the water with its tail, the fringe of which then becomes very useful to it, since the tail is thereby rendered more fit to resist the water, and cause the body to advance.

This larva appears at once to perceive the smallest insect which moves in the water, and never fails to pursue it on the instant, and seize it with its teeth. These teeth, two in number, are attached in front, on each side of the head. They are curved, and meet each other when the larva keeps them in a state of repose. They have no denticulations, but diminish by little and little into a point. Swammerdam has mentioned that the teeth of the larvæ of this genus have an aperture in the form of a cleft near their point, and that it is by this aperture that they suck the insects, the fluid substance of which passes from there into their mouth and stomach. It is known that the antlion sucks in insects in the same manner. Degeer, in confirming the observations of Swammerdam, has thought that the larva has also a second mouth, and that this mouth was placed between the two lips. What appears to prove this is, that he has seen a larva not only suck in an aquatic wood-louse, but also devour by little and little, all the solid parts of that insect; which assuredly could not have passed through the little apertures of the teeth. Two muscles have been remarked, divided into many flat and fibrous ramifications. One is attached to the external side of the tooth, and serves to draw it away from the head. The other is attached to the interior edge of the tooth, and it is by this muscle, that the insect draws it close to the head, when it has seized its prey. The head is furnished with two small antennæ placed immediately in front of the eyes.

These larvæ are exceedingly voracious. With their large teeth they seize all the insects which they meet, to suck and devour them, and especially the larvæ of the libellulæ, ephemera, tipulæ, &c.

It is not rare to find these larvæ in all the dormant waters of marshes and lakes. Rœsel informs us, that when the time of transformation is arrived, the larva quits the water and proceeds to bury itself in the earth which borders on the marshes and streams. There it digs itself a cavity in the form of an oval shell, in which it becomes changed into a nymph, and subsequently into the perfect insect. Swammerdam also tells us that these larvæ are transformed in the earth, but nevertheless avows that he speaks only from conjecture. There are strong grounds, however, for the presumption that this is the fact, and that the dytisci are purely aquatic in the larva state, that they become terrestrial in the nymph form, and, finally, that in their state of perfection they are, in some sort, amphibious, or live equally in the water or on the land.

To *Hydaticus* of *Leach* we will add a new species from Mexico, which Mr. Hope has named *marmoratus*. It is black, with large yellow spots; the antennæ and the anterior legs are yellow, beneath reddish. It is in the above named gentleman's cabinet. Also a species of *Calymbetes*, which we name *Lowei*, and is black, varied with yellow; the elytra with punctured striæ; the head and thorax black, the latter with the margin and the legs reddish. Thirteen lines in length. This species is from Madeira, and is more elongate than the other species. The thorax is margined; it has a palette on the middle tarsi as well as on the anterior.

We now come to the last genus of this family, the GYRINI, which are found on the surface of the stagnant waters of marshes, lakes, and ditches, where they are seen to swim, and as it were, to run, usually assembled in troops, and describing circles and pirouettes, with the most surprising swiftness and agility. They very much approach the dytisci, having, like them, the feet adapted for swimming.

Pl. 32

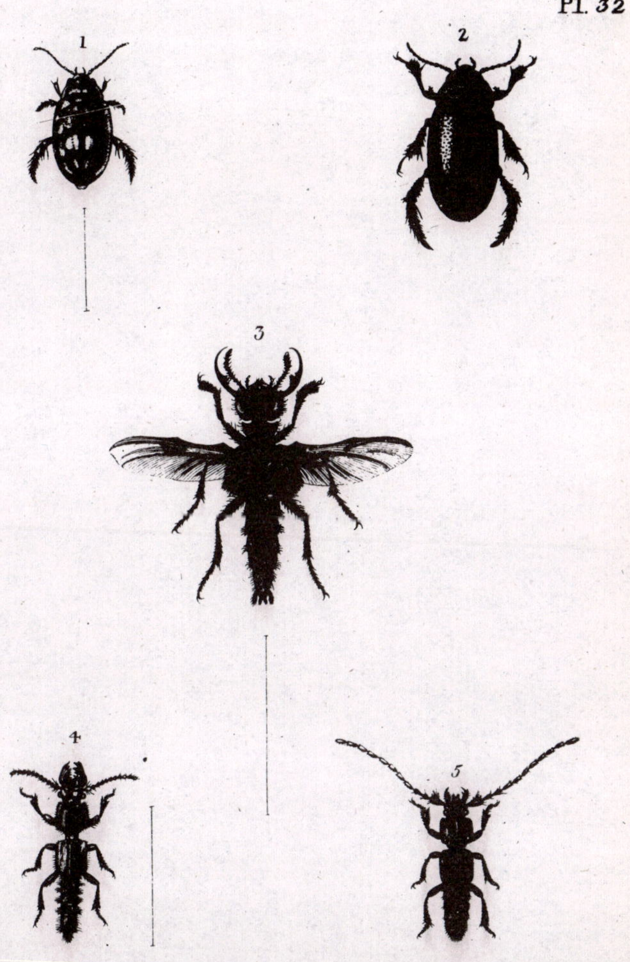

1. Colymbetes *marmoratus.* *Hope.* 2. Colymbetes *Lowei.* *G.R. Gray.* 3. Schizochilus. *Brasiliensis.* *G.R. Gray.* 4. Zirophorus *coriaceus.* *Lacord.* 5. Trichocoryne *striata.* *G.R. Gray.*

But still they are marked by some particular characters which constitute them a very distinct genus.

The gyrini may, in general, be considered as small insects, the species which is most known not exceeding the dimensions of the common fly. Some species, however, are known which are of a larger size. No insects exhibit, like them, what may be termed four net-work eyes. It is very easy to perceive them in looking at the animal from above, and then examining it underneath: for the two lower eyes are concealed underneath the head. The insect can see at the same time, and by different eyes, the objects which are above its head, and those which are below it. Accordingly its sight is very good and piercing, the proof of which may be very easily seen by placing it in a glass of water. After having made at first a few turns in swimming, it remains at last tranquilly on the surface of the water. The instant the hand is approached towards it, or any movement is made, without even touching the glass, it is seen of a sudden to put itself into a state of agitation, and usually sink into the water.

These insects can make use of their wings, and do, sometimes, elevate themselves into the air, but the particular conformation of the feet sufficiently indicate that they are also destined to live in the water.

The swiftness with which the gyrini swim on the surface of the water, or in the water itself, is surprising. The eye cannot follow the rapidity of their circumvolutions, and various motions in all directions.. As the coverings of these insects are very smooth and shining, when the sun casts its rays upon them, they seem like so many brilliant pearls in motion, which reflect the light in different shades, and present a very beautiful spectacle.

They are almost always assembled in troops upon the surface of the water, where they sometimes repose in a state of perfect quiescence, but the moment they are approached, they put themselves in motion, and escape by swimming or sinking into the water with the greatest quickness. Sometimes they remain at the bottom of the water, and keep themselves hooked to some aquatic plant; for, as they are lighter than the water, they come to the surface if they do not keep themselves fastened to something, and not stir their feet. The upper part of their body remains entirely dry when they are at the surface of the water; but when they dive, a sort of little silvery ball remains attached to their back, which produces a very pretty effect. A disagreeable odour exhales from their bodies when they are touched, which it is very difficult to get rid of. They are found in the waters from the time that the frosts give over, in spring, until it is far advanced in the autumn.

The females lay their eggs on the leaves of aquatic plants. Degeer kept some in a decanter filled with water, which deposited their eggs on the sides of the glass, one after the other. In about eight days the young hexapod larvæ, very small, issued forth from the eggs, and swam at first in the water; they also walked against the sides of the glass. Rœsel was acquainted with these larvæ, but was not able to bring them up to their full size. This also happened to Degeer, who never could preserve them alive for any length of time. But M. Modeer, who has given a history of these little insects in the memoirs of the Royal Academy of Sciences of Sweden, has had these larvæ in their full size, and has given a description of them, which agrees in all points with that of the young larvæ. This gentleman, however, has not informed us how these larvæ came into his possession, whether he hit upon the means of rearing them from their birth, or gained possession of them when they were fully formed in the waters. He also speaks of their transformations which took place out of the water. About the commencement of August, he says, this larva issues from the water, and gets upon the large leaves of such plants as grow there, where it fixes, and shuts itself up in a small oval shell, made of some substance derived from its own body. In this shell, having assumed the form of nymph, it gets out of it under that of the winged insect, towards the end of the same month, and jumps suddenly into the water. This author adds, that these nymphs are very subject to be devoured by the larvæ of the ichneumons, which lay their eggs near them in the same shell.

THE

SECOND FAMILY

OF THE

PENTAMEROUS COLEOPTERA.

Brachelytra, Cuv., Microptera, Gravenhorst,

Have but one palpus in the jaws, or four in all. The antennæ sometimes of equal thickness, sometimes a little thicker towards the end, and universally composed of grain-formed or lenticular articulations. The wing-cases are much shorter than the body, which is narrow and elongated, with the branches of the two anterior feet very large, and two vesicles near the anus, which the animal can put forth at pleasure.

These coleoptera compose the genus Staphylinus of Linnæus.

They have been considered as constituting the passage of the coleoptera to the forpiculæ, the first genus of the following order. Under some relations they also border on the insects of the preceding family, and under many others, on the necrophori, a genus of the fourth. They have most frequently the head large and flatted, strong mandibles, short antennæ, the corslet as broad as the abdomen, the cases truncated at their extremity, and nevertheless covering the wings, which preserve their usual extent. The semi-rings of the upper part of the abdomen are as scaly as the lower. The vesicles of the anus consist in two conical points, furnished with hair, which the animal causes to issue forth and re-enter at will. A subtile vapour thence escapes, and which in some species has a strong scent of sulphuric ether. M. Leon Dufour (*Annales des Sciences Natur*. t. viii. p. 16.) has given a description of the apparatus which produces it. The last segment of the abdomen, that in which the anus is situated, is prolonged into, and terminates in a point.

These coleoptera, when they are touched, or when they run, raise the end of their abdomen, and give it all kinds of inflexions. They also employ it to push their wings under the cases, and cause them to re-enter there. The two anterior feet have often the tarsi broad and dilated. Their branches, as well as those of the intermediate feet, are very large. They live for the most part in the earth, in dung-hills, in excrementitious matters. Others are found under mushrooms, the rotten parts of trees, holes, and under stones. Some inhabit only aquatic places. Some again are known, but very small, which attach themselves to flowers. All are voracious, run with great swiftness, and take flight very promptly.

Their larvæ very much resemble the perfect insect; they have the form of an elongated cone, the basis or the thickest part of which is occupied by the head, which is very large. The last ring is prolonged, in the manner of a tube, and is accompanied by two conical and hairy appendages. These larvæ feed upon the same matters as the insect in the perfect state. The first stomach of the Staphylinus is small and without folds. The second is very long and very hairy. The intestine is very short.

According to M. Leon Dufour, their alimentary canal does not differ essentially from that of the carnivorous coleoptera, but by the absence of a crop. Their biliary vessels are inserted on one and the same lateral point, and in some species, at least, offer, towards their middle, a knot or vescicle not remarked in any other insect. Their apparatus of generation differs much from that of the carnivorous coleoptera.

This genus is considerable. We divide it into five sections.

The first, that of Fissilabra, has the head entirely naked, and separated from the corslet, which is sometimes squared, or in a semi-oval, sometimes rounded, or in a truncated heart, by a neck, or visible strangulation. The labrum is deeply divided into two lobes. Such are:—

OXPORUS, Fab.,

Of which the maxillary palpi are filiform, and the labial terminated by a very large, and crescent-formed articulation.

The antennæ are thick, perfoliated, and compressed. The anterior tarsi are not dilated. The last articulation, and next to that the second, are the longest. They live in fungi, and agarics.

Staphylinus rufus, Lin. Panz. Faun. Insect. Germ. XVI. 19, about three lines in length, fawn-coloured, with the head, the chest, the extremity, and the internal edge of the wing-cases, as well as the anus, black.*

ASTRAPÆUS, Grav.,

In which the four palpi are terminated by a larger and almost triangular articulation. The anterior tarsi are very much dilated. The first and the last articulations are the longest.†

* Add *O. Maxillosus*, Fab. Panz. *ibid.* 20. The other Oxypori of Fabricius belong to the sub-genera of our fourth section. *See* Olivier, Encyc. Method. genus *Oxyporus;* and M. Gravenhorst, *Coleoptera microptera.*

† *Staphylinus ulmi,* Oliv. Ross. Faun. Etrusc. I v. 6. Panz. *ibid.* LXXXVIII. 4; Lat. Gener. Crust. et Insect. I. 284.

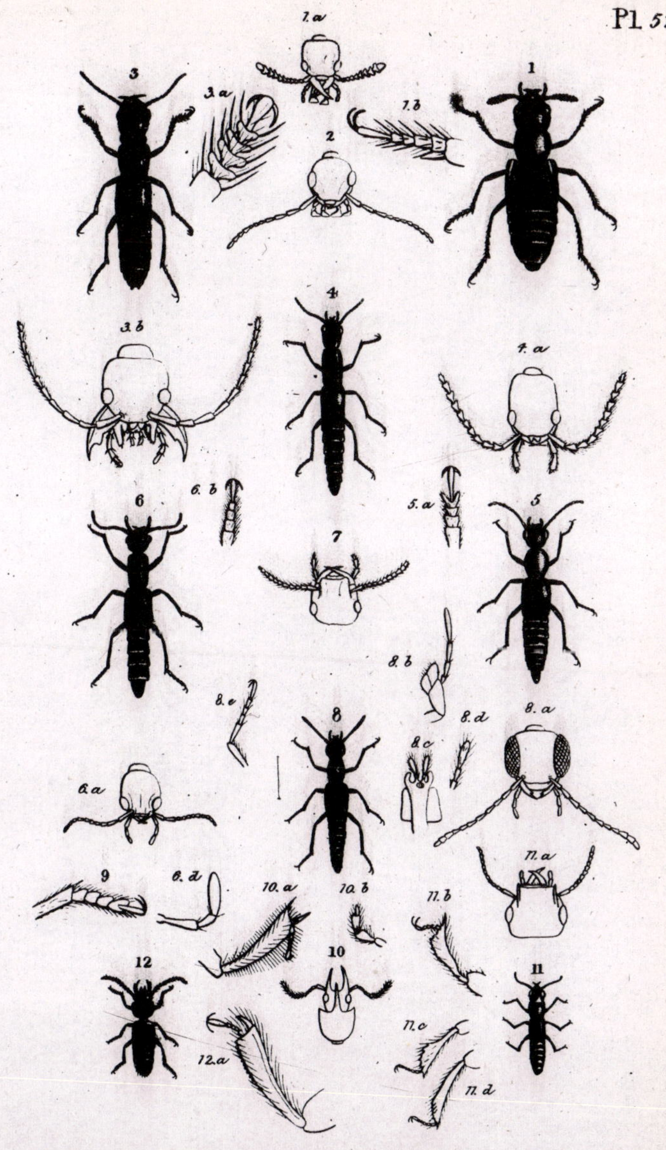

1. Oxyporus *rufus. Lin.* 2. *Head of* Astrapœus *ulmineus. Oliv.* 3. Staphylinus *tataricus. Fisc.* 4. Lathrobium *elongatum. Lin.* 5. Paederus *ruficollis. Fab.* 6. Procirrus *Lefeburi. Lat.* 7. *Head of* Evaesthetus. 8. Stenus *biguttatus. Lin.* 9. *Ant. Claw of* Stilicus. 10. Oxytelus *tricornis. Fab.* 11. Osorius *brasiliensis. Guer.* 12. Zyrophorus *striatus. Leach.*

STAPHYLINUS (Proper), Fab.,

Which have all the palpi filiform, and the antennæ inserted above the labrum, and mandibles between the eyes. Some, and especially the males, have the anterior tarsi very much dilated, the antennæ separated at their origin, and the first articulation of which equals, for the most part, in length, the fourth of the total length. The head is but little elongated. The species presenting these characters compose alone, in some methods, the genus *Staphylinus.* Authors have even separated, to form another, the *S. dilatatus*, Fab. Germ. Faun. Insect. d'Europe, VI. xiv, in consequence of its antennæ forming an elongated knob, and denticulated like a saw. According to the observations of M. Chevrolat, a very zealous entomologist, this insect feeds on caterpillars, which it goes to seek on the trees.

S. Hirtus, Lin. Panz. Faun. Insect. Germ. IV. 19. Ten lines long, black, very hairy, with the upper part of the head of the corslet, and the final rings of the abdomen covered with thick hairs, of a golden and lustrous yellow. Wing-cases of an ashen grey, with the base black. Under part of the body of a blueish black. North of Europe, France and Germany.

S. Olens, Fab. Panz. *ibid.* XXVII. i. An inch long, of a dead black, with the head larger than the corslet, and the wings reddish. Its eggs are of a remarkable thickness.

Very common in the neighbourhood of Paris, under the stones.

S. Maxillosus, Panz. *ibid.* 2, being nearly eight lines in length, black and shining ; the head broader than the corslet, a great part of the abdomen and elytra of an ashen grey, with points and black spots. In earth and dung.

S. Murinus, F. Panz. *ibid.* LXVI. 16, from four to six lines in length ; the head, corslet, and wing-cases of a deep and shining bronze, with obscure spots ; scutellum yellowish, marked with two very black spots ; the abdomen black, major part of the antennæ reddish. With the preceding.

S. Erythropterus, Lin. Panz. XXVIII. 4, from six to ten lines in length, black, with the wing-cases, the base of the antennæ, and the feet, fawn-colour.*

The others, whose form is linear, with the head and corslet elongated in the form of a long square, have the antennæ approximated at their base, strongly bent and grained, their anterior tarsi in general are not at all, or very little, dilated. The anterior legs are spinous, with a strong spine at the end. The labrum is small. These compose the genus XANTHOLINUS of some entomologists.†

PINOPHILUS, Grav.,

Which have also the palpi filiform, but whose antennæ are inserted in front of the eyes, without the labrum, and near the external base of the mandibles, (*Pinophilus latipes*) Grav. Amer. Septen. It is united to the following genus in his *mantissa*.

LATHROBIUM, Grav. *Pæderus*, Fab.,

Whose palpi are terminated abruptly by an articulation much smaller than the preceding, pointed, and often but little distinct. The maxillary are much longer than the

* *See* the monograph of this family (*Coleoptera microptera*,) of M. Gravenhorst, Panz. Index Entom. Part I. p. 208, et seq. Latr. *ibid.* I. 285. Refer to this genus the following species of Olivier: *aureus, æneus, hæmorrhoidalis, oculatus, erythrocephalus, similis, cyaneus, pubescens, cupreus, stercorarius, brunnipes, pilosus, politus, amœnus,* besides the five of which we give the description here.

† The Staphylini, *fulgidus, fulmineus, pyropterus, elegans, elongatus, ochraceus, alternans, melanocephalus,* of M. Gravenhorst.

labial, and the insertion of the antennæ is the same as in the preceding genus. The anterior tarsi are very much dilated in both sexes. The length of the last articulation of the four posterior ones, equals almost that of the four preceding articulations united.*

The second section, LONGIPALPI, which have also the head entirely uncovered, but whose labrum is entire, and whose maxillary palpi are almost as long as the head, terminating in a club-form, composed by the third articulation, with the fourth concealed or but little distinct, and in the figure of a little point, terminating the club when visible; the preceding one is very much swelled. These insects live on the banks of waters.

PŒDERUS, Fabri.,

In which the antennæ inserted in front of the eyes are either filiform, or grow thick insensibly, and are longer than the head, whose body is long and narrow, with the mandibles denticulated at the internal side, and terminating in a simple point.

Some (*Pæderes*, Lat.) have the penultimate articulation of the tarsi bifid.†

* *See* Gravenhorst, Coleop. Microp. and Latr. Gen. Crust. et Insect. I. 289. The *L. Elongatum* (*S. Elongatus*, Lin.) has been figured by Panzer, *ibid.* IX. 12; *Staphylinus linearis*, Oliv. Col. III. 2. iv. 38. *See* also Gyllen. Insect. Suec. I. Part II. p. 363, et seq. and the Catalogue of the Collection of M. le Comte Dejean, p. 24.

† M. Lefevre has brought from Sicily an insect neighbouring on the Pœderus, but evidently forming a new genus. The fourth and last articulation of the maxillary palpi is here very distinct, and terminates them in the manner of a knob. The last of the antennæ is larger than the preceding, and ovoido-conical. The head attaches to the corslet by an elongated pedicle, and on a level at its origin with the head. The corslet is narrow and elongated. The two anterior tarsi are very much dilated. The first articulation of the others is very long, and their penultimate

Staphylinus riparius, Panz. Faun. Insect. Germ. IX. ii, about three lines in length, very narrow, and very elongated, fawn-coloured, with the head, chest, upper extremity of the abdomen and knees black; elytra blue. Very common in the humid sand, under stones, at the root of trees, &c.

The others, Stilichus, Lat., have all the articulations of the tarsi entire. (*See* Lat. Gen. Crust. et Insect. I, page 290, et seq., and Gyllen. Insect. Suec. I. Part 2, p. 372.)

Evæsthetus, Grav.,

Whose antennæ are likewise inserted in front of the eyes, but little longer than the head, and almost entirely moniliform. The body is but little elongated, with the head as broad as the corslet. (*Evæsthetus Scaber*, Grav. Germ. Faun. Insect. Europ. VII. 13. Gyllen. Insect. Suec. I, Part II., p. 461. M. Blondel (the son), of Versailles, has discovered a new species in the environs of that town.)

Stenus, Lat.,

In which the antennæ inserted near the internal edge of the eyes are terminated by a club of three articulations. They have the extremity of the mandibles forked, and large eyes.

Staphylinus—2-guttatus, Lin. Panz. Faun. Insect. Germ. XI. 18. Two lines in length, altogether black, with a reddish point on each case.*

The third section, that of Denticrura, differs from the preceding in the maxillary palpi, which are much shorter

has appeared to me emarginated or bifid. I shall designate this genus by the denomination of *procirrus*, and this species shall be consecrated to the zealous naturalist (*Lefeburi*) who has discovered it.

* Add *Staphylinus Juno*, Payk; *Pæderus proboscideus*, Oliv. Col. III. 44. 1. 5; *Staphylinus clavicornis*, Panz. Faun. Insect. Germ. XXVII. 2. See Gravenhorst, *Coleopt. microp.* Latr. Gener. Crust. et Insect., genus *Stenus*, et Gyll. *ibid.* p. 465.

than the head, and always with four distinct articulations. The anterior has at least one denticulated or spinous at the interior side. The tarsi, which, in the majority, fold back on the limbs, have the last articulation as long, or longer than the preceding taken together. The first, or the first two, are usually so small, or so concealed, that their total number does not appear to be but two or three.

The front of the head, and sometimes even the corslet, is armed with horns in many of the males. The antennæ are inserted in front of the eyes.

Some whose palpi terminate in an oval form, whose antennæ are for the most part grained, and go on thickening, present distinctly but three articulations in the tarsi. With the exception of the *Trachini*, the anterior tarsi are not more notably dilated.

Oxytelus, Grav.*

The others have the filiform palpi and four articulations, at least, very apparent in the tarsi.

Osorius, Leach, Dej.

Have the body cylindrical, all the limbs widened and dentated, the head as long as broad, the corslet almost in the form of a heart, narrowed and truncated posteriorly, and the antennæ for the most part grained, thickening insensibly towards the end, and shorter than the head and corslet, the mandibles much shorter than the head, very much crossed, and terminated in a single point, and the mentum large and buckler-formed.

* See the article Oxytele of the Encyc. Method. The fore-cited Monograph of M. Gravenhorst, and Gyllenhall, Insect. Suec. I. Part II. p. 144.

We know but a small number of species which have not been yet described, and which inhabit French Guiana and Brazil.

ZYROPHORUS, Dalm. *Leptochire*, Germ. *Irenæus*, Leach. *Oxytelus*, Oliv. *Piestus*, Grav.

Whose body is depressed; whose anterior limbs, broader than the others, are alone denticulated externally; which have the head transverse, the corslet squared, the antennæ of the same thickness throughout, as long at least as the head and corslet, composed of articulations for the most part ovaliform, or cylindrical, and rounded at the two ends, and the mandibles as long as the head, and denticulated at their extremity.*

PROGNATHA, Lat. Blond. *Siagona*, Kirby,

Which differ but little from Zyrophorus, but in their filiform antennæ, comprised of elongated articulations. (*Siagonum quadricorne*, Kirb. and Spenc. Introd. Entom. I. i. 5. Blondel, Annal. des Sc. Nat. April 1817, XVII. 14-17.)

COPROPHILUS, Lat. *Omalium*, Grav. Oliv. Gyll.,

In which the body is still flatted, but all the limbs are denticulated, or spinous externally. The antennæ much longer than the head, are grained, and thicken insensibly towards the end, and the mandibles arched externally, almost into a crescent, are not sensibly denticulated, and are pro-

* See Dalman, Anal. en tom. p. 23, his *Z. Fronticornis*, iv. fig. 1. appears to be the *Oxytelus bicornis* of Olivier. (Encyc. Method.) That which he names, *penicillatus*, ibid. fig. 2, appears to have great relations with the *piestus sulcatus* of M. Gravenhorst. The *leptochirus scoriaceus* of M. Germar, (Insect. Spec. Nov. I. i.) is a species very distinct from the preceding.

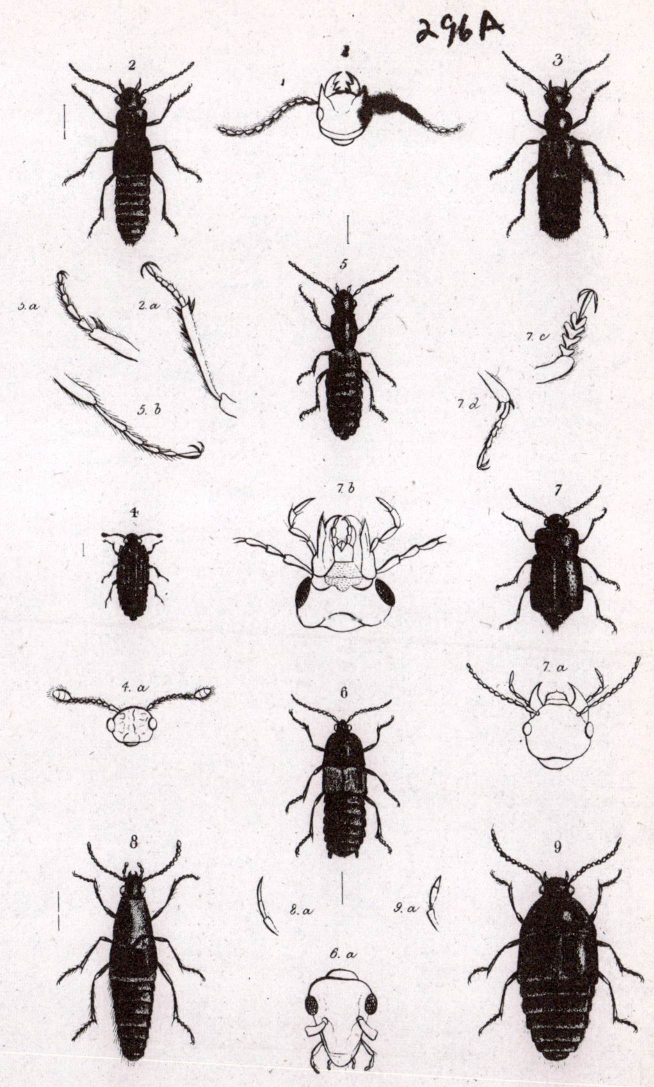

1. *Head of* Prognatha *rufipenne. Lat.* 2. Coprophilus *rugosus. Grav.* 3. Lesteva *dichroa. Lat.*
4. Micropeplus *maillei. Dej.* 5. Aleochara *canaliculata. Fab.* 6. Lomechusa *paradoxa. Grav.*
7. Omalium *blattoides. Grav.* 8. Tachinus *atricapillus. Fab.* 9. Tachiporus *marginatus. Grav.*

longed but little at their extremity. (*Omalium rugosum*, Gravenhorst, and other species with short elytra.)

The fourth section, that of Depressa, presents us, as well as the preceding, with a disengaged head, an entire labrum, short maxillary palpi, and with four distinct articulations, but the legs are simple, or without teeth or spines on the external side, and the tarsi have manifestly six articulations.

Here the palpi are filiform.

Omalium,

In which the corslet is of the breadth of the elytra, broader than the head, almost in a transverse square (with the angles, or at least the anterior ones, rounded), and often bordered laterally, and of which the antennæ go on thickening towards their extremity.*

Lesteva, Lat. *Anthophagus*. Grav.,

Which have the corslet in the form of a heart, narrowed, and truncated posteriorly almost isometrical, of the breadth of the head, more narrow than the elytra, and the antennæ generally filiform, and with elongated articulations.†

Micropeplus, Lat.,

Distinguished by their antennæ finishing in a solid knob, and lodging in the fossets of the corslet.‡

Proteinus, Lat.,

In which the antennæ being grained, a little perfoliated, and

* *See* Gravenhorst, article *omalie* of the Encyc. Meth., and Gyllenhall, *ibid.* p. 198.

† See Latreille, Gen. Crust. Insect. I. p. 296-297; Gravenhorst and Gyllenhall, genus *Anthophagus*.

‡ See Latreille, Gen. Crust. et Insect. IV. p. 377; *Omalium porcatum*, Gyll. Insect. Suec. I. Part II. p 211. *Micropeplus porcatus*, Charp. Horæ Entom. VIII. 9; Gyll. *ibid.* *O. Staphylinoides*, p. 213.

thicker towards the end, but under the form of a club, and always uncovered, are inserted in front of the eyes: in which the corslet is short, and the elytra covers the major part of the abdomen. (*See* Latr. *ibid.* I. p. 298, and the *omalium ovatum*, and *macropterum* of Gravenhorst.)

Aleochara, Grav.,

In which the antennæ are inserted between the eyes, or near their inferior edge, and naked at their origin, with the first three articulations sensibly longer than the following, these last perfoliated, and the last elongated and conical. The corslet is almost oval, or in a square rounded at the angles.*

The fifth section, the Microcephala, have the head sunk posteriorly as far as near the eyes, in the corslet. It is not separated by a neck, nor visible strangulation. The corslet has the form of a trapezium, and is widened from front to rear.

They have the body less elongated than the preceding, and approaching more to the elliptical form, the head much more narrow, contracted and advanced in front, the mandibles of middle size without denticulations, and simply arched at the point. The elytra, in many, cover a little more than half of the length of the upper part of the abdomen. Some live in mushrooms, on flowers, and others in dung. Fabricius has joined many of their species with the oxypori.

* *Staphylinus canaliculatus*, Fab. Panz. *ibid.* XXVII. 13; *Staphylinus impressus*, Oliv. Col. *ibid.* v. 41; *S. Boleti*, Lin. Oliv. Col. *ibid.* iii. 25; *S. collaris*, ejusd. ibid. ii. 13; *S. minutus*, ejusd. ibid. vi. 53; *S. socialis*, ejusd. ibid. iii 25, and generally the first three families of the genus *Aleochara*, of Gravenhorst, Col. Mic. tom. II. See also Gyllenhall, Insect. Suec. I. Part II. p. 377. But it will be observed that neither this author nor M. Gravenhorst have assigned to Aleochara, and Lomechusa, clear and rigorous characters. These two sub-genera require some new examination.

LOMECHUSA, *Aleochara*, Grav.,

Which have no spines on the legs, and of which the antennæ from the fourth articulation, form a perfoliated club, or like an elongated spindle, and the palpi terminate awl-like. The antennæ are often shorter than the head and corslet.*

TACHINUS, Grav.,

Which have spinous legs. The antennæ are composed of articulations like a reversed cone, or a pear, and thickening insensibly, and whose palpi are filiform.†

TACHYPORUS, Grav.,

Similar to the Tachini in the limbs and the antennæ, but having the palpi terminating awl-like.‡

* Some have the corslet smooth, and not raised over its edges. Such the Aleochara, *bipunctata, lanuginosa, nitida* (*Staphylinus bi-pustulatus*, Lin. Oliv. Col. III. 42, v. 44.) *fumata, nana*, of Gravenhorst, or his families, iii. vi. (Col. Micropt. tom. 2.) The others have the edges of the corslet raised, and form his genus *Lomechusa ; L. paradoxa ; Staphylinus emarginatus*, Oliv. *ibid.* ii. 12 ; *L. dentata*, Grav. ; *Staphylinus strumosus*, Payk. V.

† *Oxyporus subterraneus*, Fab. ; *O. bi-pustulatus*, ejusd. Panz. Faun. Insect. Germ. XVI. 21 ; *O. marginellus*, Panz. *ibid.* IX. 13 ; *Staphylinus fuscipes, ibid.* XXVII. 12 ; *Oxyporus suturalis*, ibid. XVIII. 20 ; *O. Pygmæus*, ibid. 27 ; *O. lunulatus*, ibid. XXII. 19 ; 15 ; *Staphylinus atricapillus*, F. ; *Oxyporus merdarius*, Panz. ibid. XXVI. 18 ; *Staphylinus striatus*, Oliv. ib. v. 47 ; *S. lunulatus*, Lin. See also, as well for this sub-genus as the following, the second part of the first volume of the Insects of Sweden, of M. Gyllenhall. There will be found excellent remarks on the several differences of many species, and the application of which may be very useful.

The Tachini, which, such as *atricapillus*, have the corslet almost as long as broad, the muzzle advanced, the four posterior tarsi sensibly longer than their respective legs, appear proper to form a particular section.

‡ *Oxyporus rufipes*, Fab. Panz. *ibid.* XXVII. 20 ; *O. marginatus*, F. Panz. *ibid.* 17 ; *O. chrysomelinus*, F. Panz. ibid. IX. 14 ; *O. analis*, F. Panz. *ibid.* XXII. 16 ; *O. abdominalis*, F.

The genus Callicerus of M. Gravenhorst is unknown to me. That of Stenosthetus, of M. Megerle, indicated in the catalogue of the collection of Coleoptera of M. le Comte Dejean, presents all the characters of a true Pselaphus, and ought to be suppressed. Such is also at present the opinion of this last naturalist.

SUPPLEMENT

ON THE

BRACHELYTRA.

THIS family is entirely composed of the genus STAPHYLINUS of Linnæus. It is a family of very considerable extent, comprehending about six hundred species, or perhaps more.

Fabricius made the first separation from it of *pederus* and *oxyporus*. Subsequently M. Latreille introduced two other genera, or, as they now stand in "The Animal Kingdom," sub-genera, those of *lestera* and *stenus*. Nine others were instituted by M. Gravenhorst, and one by Mr. Kirby, and from the communications of Dr. Leach to M. Latreille, of certain exotic species, which form so many types of new generic sections, some other divisions have been formed.

The STAPHYLINI, taken in general, are to be found in fresh and humid places under stones, under the bark of dead or rotten trees. They may be seen running in the fields, gardens, and on the highways, when in search of their prey. They are very courageous and extremely agile; provided equally with strong feet and good wings, they run with great swiftness, and fly with the utmost facility. They destroy the insects which live in dung-hills, cow-dung, on the branches of trees, &c., carrying on a continual war upon them; sometimes they surprise them in their retreats, sometimes they pursue them in the fields, attack them with advantage, and combat them with bitterness. But they appear to have been more especially designed for the reduction of carcasses to their final state of dissolution. Among these the majority of them establish their dwelling, and there it is (however disgusting the task) that we must resolve to seek for them. This kind of habitation, but little calculated to attract the attention of naturalists, must for a long period have prevented their acquaintance with a great number of these insects, and no doubt must still keep many of them undiscovered. Nevertheless it must in fine be acknowledged, that the objects which false delicacy repels with the utmost degree of disdain, are those which are often most calculated to repay the zeal of the curious observers of the productions of nature.

The head of the Staphylini, usually very broad, is furnished in front with two teeth or mandibles, which at once announces the instinct or habitual character of the animal. These teeth are large, curved and pointed. In a state of inaction they repose one upon the other, and cross by their points; but when the insect opens them they give an air which might well appear terrible if the animal was capable of imposing on the observer by its general bulk. Their substance is extremely hard, and their colour, in general, is black. The mere aspect of these redoubtable teeth would indicate that the staphylinus employed them to seize and devour whatever insects it could catch. Nor would this opinion be unfounded; it is an animal of the most extreme voracity, which often does not even spare its consimilars. Degeer relates, that having presented a fly to a large species of staphylinus, the latter seized it at once, plunged its murderous teeth into its body, and finally tore it in pieces with its denticulations. The nature of this insect is discovered the moment one desires to lay hold of it; it endeavours to defend itself, and to bite the fingers which detain it.

The two elytra, which are very short, and as it were truncated at the middle of the back, are one of those characters which strike us most when we consider the staphylini. They cover but one half of the body, and often scarcely more than a third, and usually leave the entire abdomen naked. Nevertheless the two large wings with which the insect is provided, longer than the body when unfolded, are entirely concealed in a state of inaction, under these same elytra. They are folded there with much art, three or four times, so that they are reduced to a small twisted packet. When the insect wishes to employ them it extends them rapidly, they are then admirably adapted for flight, and it is difficult to conceive how they can find room under cases of so small a volume.

The abdomen, which is left uncovered by the elytra, possesses much suppleness and flexibility. When the staphylinus is touched it raises the tail or hinder part, and gives it all kinds of inflexions. It appears desirous of defending a naked and feeble part, not protected by the elytra, as in the other coleoptera. The position which it gives to the abdomen renders it more difficult to be seized, and less opportunity is thus given to the danger which threatens it. It is also by means of the abdomen, that the insect after having ceased to fly, desiring to replace the wings under the elytra, pushes them in there, and makes them fold up.

This may easily be observed every time that a staphylinus in flying, comes to rest itself on the ground. Two small hairy and moveable points, between which a conical part equally moveable is observed, in which is placed the aperture of the anus, terminate the extremity of the body. The staphylinus can withdraw these three parts into its body, and thus make them entirely disappear. The anterior tarsi are thick and broad, and this confirmation appears well adapted to enable the staphylinus to dig in earth and dung, for the purpose of depositing its eggs.

Many staphylini have both body and elytra well furnished with hair; but there are others which have them smooth, and provided with very few hairs. Ray, and other writers, have remarked that all the staphylini, when touched a little roughly, suddenly send forth from the last ring of the abdomen two small curved vesicles, generally yellow or white.

The larvæ of the staphylini are long, and resemble in some measure the perfect insect. They live in the earth and dung, where they are transformed into nymphs, which, generally speaking, are similar to those of the rest of the coleoptera. They are rather rare and difficult to be found.

We have established a new genus on one from the Brazil, which we name *Schizochilus*, of which the labrum is rather crescent-shape, strongly divided, and lobed, with two hooks curved in; the mandibles very long, slender, with a tooth in the middle, beneath which there are two small filaments; the maxillary palpi long, with the last joint very acute; the labial palpi short, the last joint triangular; the antennæ short, with the first joint long, the others enlarging toward their extremities, the last joints rather serrated, the tarsi very dilated; the head very flat; the thorax truncate, heart-shape; the elytra short, quadrate; the body long, depressed. This species we call *Brasiliensis*. Its specific characters are bronze, varied with yellowish and black hair; the mandibles black; the eyes white; the body black, with the two last segments golden yellow. It is twelve lines and a half long, and is from Brazil.

The Oxypori are small insects, whose body is proportionally less elongated than those of the other brachelytra, and smooth. The elytra are very hard, and scarcely cover one third of the abdomen. Like the staphylini in general, the wings are curiously folded underneath them.

The habitudes of the oxypori differ a little from those of

the insects we have last treated of, though the form of the body is very nearly similar. Instead of frequenting dunghills, ordure, and decomposed animal matter, the oxypori inhabit agarics, &c. There the larva lives and acquires its growth; there the perfect insects are found in more or less numerous parties, and there the sexes unite.

The LATHROBIA have great relations with the staphylini proper, to which our introductory observations, though generally applicable to all the sub-genera, must yet be more particularly understood to refer, and seem to constitute a proper link between them and pœderus. They are found under stones, in the debris of animal and vegetable matters, and often in fresh and humid places.

The PÆDERI, *Evæstheti* and *Steni*, are distinguished from the other brachelytra, by the length of their maxillary palpi. In this respect they approach Lathrobium, but are nevertheless separated from it by characters of sufficient importance. The pæderi are distinguished from stenus and evæsthetus, by the filiform character of the antennæ.

The majority of them frequent the sandy borders of rivers, rivulets, and marshes. The others live under stones, mosses, in ordure, &c. Their habits are otherwise similar to those of the Staphylini, and the same, in all probability, is the case with their metamorphoses.

The STENI are insects of a very small size, which, in the same way as the last, live habitually in aquatic places. Respecting their habits there is nothing more to be said.

The OXYTELI are also very small coleoptera, which appear to have the habits of the other brachelytra, but which form, nevertheless, a very distinct genus. The species of which it is composed present some differences in the mandibles, the antennæ, &c., so that, according to the remark of Olivier, this group is susceptible of simplification. This has been performed by Dr. Leach, in a new monograph of the coleoptera of this family. Many oxyteli are more particularly found in the dung of animals, and human excrements. Some others habitually frequent humid or aquatic situations. Others remain under moss, stones, &c. The males of some are remarkable for two prominences in the shape of horns, which they have in front of the head. Even some among them have a strong and advanced point upon the corslet.

We have figured a species of *Zirophorus*, which M. Lacordaire has called *coriaceus*. It is smooth, and of a greyish black, with the antennæ and legs black, and is nine lines and a half in length ; also from Brazil.

Next to the Zirophorus we will add a new genus, which we call *Trichocoryne*. The characters are as follows: the mandibles slightly dentated ; the antennæ with the first joint club-shape, and stuffed—the second and third slender, rather clubbed, the other joints oblong-square ; the labrum small, bilobed ; the labium large, crescent-shaped ; the maxillary palpi with the last joint large and sub-oval ; the labial palpi small : it is readily distinguished from the Zirophori, by the length of the antennæ, and its peculiar form, which is easily seen by referring to the plate of dissections. This species is from the West Indies. It appears to be the same as M. Dalman described in his Anal. Ent., under the name of Zirophorus penicillatus. The specific characters are, piceous, with the antennæ and legs reddish ; the elytra strongly striated ; the length is three lines and a half. The type is in Mr. Children's cabinet.

The Omalia are also very small coleoptera, which in their manner of living differ but little from the Staphylini, from which they have been distinguished generically by M. Gravenhorst. Some are found under moss and flowers ; others delight in cow-dung and other excrements. Some few live in agarics.

The TACHYPORI have been associated by Fabricius with the oxypori. The species most known, *tachyporus chrysomelinus*, is very common under stones, mosses, leaves fallen to the earth, and even on flowers and the leaves of fruit-trees. It runs with remarkable swiftness.

Respecting the sub-divisions of this family, which we have passed unnoticed, either their habits are unknown, or do not differ from those we have already described.

THE

SERRICORNES.

THE third* family of the PENTAMEROUS COLEOPTERA present us, as well as the preceding family and the following of the same order, but four palpi. Their elytra cover the abdomen, which distinguishes them, with some other characters, from the Brachelytra just described. The antennæ, with some few exceptions, are of the same thickness throughout, or more slender at their extremity, denticulated, either like a saw or a comb, or even forming a fan, and more developed in this point of view in the males. The penultimate articulation of the tarsi is often bilobate or bifid. These characters present themselves very rarely in the following family, that of Clavicornes, and to which we arrive

* The *Silpha* are the only pentamerous coleoptera which present, as well as the preceding, an excrementitious apparatus, at the same time it is not binary as in the latter, and the external conduit disgorges itself directly into the rectum, like the urethra of birds. It would appear then, from these relations, that Silpha should come, as well as the other clavicorni, immediately after the brachelytra. Other considerations had conducted me to the same approximation. (*See* the Preface to my Work, entitled, General Considerations on the Natural Order of the Crustacea, &c.) According to M. Leon Dufour, who has furnished me with these anatomical observations, the hepatic conduits of the Buprestides and Enterides, or of my Sternoxi, resemble in their number, length, and mode of insertion, those of the carabici. The Lampyri and Melyrides have also but two hepatic vessels. But there are four in the Telephori, Lycus and Ptiniori.

by transitions so gentle, that it is extremely difficult to assign its limits rigorously.

Some, whose body is always of a firm and solid consistence, most frequently oval or elliptical, with the feet in part contractile, have the head engaged vertically as far as the eyes in the corslet; and the presternum, or the medial portion of this latter part of the body elongated, dilated, or advanced in front as far as under the mouth, distinguished usually on each side by a groove, in which the antennæ (always short) are applied and prolonged posteriorly into a point, which is received in a sinking of the anterior extremity of the mesosternum. These anterior feet are remote from the anterior extremity of the corslet. These Serricornes form a first section, that of STERNOXI.

Others, having also the head engaged posteriorly in the corslet, or at least covered by it at its base, but of which the presternum is not dilated and advanced anteriorly in the manner of a chin, nor usually* terminating posteriorly in a point received in a cavity of the mesosternum, whose body is most frequently altogether, or in part, of a soft and flexible consistence, will compose a second section, that of MALACODERMI.

A third and last, that of XYLOTROGI, will comprehend those Serricornes whose præsternum is not equally prolonged at its posterior extremity, but whose head is entirely uncovered, and separated from the corslet by a strangulation, or sort of neck.

* The cebriones form an exception, and approach in this respect to *elater*; but the inferior extremity of the præsternum, does not advance on the under part of the head. The mandibles are advanced, arched, and simple; the palpi are filiform. The feet are not contractile, and the two anterior ones are but little removed, at their origin, from the anterior extremity of the corslet, and very much approximating to each other.

We shall divide the STERNOXI into two tribes.

The first, that of BUPRESTIDES, has the posterior projection of the præsternum flatted, and not terminated in a point laterally compressed, and simply received in a depression or in an emargination of the mesosternum. The mandibles often terminate in an entire point, or without emargination or fissure. The posterior angles of the corslet are either not at all, or very little prolonged. The last articulation of the palpi is most frequently almost cylindrical, a little thicker than the preceding, and globular, or ovoid in the others. The most part of those of the tarsi are commonly broad, or dilated, and furnished underneath with small pellets. These insects do not leap, a character which eminently distinguishes them from those of the following tribe.*

They compose the genus

BUPRESTIS, of Linnæus.

The generic denomination of *Richard*, given by M. Geoffroy, to these coleoptera, announces to us the beauty of their clothing. Many indigenous species, and many exotic, otherwise remarkable for the greatness of their size, and the brilliancy of polished gold on a ground of emerald; in others, azure shines over the gold, where many other metallic colours are united. Their body, in general, is oval, a little broader, and more obtuse or truncated in front, and narrowed behind from the base of the abdomen, which occupies the greatest part of its length. The eyes are oval, and the corslet is short and broad. The scutellum is small or nullified. The extremity of the elytra is more or less denticulated in a great number. The feet are short.

* The insects of this tribe differ moreover from all the others of this family, by their vesicular trachea, while they are tubular in the other serricornes. See the anatomical observations of M. Leon Dufour.

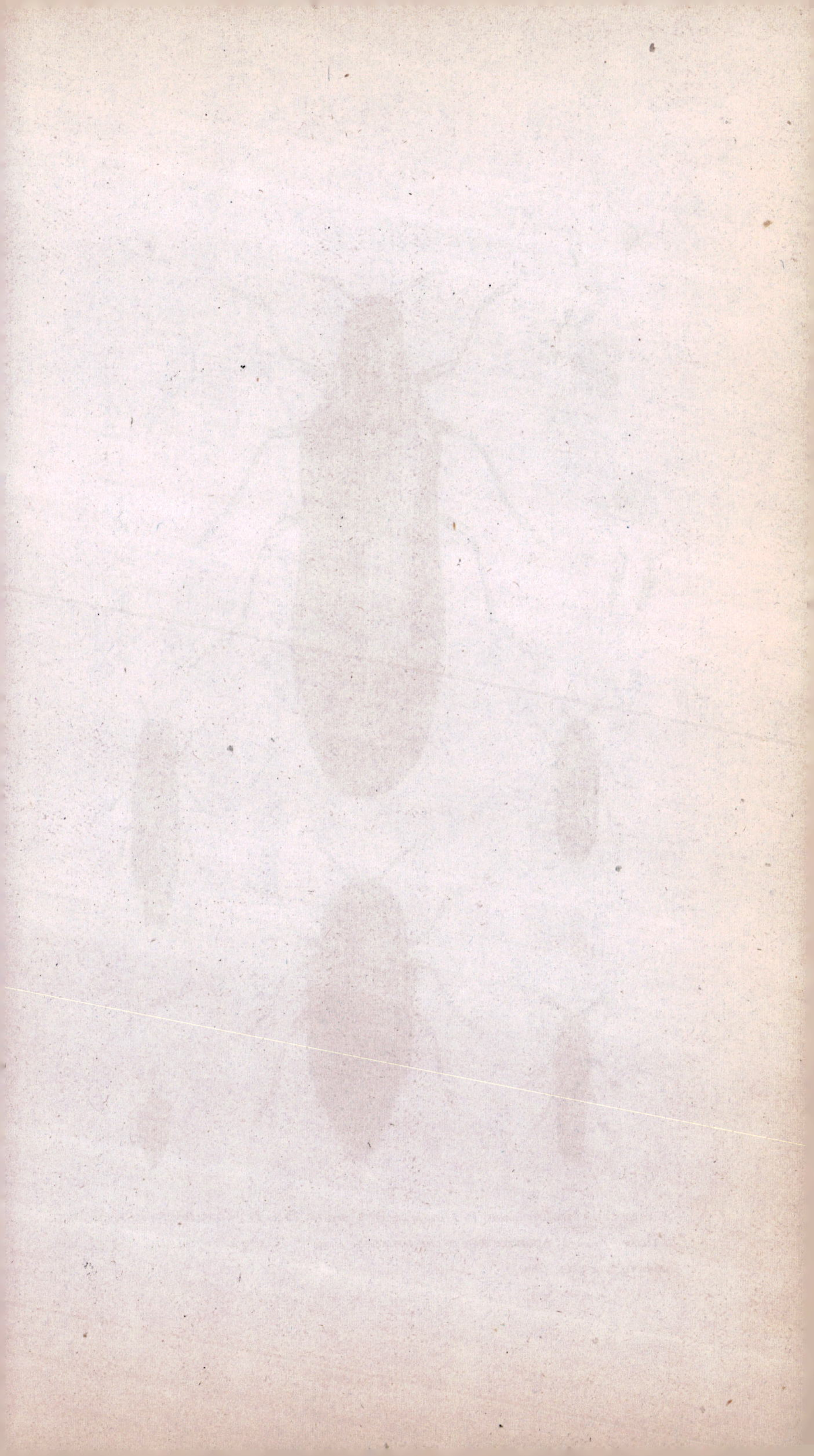

Pl. 9

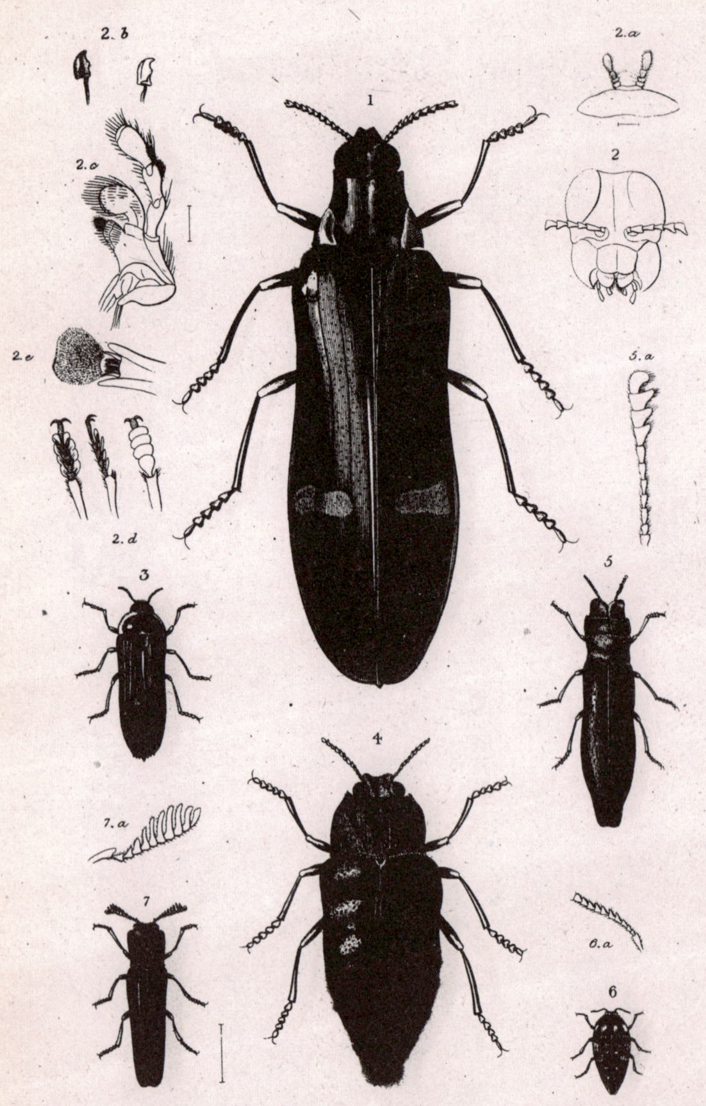

1. Buprestis *bicolor.* *Latr.* 2. *Details of* Bup. *gigas.* *Fab.* 3. Bup. *rubripennis.* 4. Bup. *Lalandii.* *Guer.* 5. Aphanisticus *emarginatus.* *Lat.* 6. Trachys *cruentata.* *Fab.* 7. Melasis *buprestoide.* *Oliv.*

They walk slowly, but their flight is very agile when the weather is hot and dry. If attempted to be seized, they let themselves fall to the ground. The females have at the posterior extremity of the abdomen, a coriaceous or corneous part in the form of a conical lamina, composed of three pieces (the last rings), and which is probably an instrument with which they deposit their eggs in the dry wood, in which their larvæ reside. Many small species are to be met with on flowers and leaves, but the others remain for the most part in forests, &c. They sometimes assume the perfect state in houses, being transported there in the larva or nymph state, along with the wood.

Sometimes the antennæ are altogether saw-like. The intermediate articulations of the tarsi are in the form of a heart reversed, and the penultimate, at least, is bifid. The palpi are filiform, or slightly more thick at the end. The jaws are bilobate.

BUPRESTIS (Proper). L.

Whose antennæ are of the same thickness throughout, and serrated from the third or fourth articulation.

Some have no scutellum. *B. fasciculata.* Lin. Oliv. Col. ii. 32., iv. 38, about an inch in length, ovoid, convex, very punctuated and wrinkled; of a golden or coppery green; sometimes obscure, with small tufts of yellowish or reddish hairs. Elytra entire.—At the Cape of Good Hope, and sometimes in such great abundance on the same shrub, that it seems altogether charged with flowers.

B. Sternicornis. Lin. Oliv. Col. *ibid*, vi. 52, a little larger; of the same form; of a green a little golden, very brilliant. Some thick sunken points, the bottom of which is furnished with whitish scales on the cases; three teeth at their extremity. The posterior sternum is advanced in the form of a horn.—East Indies.

B. Chrysis. Fab. Oliv. *ibid.* II. 8, vi. 52—b., differ from the preceding, by the cases of brown marron, and without whitish spots.

B. vittata. F. Oliv. *ibid.* III. 17, near an inch and a half in length; more narrow, and more elongated than the preceding; depressed, of a blueish green; four raised lines, and a golden and coppery band on each case, the end of which has two teeth.—Of the East Indies.

B. ocellata. F. Oliv. *ibid.* I. 3, almost similar both in size and form; has on each case a large yellow and phosphoric spot, situated between two others of a golden colour. The end of each case is terminated by three teeth.

The others have a scutellum. *B. gigas.* Lin. Oliv. *ibid.* I. i. Two inches in length; corslet coppery, mingled with brilliant green, with two large smooth spots, of the colour of browned steel. The cases are terminated by two points, coppery in their centre, of a bronzed green on their edges, with sunken points, raised lines, and wrinkles.—Of Cayenne.

We shall cite among the species of our country,

B. Affinis. Fab. *B. Chrysostigma.* Oliv. 105, *ibid.* VI. 54. Bronzed above, coppery and brilliant underneath; the elytra of which, denticulated like a saw, at their point, have three longitudinal raised lines, and two gilded impressions on each.

B. viridis. Lin. Oliv. *ibid.* XI. 127, about two lines and a half in length, with a linear form, of a green bronzed, with the cases entire, and punctuated.—On trees.

Fabricius has detached from Buprestis proper, those which have the body short, more broad proportionally, and almost triangular; the front excavated, the corslet transverse, and lobate posteriorly, and the tarsi very short, with broad pallets, the last five articulations of the antennæ form alone the teeth of a saw. The preceding, with the exception of the first two, are small, almost grained, or in a reversed cone. The first two are much thicker. These species comprehend the genus Trachys. Of this number is *B. Mumata*, Lin. Oliv. *ibid.* II. 14. Black underneath, of a coppery brown above, with the middle of the front sunk in the corslet, sinuated at its posterior edge, and some whitish stripes, undulating, formed by hairs, and transverse on the cases. Common on the hazel tree, of which it gnaws the leaves.

Aphanisticus, Lat.

Have the antennæ terminated abruptly in a club-form, oblong, compressed, slightly serrated, formed by the last four articulations. The last of the palpi is a little thicker, almost ovaliform. The interval between the two eyes is excavated, as well as in the trachis. Two or three species are known, all very small, and of a linear form.*

Sometimes the antennæ are very much pectinated (on one side only) in the males, but strongly serrated in the other sex. The articulations of the tarsi are almost cylindrical and entire. The palpi are terminated by an articulation much thicker than the preceding, and almost globular. The jaws are terminated by a single lobe.

Melasis, Oliv.

Their body is cylindrical, and the posterior angles of the corslet are prolonged into a sharp tooth, characters which, as well as those taken from the tarsi and the palpi, announce that these insects form the passage from this tribe to the following.†

* *Buprestis emarginata*, F. Oliv. *ibid.* X. 116; Germ. Faun. Insect. Eur. III. 9; *ejusd.* *Buprestis lineola*, ibid. 10.

† *Melasis Buprestoides*, Oliv. II. 30, i. i. *Melasis elateroides*, Illig., differing, according to him, from *Elater Buprestoides*, Lin.

The second tribe, that of the ELATERIDES, does not differ essentially from the preceding, but that the hinder stilet of the præsternum, terminating in a point, compressed laterally, and often a little arched, and unindentated, sinks at the will of the animal into a cavity of the breast, situated immediately above the origin of the second pair of feet, and that these insects, placed on the back, have the faculty of jumping. (See lower down.) They have for the most part mandibles, emarginated or cleft at their extremity, the palpi terminated by an articulation much larger than the preceding, in the form of a triangle or axe, and the articulations of the tarsi entire. This tribe comprehends only the genus

ELATER, of Linnæus.

Their body is generally more narrow and more elongated than that of the Buprestides, and the posterior angles of the corslet are prolonged into a sharp point, spine-formed.

They have been named in French *Scarabées à ressort*, and in Latin, *Notopeda*, *elater*. Laid upon their back, and not being able to rise in consequence of the shortness of their feet, they jump and raise themselves perpendicularly into the air, until they fall back into their natural position, or on their legs. To execute these motions they press them against the under part of the body, lower the head inferiorly, and the corslet which is very mobile, from top to bottom, then drawing together this last part of the post-pectus, they push with force the point of the præsternum against the edge of the hole situated in front of the mesosternum, where it then sinks abruptly, and as it were by a spring. The corslet with the lateral points, the head, the upper part of the elytra striking with force against the plane of position, especially if it be firm and smooth, concur, by their elasticity, to raise the body in the air. The sides of the præsternum are distinguished by a groove in which those insects lodge, in part, their antennæ, which are pectinated, or in long barbs, in many males. The females have at the anus a sort of long auger, with two lateral pieces, pointed at the end, between which is the oviductus, properly so called.

These insects remain on flowers, plants, and even on the ground, or turf; they lower the head in walking, and when they are approached they suffer themselves to fall to the ground, applying their feet under the lower part of the body.

De Geer has described the larva of a species of this genus (*undulatus*). It is long, almost cylindrical, provided with small antennæ with palpi, with six feet, and has twelve rings covered with a scaly skin, of which, that of the posterior extremity forms a plate bordered and angular on the edges, with two blunt points, and curved within; underneath is a thick, fleshy and retractile nipple, which performs the office of a foot It lives in the sort of loam formed by rotten wood; it is also found in the earth. It even appears that that of the *E. Striatus*, of Fabricius, gnaws the roots of corn, and causes much injury when it multiplies.

The stomach of the elater is long, wrinkled crosswise, and sometimes swelled at the lower part. The intestine is moderate.

We may refer to two principal divisions the various subgenera which have been formed in this tribe. Those whose antennæ can lodge entirely in the lower cavities of the corslet will compose the first.

Sometimes they are received on each side in a longitudinal groove, formed immediately under the lateral edges of the corslet, and always filiform, and simply serrated. The articulations of the tarsi are always entire, and without elongations, in the form of a palette underneath. The corslet is convex or gibbous, at least on the sides, and is dilated towards the posterior angles in the manner of a lobe, going either in a point, or triangular. These insects approximate to the buprestides.

GALBA, Lat.,

The mandibles of which terminate in a single point, the jaws present but a single lobe, the last articulation of the palpi is globular, and the body almost cylindrical.*

EUCNEMIS, Arh.,

In which the mandibles are bifid, and the jaws bilobate; in which the last articulation of the palpi is almost hatchet-formed, and the body almost elliptical.†

Occasionally the antennæ, sometimes club-formed, are lodged, at least in part, either in the longitudinal grooves of the lateral edges of the præsternum, or in the fossets situated under the posterior angles of the corslet. The tarsi are often little palettes formed by the elongation of the lower pellets; or the penultimate articulation is bifid.

Some, with filiform antennæ, have the articulations of the tarsi entire and without palettes underneath; the two ante-

* I have seen three species, and all of Brazil. One has great relations with the *Melasis tuberculata* of M. Dalman. (Anal. Entom.) The jaws terminate by a very small and pointed lobe.

† M. le Comte de Maunherheim has published a very fine monograph of this sub-genus, of which an extract has been given, and the plates re-produced in the third volume of the Annals of Natural Science. I have added there a few observations on the too great extent which this naturalist has given to this sub-genus. The species, which he names *Capucinus*, is, according to me, the only one which ought to remain in it, and such was, at first, the opinion of him who established it.

Pl. 16

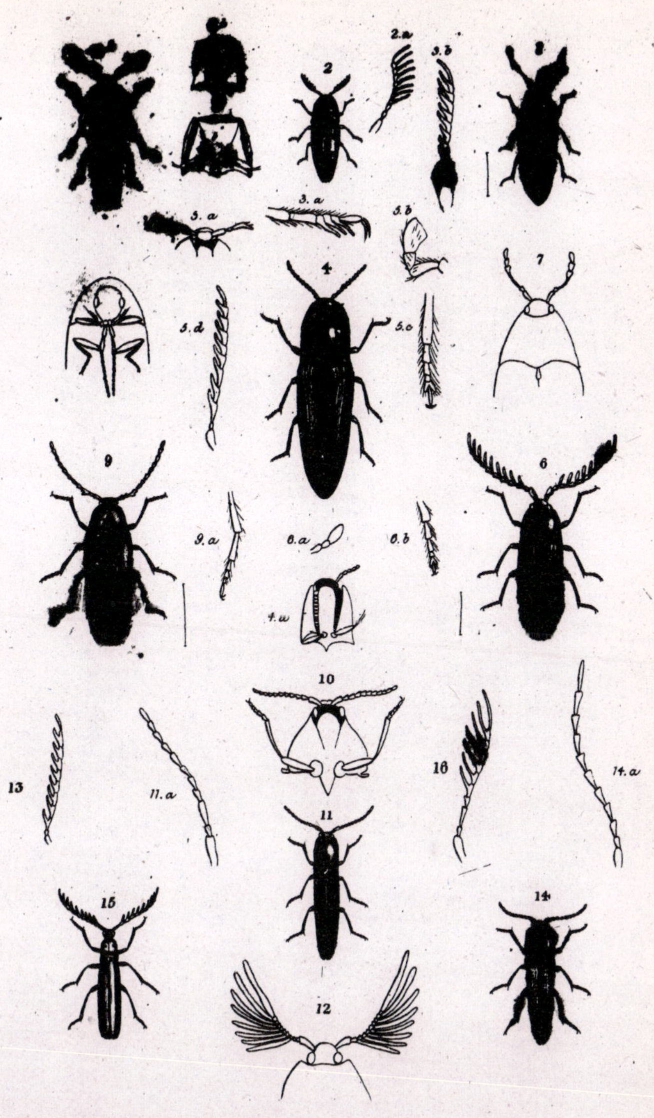

Eucnemis *capucinus*. *Man*. 2. Pterotarsus *histrio*. *Latr.* 3. *Details of* Galba *marmorata*. *Guer.* 4. Adecera *chabannii*. *Guer.* 5. Pachyderes *ruficollis*. *Guer.* 6. Cerophytum *elateroides*. *Latr.* 7. Throscus ermnestoides. *Latr.* 8. Chelonarium *undatum*. *Latr.* 9. Cryptostoma *denticornis*. *Fabr.* 10. *Head and* *corslet of* Lobaederus *monilicornis*.. *Guer.* 11 Nematodes *filum*. *Latr.* 12. *Head of* [illegible] *flabel*-*cornis*. *Latr.* 13 *Antenna of* Ctenicera *haematodes*. *Lat.* 14. Elater *plagiatus*. *Germ.* [illegible] *denti*-*collis*. *Fischer* 16. *Antenna of* Phyllocerus *flavipennis*. *Dej.*

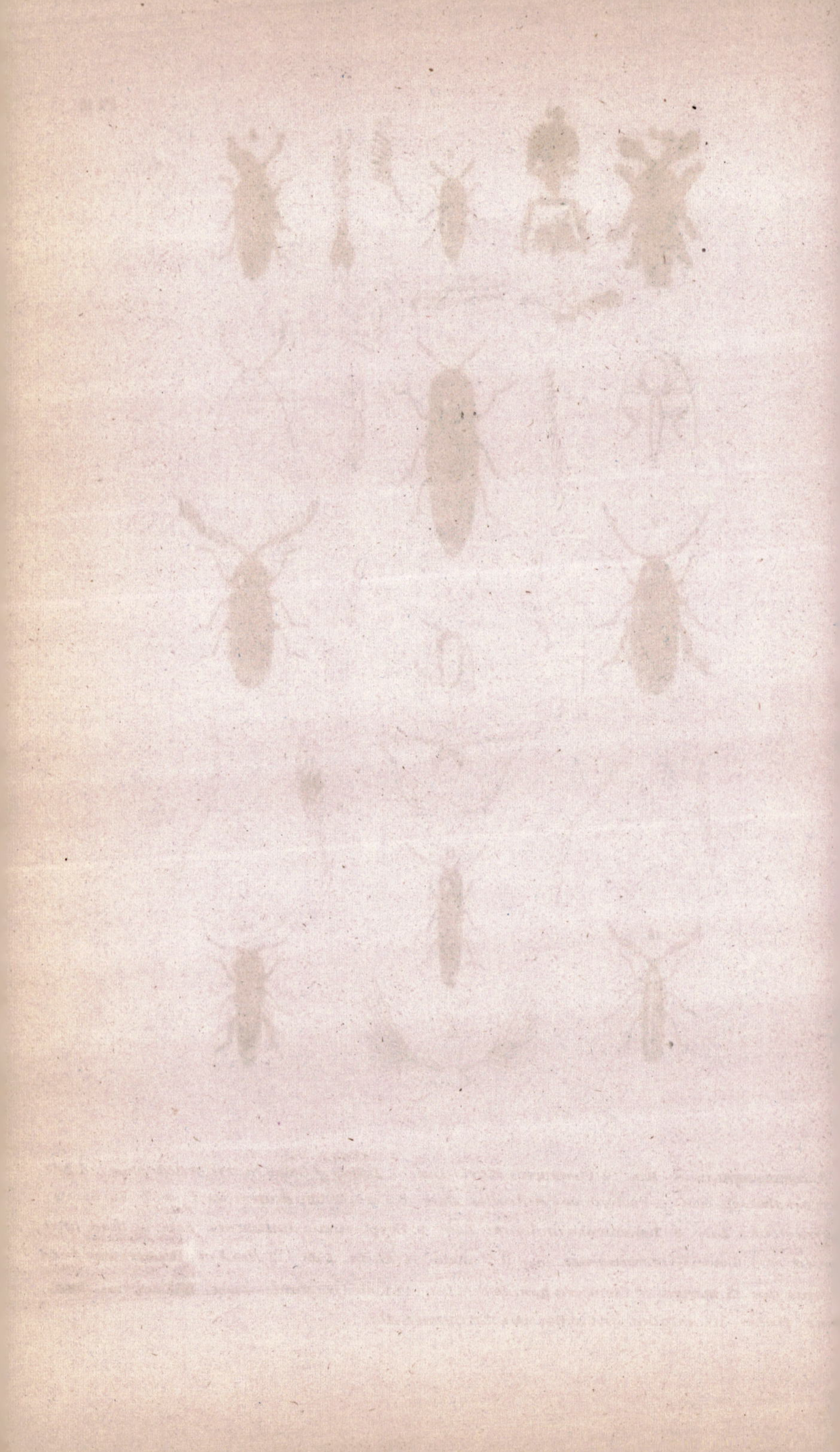

rior feet are lodged, when in a state of contraction, in the lateral sinkings of the under-part of the corslet. Such are,

ADELOCERA, Lat.*

Others, with antennæ equally of the same thickness throughout, have the articulations of the tarsi entire, but with under pellets prolonged and advanced, like little palettes or lobes. Their head is uncovered. These are,

LISSOMUS, Dalm. *Lissodes*, Lat. *Drapetes*, Meg. Dej.†

Others have the antennæ equally filiform, but the second and third articulations, which are larger than the following and flatted, alone lodge in the sternal grooves; the tarsi resemble those of lissomus; the head is concealed underneath, and, as it were, covered by a semi-circular corslet, in which it is sunk. Such are,

CHELONARIUM, Fab.

The antennæ in a state of repose, extend parallelly the length of the breast. The first and fourth articulations are the smallest of all, the following seven are of the same size, and with the exception of the last, which is ovoid, are almost in the form of a reversed cone, and equal. The body is ovoid, with the anterior legs broader than the others. All the known species are of South America.

The last sub-genus of this first division, that of

* *Elater ovalis*, Germ.; *Elater fuscus*, Fab., and some others from the East Indies, brought by M. Labillardière.

† Dalm. Epen. Entom. 1824. It is *Lissomus punctulatus*—has great relations with the *Drapetes Castaceus* of M. le Comte Dejean, and the *Elater lævigatus* of Fabricius.

Europe has one species of this sub-genus, *Elater equestris*, of the latter, figured, by Panzer, Faun. Insect. Germ. XXXI. 21.

THROSCUS, Lat. *Trixagus*, Kugel. Gyllenh. *Elater*, Lin.,

Is distinguished from all those of this tribe by its antennæ terminating in a club of three articulations, and lodged in a lateral and inferior cavity of the corslet. The penultimate articulation of the tarsi is bifid. The point of the mandibles is entire.*

Our second division of this tribe will comprehend all the elaterides whose antennæ are always uncovered or external.

We shall detach from them at first, those whose last articulation of the palpi, especially of the maxillary, is much larger than the preceding, almost in the form of an axe.

A single sub-genus, that of

CEROPHYTUM, Lat.,

Is removed from the following by its tarsi, the first four articulations of which are short, in the form of a triangle, and the penultimate articulation of which is bifid.

The antennæ of the males are branched on the internal side, the base of the third articulation, and of the following, being prolonged into a broad branch, rounded at the end. Those of the females are serrated.†

In all the other sub genera, the articulations of the tarsi are almost cylindrical, and entire.

Sometimes the head sinks up to the eyes in the corslet.

* *Elater dermestoides*, Linn.; *E. Clavicornis*, Oliv. Col. II. 31, VIII. 85, a. b.; *Dermestes adstrictor*, Fab. Panz. Faun. Insect. Germ. LXXV. 15. Its larva lives in the wood of the oak.

† Lat. Gen. Crust. et Insect. IV. 375. The *Melasis Sphondyloides* of Germar, Faun. Insect. Europ. XI. 5, has a great affinity with the female of the species, which serves as type. The *Melasis picoa* of Paliset de Beauvois, Insect. d'Afr. et d'Amer. VII. 1, has also some analogy with *Cerophytum*.

The anterior extremity of the præsternum advances over the under part of the head, and its edge is arched.

Some have the labrum and mandibles concealed by the anterior extremity of the præsternum; the chaperon, or epistomer, being widened and applying itself to this part. Such are

CRYPTOSOMA, Dej. *Elater*, Fab.,

Which have the internal angle of the third articulation of the antennæ, and the seven following, prolonged in the manner of a tooth. The second and fourth articulations are shorter, the last long and narrow, and a straight and linear branch at the internal side of the third, near its origin.

The mandibles are unindented under the point. The jaws present but a single lobe. They, as well as the tongue, are small and membranous. The palpi are very short. The tarsi are small, slight, and almost setaceous.

The only known species (*Elater denticornis*, Fab.) is found at Cayenne, from which it has been sent to the Museum of Natural History by M. Banon.

NEMATODES,

In which the antennæ have the first articulation elongated, the following five in a reversed cone, equal, with the exception of the first between them, or the second, which is a little shorter, and the last five thicker, almost perfoliated, and that of the top ovoid.

The body is almost linear.*

The labrum and mandibles are now discovered.

Here the antennæ of the males are terminated fan-like. These are

* *Eunemis filum*, Manner.

Hemirhipus, Lat.

The species are all exotic.*

Here these organs in the same sex, are pectinated in all their length.

Ctenicera, Lat.†

In the following sub-genus, or the

Elater (Proper),

The antennæ of the males are simply serrated.

The anterior extremity of the head is sometimes on a level with the labrum, or on the same horizontal, sometimes more raised and terminated abruptly; but these differences, often not to be well appreciated, are not of sufficient consequence to establish generic sections, and the genus, which I named *Lædie*, requires a further examination.

E. Noctilucus, Lin. Oliv. Col. II. 31, ix. 14, a., rather more than an inch long, of an obscure brown, with an ash-coloured down. There is a yellow. round, convex, shining, spot, on each side of the corslet, near its posterior angles; some lines with small points sunken on the cases.—Of South America.

Its spots, during the night, send forth a very strong light, which is sufficient to enable one to read the finest writing, especially if several of these insects be collected in one vessel. It is by this light that women perform their work. They also place this insect as an ornament in their head-dress, for their evening promenade. The Indians attach it

* *Elater flabellicornis*, Fab. *ejusd. E. fascicularis*, &c.

† His Elater, *pectinicornis, cupreus, hæmatodes; Taupin double-croix*, Cuv. Regn. An. IV. xiv. 3.

to their boots, for the purpose of giving them light in their nocturnal journeys. Brown will have it, that all the internal parts of this insect are luminous, and that it can suspend at will its phosphoric property. But M. de la Cordaire, who has observed this insect in its living state, has informed me that the principal reservoir of the phosphoric matter is situated underneath, at the junction of the abdomen with the thorax. Our colonists term it "*mouche lumineuse*," (fire-fly,) and the savages *Cacuyos*, *Coyouyou*, from which comes the Spanish name Cucujo. An individual of this species, brought to Paris in some wood, in the larva or nymph state, there underwent its metamorphosis, and by the light which it threw out excited the surprise of many of the inhabitants of the Faubourg St. Antoine, to whom such a phenomenon had hitherto been unknown.

E. æneus, Lin. Oliv. Col. *ibid.* VIII. 83. Six lines in length, of a bronzed green, shining, with striated elytra, and fawn-coloured feet.—Found in Germany and towards the North of Europe.

E. germanus, Lin. Oliv. *ibid.* ii. 12. Very common in the neighbourhood of Paris, not differing from the preceding except in the colour of the feet.

E. cruciatus, Oliv. *ibid.* IV. 40. A handsome European species, having the appearance of *E. æneus*, but smaller, black, with two red longitudinal bands on the corslet, near the lateral edges; the elytra are of a yellowish red, and have near the anterior angles of their base a black line, and two bands of this colour, forming a cross at the suture. It is rare in the neighbourhood of Paris.

E. castaneus, Lin. Oliv. *ibid.* III. 25; v. 51, black; corslet covered with a reddish down; elytra yellowish, with the extremity black; antennæ of the male pectinate. Of Europe. *E. ruficollis*, Lin. Oliv. *ibid.*

E. ferrugineus, Lin. Oliv. *ibid.* III. 35, ten lines long, the corslet black, with the exception of its posterior edge, and the cases of a deep blood red. Found on willows. The largest species in Europe.*

Sometimes the head is disengaged posteriorly, or is not sunk as far as the eyes, which are prominent and globular. The antennæ are inserted under the edges of a frontal projection, depressed, and arched before. The body is long and narrow, or almost linear. Such are

CAMPYLUS, Fischer. *Exophthalmus*, Latreille. *Hammionus*, Mühfeld.†

Some elaterides, with filiform palpi and antennæ pectinated from the fourth articulation, will compose a final sub-genus; that of

PHYLLOCERUS.‡

Our second section, that of the MALACODERMÆ, shall be divided into five tribes.

* See for the other species, Oliv. *ibid.* Panz. Faun. Insect. Germ., and his Ind. Entom., as well as Herbst. Col. and M. Palisot de Beauvois, Insect. d'Afri. et d'Amer. The genus DINA, of M. Ziegler, and of which the species named *elateroides*, has been figured by M. Charpentier, in his work entitled, Horæ Entomolog. VI. 8, presents to me no character to distinguish it precisely from the foregoing.

† See Fischer Entom. de la Russie, tom. II. page 153. This sub-genus comprehends the *Elater linearis* of Linnæus, of which his *masomelas* is only a variety; *E. borealis*, of Gyllenhal, and his *E. cinctus*.

‡ M. le Comte Dejean having gained but a single individual, I could not sacrifice it for the purpose of studying the characters in detail. Two insects of Java have somewhat of a similar appearance. In the subject in question, the antennæ (and most probably those of the females) are simply serrated. The mandibles appear to terminate in an entire point, or without tooth. The last articulation of the palpi is a little larger, and almost oblong. Supposing that the mandibles of the Phyllocesi are similar, those exotic species will be congenerous.

The first (CEBRIONITES), thus named from the genus CEBRIO, of Olivier, to which the others are attached, has the mandibles terminated in a simple or entire point, the palpi of the same thickness, or more slender at their extremity, the body rounded and gibbous, in some oval or oblong, but arched above, and inclined underneath in others. It is most frequently soft and flexible, with the corslet transverse, broader at its base, and its lateral angles sharp, or even prolonged in many species, in the form of a spine. The antennæ are usually longer than the head and corslet. The feet are not contractile.

Their habits are unknown. Many sojourn on plants, and in aquatic situations. These insects may be united into a single genus, that of

CEBRIO, Oliv. Fab.

Some, establishing a connection of this tribe with the preceding, whose consistence is even equally solid with that of Sternoxus, whose feet are never adapted for jumping, and whose body is generally an oblong oval, with the antennæ either pectinate or serrated in the males, the palpi filiform, or a little thicker at their extremity, and the posterior angles of the corslet prolonged into a sharp point, present us with mandibles advancing beyond the labrum, narrow and greatly arched, or in the form of crotchets. The labrum is usually very short, emarginated or bilobate.

In others, as well as in the elaterides, the præsternum terminates posteriorly in a point received in a depression of the mesosternum.

The antennæ, long in the males of some species, are composed of eleven articulations, pectinate or serrated. The last articulation of the palpi is almost cylindrical, or in the form of an inverted cone.

PHYSODACTYLUS, Fisch.

In which the three intermediate articulations of the tarsi present underneath, a pellet, which is membranous and orbicular. The hinder thighs are swelled, and the antennæ, at least in one of the sexes, are very short, serrated, and insensibly attenuated towards the extremity.

This sub-genus has been established by the celebrated author of the Entomography of Russia, on a North American insect (*P. Henningii*). (Letter on the Physodactylus, Moscow, 1824, Annales des Science, Dec. 1824, XXVII. B.)

CEBRIO, (Proper.) Oliv. Fab.,

In which all the articulations of the tarsi are entire and without pellets, and the posterior thighs are but little thicker than the others.

The species proper to Europe appear in great quantities after rain storms. The* female of the species the most known (*Gigas*, Fab.; *C. longicornis*, Oliv. Col. II. 30, bis. I. i. a, b, c; *Taupin*, I. I, a, b, c,) differs singularly from the male. Its antennæ are but little longer than the head; the first articulation is much longer than the others. The fourth and the following compose, when united, a small oblong and almost perfoliated club. The wings are partly rudimental. The feet are shorter, but proportionally more robust than those of the males. The larva probably lives in

* *Cebrio brevicornis*, Oliv. Col. II. 30. *bis*. I. 2. a. b. c.; *Tenebrio dubius*, Rossi. Faun. Etrusc. I. 1. 2. It appeared to me that this female, in consequence of her antennæ, ought to form a new genus, which I have named *Hammonie*. We find at the Cape of Good Hope a species in which the articulations of the antennæ throw out, each of them at the base of their internal side, a long and linear branch, and the palpi terminate in an ovoid articulation, and not in the form of a reversed cone, as the other species. This, therefore, might be separated from them.

Pl. [illegible]0

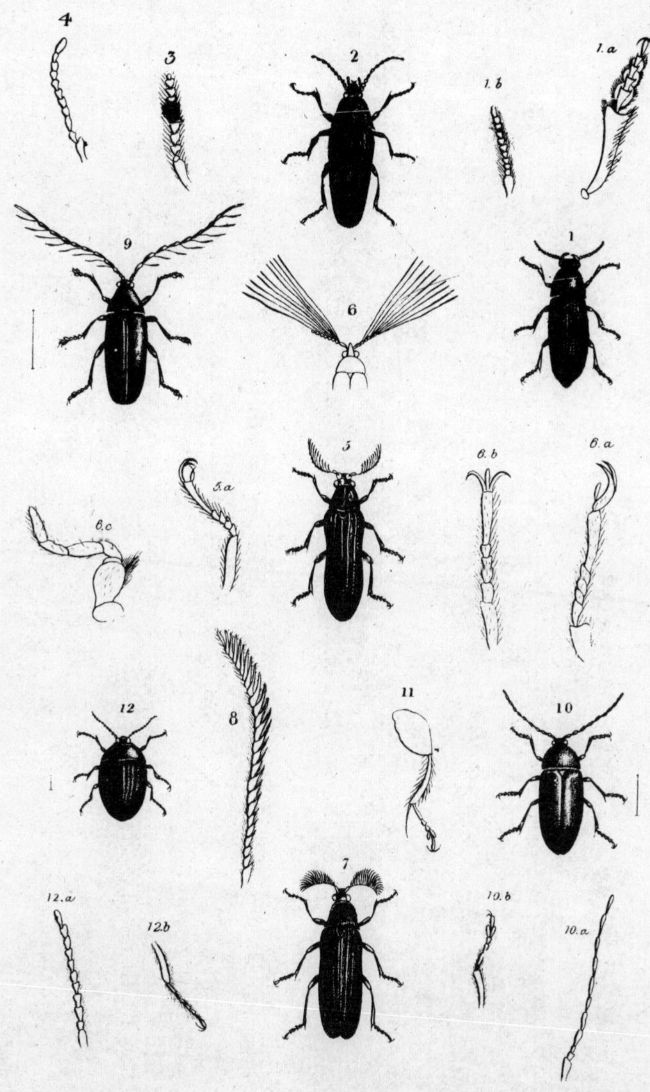

hysodactylus *Henningii.* *Fisch.* 2. Cebrio *fuscus.* *Gory.* 3. *Ant. of* Cebrio [illegible] *Fem.* 4. *Ant. of* clastes *Drurii.* *Kirby.* 5. Callirhipis *Goryi.* *Guérin.* 6. *Details of* Call. *D*[illegible] *Lat.* 7. Rhi-cera *cayanea.* *Guyer.* 8. *Ant. of* Rhip. *fem.* 9. Ptilodactyla *elaterina.* *Illig.* 10. [illegible]s. t. 11. *Post. tarsus of* Scyrtes. 12. Eubria *palustris.* *Germ.*

the earth. The *C. bicolor* of Fabricius, and some other American species whose body is elongated, less arched above, or almost straight, with the antennæ shorter, have appeared to Dr. Leach proper to compose a new generic section.

N. B. The *cebriones fuscus* and *ruficollis*, of Fabricius, have the form of the species which he calls *gigas*. M. Lefevre refers the second to Sicily. The *cebrio femoratus* of M. Germar, does not belong to the genus *Anelastes* of Mr. Kirby, as I had formerly suspected.

Here the præsternum is not prolonged remarkably into a point, and the mesosternum presents anteriorly no cavity.

Sometimes all the articulations of the tarsi are entire, and without any membranous palette advanced underneath.

ANELASTES, Kirby,

Whose antennæ are separated at their origin, short, almost grained, with the last articulation almost in the form of a crescent. The third is longer than the preceding and following, while in the cebriones this articulation and the second are shorter than the fourth and following ones. These organs, as well as in the elaterides, seem to have twelve articulations, the eleventh being abruptly attenuated towards the end, and terminating in a point, having the appearance of a small conical and triangular articulation. The last articulation of the palpi in the Anelastes, is almost in the form of an inverted cone. M. Kirby mentions but a single species. (*A. Drurii*, Lin. Trans. XII. xxi. 2.)

CALLIRHIPIS, Lat.,

Whose antennæ are very approximated at their origin, inserted on an eminence, and from the third articulation, form in the males a large fan. The last of the palpi is ovoid. The same of the tarsi is almost as long as the others taken together, and presents between its crotchets, a small linear and silken appendage.

The species which serves as type *C. Dejeunii*, is found in Java, and has been sent to the Museum of Natural History, by M. Diard and the late M. Duvaucel. The antennæ have but four articulations, and differ in that respect from those of the rhipicera, which have very much the same figure, but whose articulations are much more numerous in individuals of the same sex, or males.

Sometimes the tarsi have underneath membranous palettes, or their penultimate articulation is deeply bilobate.

In the two following subgenera, the first four articulations of the tarsi present, each underneath, two membranous and advanced lobes. The last is long, and terminated between the crotchets by a small silken appendage. The antennæ of some are composed of more than eleven articulations, disposed like a fan; those of the others have also eleven, toothed like a saw, and of which the last four being thicker, form a club or knob.

SANDALUS, Knock.

The antennæ, at least those of the females, are simply a little longer than the head, composed of eleven articulations, of which the third and the following, the last excepted, formed like the teeth of a saw, and the last four, a little more dilated, compose a knob. The terminal one is almost oval, rounded, or very obtuse at the end. (*Sandalus petrophya*, Knock. N. Beyt. I. p. 131, v. 5; *S. niger*, ejusd, ibid.)

RHIPICERA, Lat. Kirb. *Ptyocerus*, Hoffmans. *Polytomus*, Dalm.

The antennæ form in the two sexes a fan, and are composed of a great number of articulations (20—40), but in less quantity in the females.

This sub-genus is composed of from five to six species, two of which are of New Holland, and the others of America.*

The first three articulations of the tarsi of the two following sub-genera, are in the form of a reversed heart, without membranaceous elongations underneath. The fourth is profoundly bilobate; the last, but little elongated, presents between its crotchets no projecting and silky appendage. The antennæ are filiform, simple, or at most pectinate, and have never above eleven articulations.

PTILODACTYLA, Ilig. *Pyrochroa*, De G.

Are distinguished by their antennæ semi-pectinated, or toothed like a saw, in the males.

This sub-genus is composed of species proper to America. (*Ptylodactyla elaterina*, Ilig.; *Pyrochroa nitida*, De G. Insect. V. xiii. 6—17.)

DASCILLUS, Lat. *Atopa*, Fab.

Differ only in their antennæ being simple in the two sexes. (*Atopa cervina*, Fab. ejusd. A. Cinerea, var.; *Ptinus testaceo-villosus*, De G. IV. ix, 8; *Cistela cervina*, Oliv. Col. III. 5, 4, 1, 2, a)

The other cebrionites have small mandibles, little or not at all projecting beyond the labrum, the body generally soft, almost hemispherical or ovoid, and the palpi terminating in a point. The antennæ are simple, or but slightly toothed. In many the hinder feet answer the purposes of leaping.

* *Rhipicera marginata*, Lat. Cuv. Regn. Anim. III. p. 285, Kirb. Lin. Trans. XII. mas.; *polytomus marginatus*, Dalm. Anal. Entom. p. 22; ejusd. *P. femoratus*, ibid. 21; ejusd. *P. mystacinus*, p. 22; *Hispa mystacina*, Fab. Drur. Ins. III. viii. 7. I have seen, in the Collection of M. le Comte Dejean, another species, altogether fawn-colour, obtained in America by M. Leconte.

These insects inhabit plants and aquatic situations.

These have the penultimate articulation of the tarsi bilobate. The second and third of the antennæ are shorter than the following.

ELODES, Latreille. *Cyphon*, Fab. Dej.,

In which the hinder thighs differ little in thickness from the preceding. They constitute the first division of the *Cyphons* of Fabricius.

SCYRTES, Lat. *Cyphon*, Fab.

The hinder feet of which have the thighs very bulky, and the legs terminated by two strong spurs, one of which being very long, gives to these insects the faculty of leaping.

The labial palpi are forked. The first articulation of the posterior tarsi is as long as the others taken together. This is his second division. *See* the catalogue of the collection of M. Dejean.

Those have the articulations of the tarsi entire.

NYCTEUS, Lat. *Hamaxobium*, Ziegl. *Encynetus*, Schüpel.

In which the third articulation of the antennæ is very small, and much shorter than the second and the following, and in which the last are almost grained. The four feet have the legs terminated by two very distinct spurs, with the tarsi long, and more slender towards the end. (*Encynetus hæmorrhoidalis*, Germ. Faun. Insect. Europ. V. ii.)

EUBRIA, Zieg. Dej.,

Which have the antennæ a little denticulated, like a saw, with the second articulation very small, the following two the largest of all, and the last a little emarginated at the end, and going in a point. The spurs of the legs are very small, or almost nothing. The tarsi are filiform. (*Cyphon palustris*, Germ. *ibid.* IV. 3.)

The second tribe of *Malacodermi*, that of LAMPYRIDES, is distinguished from the preceding by the swelling which terminates their palpi, or at least the maxillary ones; by their body, which is always soft, straight, depressed, or but little convex, and the corslet, sometimes semi-circular, sometimes almost square, or in the form of a trapezium, advances on the head, which it covers entirely, or in part. The mandibles are generally small, terminating in a slender, arched, and very acute point, entire at the end in the majority. The penultimate articulation of the tarsi is always bilobate, and the crotchets of the last are neither denticulated, nor have they any appendages.

The females of some of them are deprived of wings, or have only very short elytra.

When these insects are seized they fold back their antennæ and feet against their body, and make no motion, just as if they were dead. Many, in this case, curve the abdomen underneath. They comprehend the genus

LAMPYRIS, Lin.

Antennæ very much approximated at their base; head either uncovered and prolonged interiorly, in the manner of a muzzle, or concealed entirely, or far the greater part under the corslet, with the eyes large and globular in the males, and the mouth small; such are the essential characteristics of a first division of this tribe, which we shall divide into those of which neither of the sexes is phosphorescent, and those whose females at least enjoy this property. All the individuals of the first are winged, have the head uncovered, often narrowed and advanced in front, or in the form of a muzzle, and the corslet widened posteriorly, with the lateral angles painted. The two or three last rings of the abdomen do not present that tint of pale yellow, or whitish, which affects this part of the body in Lampyris (proper), and announces their phosphorescence. The elytra, in many, go on widening, and are even sometimes very much dilated and rounded posteriorly, particularly in the females. They are very much punctuated, and often reticulated.

LYCUS, Fab. Oliv. *Cantharis*, Lin.

We shall confine this sub-genus to the species of Fabricius, in which the muzzle is as long, or longer, than the portion of the head which precedes it, and the antennæ are serrated. The elytra are most frequently dilated, either laterally, or at their posterior extremity, and the two sexes differ much in this respect, particularly in some species peculiar to Africa.*

Other species of the same writer, but with a very short muzzle, and whose compressed antennæ, sometimes simple, and sometimes serrated or pectinated, have their third articulation longer than the preceding, and in which the intermediate articulations of the tarsi are in the form of a reversed heart, will compose another sub-genus, that of

DYCTIOPTERA, Lat.,

Which we find in some woods in the neighbourhood of Paris, in the flowers of the millefolium, and others, and sometimes in abundance.

The *Lampyris Sanguinea*, Lin. Panz. Faun. Insect Germ. XLI. 9. It is almost three lines in length, black, with the sides of the corslet, and the elytra of a blood-red. These elytra are silken, and slightly striated. Its larva lives under the bark of the oak. It is linear, flatted, black, with the last

* The Lycus, *latissimus*, *rostratus*, *proboscideus*, &c. of Fabricius. See for the other species the appendix of the third part of the first volume of the synonimy of the insects of M. Schœnherr, in which he has described and figured several.

Pl. 36

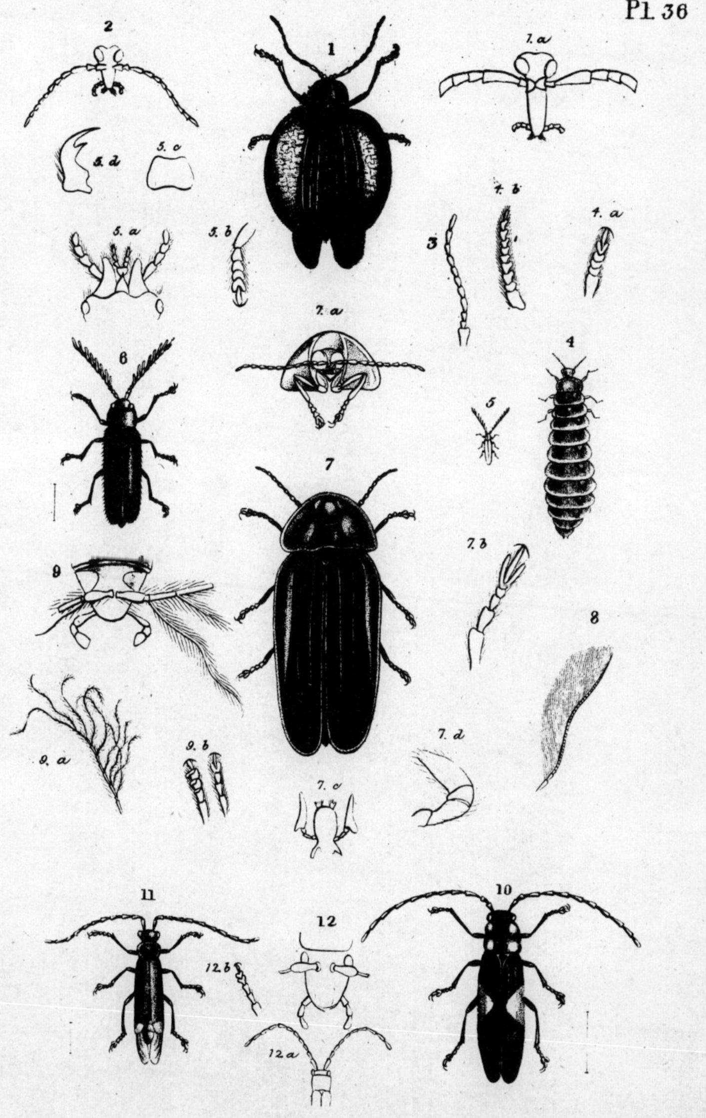

1. Lycus *latissimus. Fabr.* 2. *Head of* Dictyoptera *sanguinea. Lat.* 3. *Antenna of* Omalisus *saturalis Geoff.* 4. Drilus *flavescens. fem.* 5. *Id. male.* 6. Drilus *rufi-collis. Dej.* 7. Lampyris *Savignyi. Kirby.* 8. *Antenna of* Armydetes. *Hoff.* 9. *Details of* Cladophorus *ruficollis. Guer.* 10. Silis *tricolor. Guer.* 11. Malthinus *bi-guttatus. Oliv.* 12. Cordylocera *antennata Guer.*

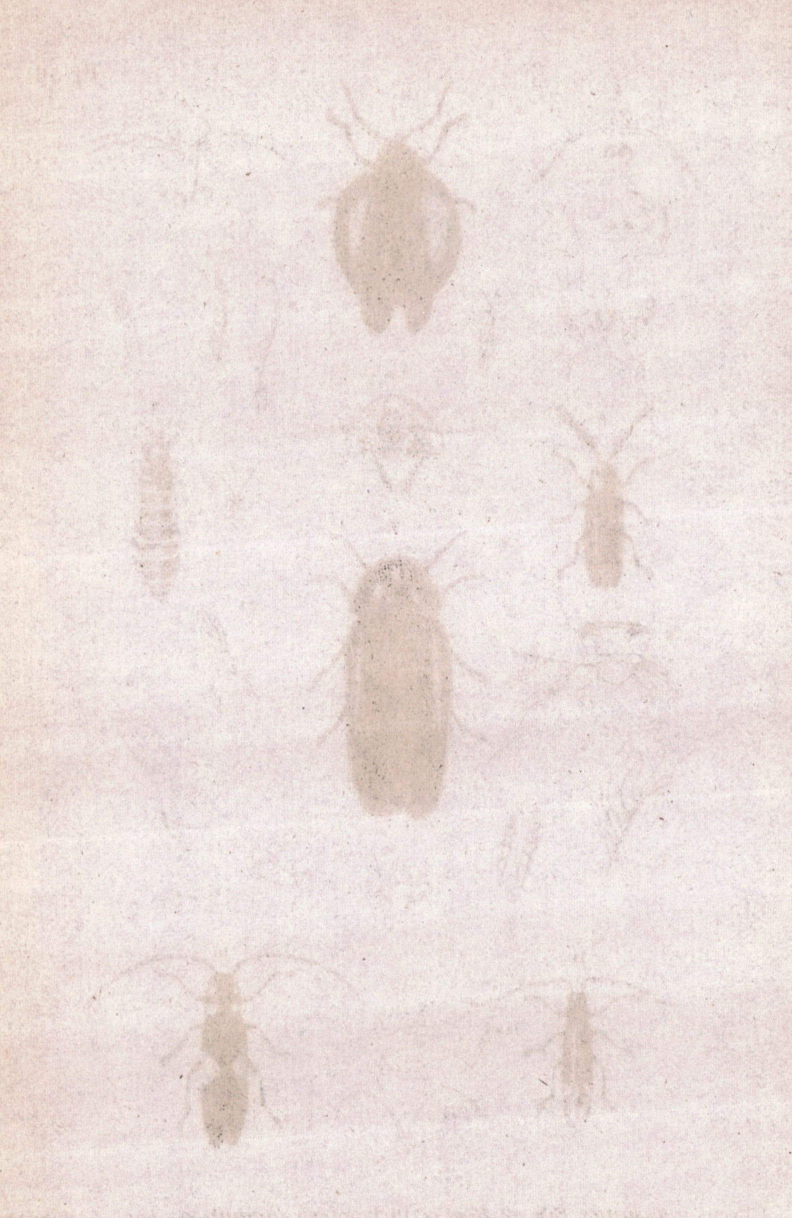

ring red, in the form of a plate, having two sorts of cylindrical horns, as it were annulated or articulated, and arched within. It has six small feet.

Another species, but smaller, altogether black, with the exception of the elytra, which are red, and the end of the antennæ, which is reddish (*Lycus minutus*, Fab. Panz. Faun. Insect. Germ. XLI. 2), is also found in France, but in fir-woods of the mountains.*

Omalisus, Geoff. Oliv. Fab.

Have no very perceptible muzzle. The articulations of their antennæ are almost cylindrical, a little attenuated at their base, and the second and third are much shorter than the following. The penultimate of the tarsi alone is in the form of a reversed heart. The others are elongated and cylindrical. The elytra are of a tolerably solid consistence.

O. Suturalis, Fab. Oliv. Col. II. 24, 1, 2. A little more than two lines in length, black, with the cases, their anterior portion or suture excepted, of a blood-red. In the woods of the environs of Paris, and particularly of the forest of St. Germain, on oak-trees in spring. (See the article *Omalise* of the Encyc. Method.)

The other lampyrides of our first division are distinguished from the preceding, not only by none of them having any muzzle, and that their head, almost entirely occupied by the eyes in the males, is concealed totally, or for the greater part, under a semi-circular or square corslet, but still more by a very remarkable character, either common to the two sexes, or peculiar to the females, that of being phosphorescent. From this comes the names of *glow-worms*, *luminous-flies*, *fire-flies*, given to these insects.

The body of these insects is very soft, especially the ab-

* Lycus *reticulatus*, *bicolor*, *serraticornis*, *fasciatus*, *aurora*, &c.

domen, which is, as it were, folded. The luminous matter occupies the underpart of the two or three last rings of this portion of the body, which are otherwise coloured, and usually yellowish or whitish. The light which they spread is more or less vivid, of a greenish white, or whitish, like that of the different phosphori. It appears that these insects can vary its action at will, which is observed, especially when they are seized or held in the hand. They live a very long time in vacuo, or in the different gases, except in nitrous acid, muriatic and sulphureous gas, in which they die in a few minutes. Their living in hydrogen gas renders it, at least sometimes, detonating. Deprived by mutilation of this luminous part of the body, they still continue to live, and the same part, thus detached, preserves during some time its luminous property, whether it be submitted to the action of the different gases, or whether in vacuo, or in the free air. The phosphorescence depends more on the degree of softness of the matter than on the life of the insect. It may be reproduced by softening this matter in water. In lukewarm water these insects shine with vivacity, and are extinguished in cold water. It appears that this fluid is the sole dissolving agent of the phosphoric matter.*

These insects are nocturnal. The males are often seen to fly, like the Phalænæ, around lights, from which we may conjecture that the phosphoric brilliancy emitted chiefly by the females, is intended to attract individuals of the other sex ; and if the larvæ and nymphs of the species of our own country be, as Degeer supposes, luminous, we should only conclude from this, that the phosphoric substance is developed from their very first age. It has been said that some males

* Besides the experiments reported in the " Annales de Chimie," consult the " Annales Generales des Sciences Physiques," by MM. Bory de St. Vincent, Drapier, and Van Mons, t. VIII. p. 31, where the researches of M. Grotthus on the phosphorescence of the *Lampyris Italica* are given.

did not possess this property; it appears, however, that they do possess it, but in a very feeble degree. Almost all the lampyrides of warm climates, both males, as well as females, being winged, and in great abundance, present to the inhabitants, after the setting of the sun, and during the night, an amusing spectacle, a natural illumination, which, like sparks, or little stars, wander through the air. Light may be obtained by assembling many of these insects together.

According to M. Dufour, (Annal. des Sciences Natur. III. p. 225,) the alimentary canal of the female of our common Lampyris (*splendidula*) is about as long again as the body Its œsophagus is extremely short, and dilates itself immediately into a short crop, separated from the chylific ventricle, by a valvular strangulation. This last part is very long, smooth, inflated and cylindrical for two-thirds of its length, and then intestini form. The slender intestine is very short, flexuous, and presents a swelling representing the cœcum, but perhaps not constantly so, and which terminates in an elongated rectum.

From the genus *Lampyris*, of Linnæus, some Brazilian species have been separated, whose males have the antennæ composed of more than eleven articulations, in the form of barbs of feathers. These species form the genus AMYDETES, Hoffm. Germ.*

Other lampyrides, and also proper to South America, having but eleven articulations to the antennæ, present some peculiar characters, which have obtained for them the same generic distinction. (PHENGODES, Hoffm.) The third articulation of these organs, and the subsequent ones, throw out each, on the internal side, two long ciliated threads, appearing articulated, and rolled back upon themselves. The elytra

* *Lampyris plumicornis*, Lat. Voy. de MM. Humb. et Bonpl. Zool. XVI. 4; *Amydetes apicalis*, Germ. Insect. Sp. Nov. p. 67.

are abruptly contracted into a point. The wings are extended in all their length, and simply folded longitudinally. The maxillary palpi are very prominent, and almost filiform. The corslet is transverse. The tarsi are filiform, with the penultimate articulation very short, and scarcely bilobate. The body is narrow and elongated, with the head uncovered. (Ilig. Mag. VI. p. 342.)

The other species, at present, compose the genus

LAMPYRIS (properly so called),

Which, in consequence of the form of the antennæ, the presence or absence of elytra, wings, &c., is susceptible of many divisions.

L. noctiluca, Lin. Panz. Faun. Insect. Germ. XLI. 7. Male four lines long, blackish; antennæ simple; corslet semi-circular, receiving the head entirely, with two transparent spots, crescent-formed; belly black; last rings of a pale yellowish.

L. splendidula, Lin. Panz. *ibid.* 8, very much approximating to the preceding, but a little larger. The corslet yellowish, with the disk blackish, and two transparent spots in front; elytra blackish; under part of the body and feet of a livid yellowish; first rings of the belly sometimes of this colour, sometimes more obscure.

Female destitute of elytra and wings, blackish above, with the circumferential edge of the corslet, and the last ring yellowish; the lateral angles of the second and third rings, flesh colour. Under part of the body yellowish, with the last three rings of the colour of sulphur.

It is particularly to these individuals that the name of *glow-worms* has been given. They are found every where in the country, on the borders of roads, in hedges, meadows, &c., in the months of June, July and August. They lay a great number of eggs, which are thick, spherical, and of a citron yellow, in the earth, or on plants. They are fixed by means of a viscous matter which invests them.

The larva very much resembles the female, but it is black, with a reddish spot on the posterior angles of the rings. Its antennæ and feet are short. It walks very slowly, can elongate, shorten, or curve underneath its body. It is probably carnivorous.

L. Italica, Lin. Oliv. Col. II. 28, 11, 12, named by the inhabitants *Lucciola*. The corslet not covering the entire head, transverse, reddish, as well as the scutel, the breast, and a part of the feet; the head, cases and abdomen are black. The last two rings of the body are yellowish. The two sexes are winged. (*See* Fabricius and Olivier, Col. II. No. 28.)

In our second division of Lampyrides the antennæ are remarkably separated one from the other at their origin. The head is neither prolonged nor contracted anteriorly in the form of a muzzle, and the eyes are of the usual size in the two sexes.

Drilus, Oliv. *Ptilinus*, Geoff. Fab.

The males are winged, and the internal side of their antennæ beginning at the fourth articulation, is prolonged like the tooth of a comb. Those of the female are shorter, a little perfoliated, and slightly serrated. In both sexes the maxillary palpi are thicker towards the end, and terminate in a point. The internal side of the mandibles presents a tooth.

The female of the species which serves as a type to the genus, and the male of which is very common, had been unknown until these later times, as well as the metamorphoses of the two sexes. The observations made at Geneva, by M. le Comte Mielzinsky, on the larva of this insect, and on the female, in a perfect state, excited the attention of two

French naturalists, who had already given proofs of their talents—M. Desmarest, Professor at the Veterinary School at Alfort, and M. Victor Audouin; this last had received, from the author of this discovery, some larvæ in a living state. They were found in the interior of the shell *Helix nemoralis* of Linnæus. M. Mielzinsky had made them known as well as the female, when she had attained the last state of transformation, the only sort of individuals which he had obtained in the perfect state. But he was deceived in considering as nymphs, the larvæ arrived at their full growth, and which pass the winter in the interior of these shells. Under this form these insects have a sufficient resemblance to the larvæ of our lampyris, but the sides of their abdomen present a range of conical mammellæ, and two series of tufts of hairs, placed on other mammellæ, or dermic elongations. The posterior extremity of the body is forked, and the anus assists the animal in progression. It devours, and that quickly, the natural inhabitant of the shell, and hence is derived its generic name of Cochleoctonus, given to this insect by this naturalist. M. Desmarest presumed, with reason, that since these larvæ were common enough in the neighbourhood of Genoa, they might also be discovered in the environs of Paris. Assisted by his pupils, he did, in fact, procure a great number of these individuals, which enabled him to give a complete history of this insect, and to discover that the individuals in a perfect state, described by M. Mielzinsky, were females of the *Drilus subflavus*, or *panache jaunatre* of Geoffroy (I. 1, 2, Oliv. Col. II. 23, 1.), whose body is about three lines in length, black, with the elytra yellowish. The female is almost three times as large, of an orange or reddish yellow, and resembles those of lampyris, but without being phosphorescent. M. Audouin has published the anatomy of it. He has remarked that the old skin of the larva exactly stops the entrance of the shell, and forms to it a sort of opercle. While the insect is in the larva state, if it retires to the bottom of its habitation, it places itself there in such a manner that the posterior extremity of its body faces the aperture of the shell, but having passed into the nymph state it is turned about in the opposite direction. This observation is owing to M. Desmarest (See "Annales des Sciences Naturelles," January, July and August, 1824, and the "Bulletin de la Soc. Philm.," April of the same year). M. Leon Dufour has also published some anatomical observations made upon the male of this species.

There is found in Germany another of them (*Ater*, Dej.), altogether black, and with antennæ less pectinated. It has been figured, as well as a third (*ruficollis*), discovered by M. le Comte Dejean, in Dalmatia, in a memoir of M. Audouin (Annal. des Sciences Nat., August, 1824), which, under the title of Anatomical Researches, on the female of the yellowish drilus, and on the male of this species, forms a complete monograph of this genus, enriched with excellent figures.

All the individuals of the other lampyrides of this second division are winged, and their maxillary palpi are not much longer than the labial. They embrace a great portion of the genus *Cantharis* of Linnæus, or that of *Cicindela* of Geoffroy.

TELEPHORUS, Schœff. *Cantharis*, Linnæus,

In which the palpi are terminated by an articulation in the form of an axe, and the corslet presents no lateral emarginations. They are carnivorous, and run over plants. Their stomach is long, and wrinkled crosswise. Their intestines very short.

Cantharis fusca, Lin. Oliv. Col. II. 26, 1, 1. Five or six lines in length. The hinder part of the head, wing-cases, breast, and a great portion of the feet, of a slate black, the other parts of a yellowish red. One black spot on the corslet. It is found frequently in Europe in spring. Its larva is almost cylindrical, elongated, soft, of a dead black, and hairy, with the antennæ, the palpi, and the feet of yellowish red. The head is scaly, with strong mandibles. Under the twelfth and last ring is a nipple, which it employs in walking. It lives in humid earth, and feeds on prey.

There has been seen, some years, during the winter, in Sweden, and even in the mountainous parts of France, a considerable extent of ground covered with an infinite quantity of these larvæ, as well as with different other species of living insects. It has been suspected, and with some foundation, that they had been carried off and transported thither by gusts of wind, which deracinate and lay prostrate a great number of trees, particularly pine trees and firs. Such is the origin of what has been named *insect rains.* The species which are then found, and sometimes even on frozen lakes, are probably of the number of those which appear early in the season.

Cantharis livida, Lin. Oliv. *ibid.* II. 28. Size and form of the preceding; corslet reddish, without spot; cases of an ochreaceous yellow, and end of the hinder thighs black. On flowers. (Consult for the other species, Schænherr, Synon., Insect. II. p 60, and Panz. Ind. Entom. p. 91.)

Silis, Meg. Dej. Charp.

Do not differ from the telephori, but that their corslet is emarginated, on each side, posteriorly, and that we see underneath (at least in the *S. Spinicollis*) a small coriaceous appendage, terminating in a knob, and the extremity of which, probably more membranous, forms in dried individuals, the appearance of an articulation. M. Touissant de Charpentier has figured one species, *rubricollis*, in his Horæ Entom., p. 194, 195, vi. 7.

Malthinus, Lat. Schœnh. *Necydalis*, Geoff.

The palpi of which are terminated by an ovoid articulation.

The head is narrowed behind; the cases are shorter in the abdomen in many of them.

On plants, and particularly on trees. (Lat. Gen. Crust. et Insect, I. 261. Schœnh. Synon. Insect. II. p. 73. The telephori, *biguttatus* and *minimus* of Olivier, are of this genus.

The third tribe of Malacodermi, the Melyrides, exhibits palpi, most frequently filiform and short; mandibles emarginated at the point; a body most frequently narrow and elongated, with the head alone covered at its base, by a flat, or but triflingly convex corslet, usually square, or quadrilaterally elongated, and the articulations of the tarsi entire. The crotchets of the last are unindentated, or bordered by a membrane. The antennæ are usually serrated, and even pectinated in the males of some species.

The majority are very agile, and are found on flowers and leaves.

This tribe, which is only a dismemberment of the genera *Cantharis* and *Dermestes* of Linnæus, will compose that of

Melyris, Fab.

Some have the palpi of the same thickness throughout.

Here we discover under each anterior angle of the corslet, and on each side of the base of the abdomen, a vesicle in the form of a horn or cockade, retractile and susceptible of dilatation, which the animal causes to issue forth when it is frightened, and the use of which is unknown. The body is proportionally shorter than in the following sub-genus, broader and more depressed, with the corslet broader than long.

Under each crotchet of the end of the tarsi is seen a membranaceous appendage, in the form of a tooth.

MALACHIUS, Fab. Oliv. *Cantharis*, Lin.

One of the sexes has, in some species, an appendage in the form of a crotchet, at the end of each case, which the individual of the other sex seizes with its mandibles, when it flies or runs too fast. The first articulations of the antennæ are often dilated and irregular in the males. These insects have agreeable colours.

Cantharis ænea, Lin. Panz. *ibid.* X. 2, three lines in length, of a shining green, with the cases red at the edge, and the front of the head yellow. (*See* the same works, and Schœnh. Sym. Insect. I. I. p. 67.)

Cantharis bipustulata, Panz. *ibid.* 3, a little smaller, of a shining green, with the end of the cases red. (*See* as above.)

Amongst the following melyrides, with filiform palpi, and whose corslet and abdomen are without retractile vesicles, we shall place at first those whose antennæ are, at least, of the length of the head and corslet, whose body is generally narrow, elongated, and sometimes linear, and the crotchets of the tarsi usually, as well as those of Malachius, bordered below, by a membranous appendage.

DASYTES, Payk. Fab. *Dermestes*, Lin.

D. Cæruleus, F. Panz. Faun. Insect. Germ. XCVI. 10, three lines long, elongated, green or blueish, shining and hairy; very common in the environs of Paris, on flowers, in the fields.

Dermestes hirtus, Lin. Oliv. Col. II. 21, 11, 28: a little larger, less oblong, altogether black, and very hairy A spine at the base of the anterior tarsi, much stronger and more crooked in one of the sexes. On grasses.*

Other melyrides with crotchets of the tarsi unindentated, as well as in Dasytes, to which they are much approximating, and with which Olivier confounds them, are removed from them by antennæ shorter than the head and corslet, and the third articulation of which is at least as long again as the second. Their body is less elongated, of a more solid consistence, with the head a little prolonged, and narrowed in front, the corslet almost semi-orbicular, and truncated in front. They have a certain resemblance with the coleoptera of the genus *Silpha* of Linnæus. Such are

ZYGIA, Fab.

The fourth articulation of the antennæ and the following ones, form almost an elongated knob, compressed, denticulated like a saw, and most part of these articulations are transverse. The corslet is very convex.

Z. Oblonga, Fab. is found in Spain, and in Egypt, in the interior of houses, and more especially, as M. le Comte Dejean has informed me, in garrets. It appears that it is also met with in France, in the department of the Eastern Pyrenees. Another species has been discovered in Nubia.

MELYRIS, Fab.,

Whose antennæ thicken insensibly without forming a knob, and whose articulations are less dilated laterally, and almost isometrical. The corslet is less convex.†

* See, for the other species, Fabricius. The *melyris* of Olivier, Nos. 6, 17. Panz. Ind. Entom. p. 143, Lat. Gen. Crust. et Ins. I. p. 264. Germ. Insect. Spec. Nov. Brazil, supplies some tolerably large, some of which constitute a particular division.

† *M. viridis*, Fab. Oliv. Col. II. 21. 1. 1; *M. abdominalis*, Fab. Oliv. *ibid.* I. 7; *Opatrum granulatum*, Fab. Coqueb. Illust. icon. Insect. III. xxx. 7.

The other, and the last of the melyrides, have the maxillary palpi terminated by a larger and hatchet-formed articulation. This character, the shortness of the first articulation of the tarsi, and some other considerations, seem to approximate them to the Insects of the following tribe. These are

PELOCOPHORUS

Of M. le Comte Dejean, who places them with the tetramerous Coleoptera.*

The fourth tribe of the Malacodermi, that of Clerii, derived from the principal genus *Clerus*, is distinguished by the assemblage of the following characters. Two, at least, of their palpi are advanced, and terminate in a knob. The mandibles are denticulated. The penultimate articulation of the tarsi is bilobate, and the first is very short, or hardly visible in many. The antennæ are sometimes almost filiform and toothed like a saw, and sometimes terminating in a knob, or thickening insensibly towards the end. The body is usually almost cylindrical, with the head and corslet more narrow than the abdomen, and the eyes emarginate.

The majority of these insects are found on flowers, the others on the trunks of old trees, or in dry wood. Such of the larvæ as have been observed are carnivorous.

This tribe will comprehend the genus

CLERUS, Geoff.

There are some of them whose tarsi, seen at their two faces,

* Catalogue of the Collection of the Coleoptera, of M. le Comte Dejean, p. 115, *Notoxus Illigeri.* Schœnh. Synon. Insect. I. p. 55. 4. 7. a. I shall refer to the same sub-division of the Melyrides, a new sub-genus, which I shall name *Diglobicerus.* The antennæ have but ten distinct articulations, the last two of which are thicker, and globular. It is established on an insect, which was sent to me by M. Lefebure de Cérisy.

present distinctly five articulations. Their antennæ are always denticulated, for the greater part, in the manner of a saw.

Some, among them, have the maxillary palpi filiform, or slightly thicker towards the end.

Cylidrus, Lat.

Have the mandibles long, very much crossed, terminated in a simple point with two teeth at the internal edge. The first four articulations of the antennæ are cylindrical and elongated. The following six are serrated, and the last is oblong. The palpi are terminated by an elongated articulation; that of the maxillary is cylindrical, and the same of the labial is a little thicker, and in the form of an inverted cone. The penultimate articulation of the tarsi is formed of two distinct lobes. The head is elongated.

The sole species known (*Tricodes cyaneus*, Fab.) is found in the Isle of France.

Tillus, Oliv. Fab.

Have the mandibles of middle size, and bidenticulated at the end; antennæ sometimes toothed like a saw, from the fourth articulation to the tenth inclusively, with the last one ovoid; sometimes terminated abruptly from the sixth, in a serrated knob. The last articulation of the labial palpi is very large, and hatchet-formed. The head is short, and rounded. The third and fourth articulations of the tarsi are dilated, and in the form of a reversed triangle.

These insects are found in old wood, or on the trunks of trees.*

* *Tillus elongatus*, Oliv. Col. II. 22. i. 1.; *Chrysomela elongata*, Lin. *Clerus unifasiatus*, Fab. Oliv. *ibid.* IV. 76. ii. 21. The first has the antennæ serrated from the fourth articulation, and the corslet cylindrical. In the second, the antennæ are terminated, beginning from the sixth articula-

The other insects of this tribe, and always distinctly pentamerous, have the four palpi terminated in a knob. The last articulation of the labial is almost always hatchet-formed.

Here, the first four articulations of the tarsi are furnished underneath with membranaceous cushins, advanced, and lobate. The corslet is elongated, and almost cylindrical.

PRIOCERA, Kirb.

The body is convex, with the corslet contracted posteriorly. The last articulation of the maxillary palpi is less dilated than the same of the labials, in the form of a reversed and oblong triangle. The labrum is emarginate.

But one species is known (*Priocera variegata*, Kirb. Lin. Trans. XII. n. 389, 390. XXI. 7.).

AXINA, Kirb.

The body is depressed. The last articulation of the four palpi is very large, in the form of a hatchet.

As yet but a single species has been described (*Axina analis*, Kirb. *ibid.* fig. 6), and is found in Brazil.

There, the penultimate articulation of the tarsi is alone distinctly bilobate. The corslet is square. The body, otherwise, is depressed, as in the preceding sub-genus, and the palpi are terminated in a similar manner.

EURYPUS, Kirb.

Eurypus rubens, Kirb. *ibid.* fig. 5, also inhabits Brazil. I have seen a second species of the same country, in the fine collection of M. de la Cordaire.

At present, the tarsi, seen from above, do not appear com-

tion, in a serrated knob. The corslet is narrowed posteriorly. The last articulation of the maxillary palpi is proportionally longer than the same of the first species, and compressed.

posed of more than four articulations, the first of the five ordinary ones being very short, and concealed under the second.

The insects of this subdivision compose the genus *Clerus*, properly so called, of Geoffroy. M. Dufour admits that the posterior tarsi have six articulations, but the first of which is very short; the same articulation is only rudimentary in the intermediate tarsi, and nothing in the anterior two.

Sometimes the antennæ thicken insensibly, or terminate gradually in a knob. The intermediate articulations, proceeding from the third, are almost in the form of an elongated cone. The two, to the last four but one, are almost in the form of a reversed triangle, and the last is ovoid.

THANASIMUS, Lat. *Clerus*, Fab.,

Have the maxillary palpi filiform, and the last articulation of the labial large, and hatchet-formed.*

OPILO, Lat. *Notoxus*, Fab.,

In which the four palpi are terminated by a large articulation, hatchet-formed (*attelabus mollis*, Lin.; *Clerus mollis*, Oliv. *ibid.* 1. 10.).

Sometimes the last three articulations of the antennæ are much broader than the preceding, and form an abrupt knob, either simple, and in the form of a reversed triangle, or serrated.

Those in which this knob is simple, or not serrated, compose two sub-genera.

CLERUS (Proper), Geoff. *Trichodes*, Fab.

Their maxillary palpi are terminated by an articulation in the form of a reversed and compressed triangle; the last of the

* *Attelabus formicarius*, Lin.; *Clerus formicarius*, Oliv. Col. IV. 76. i. 13; *Clerus mutillarius*, Fab. Oliv. *ibid.* i. 12.

labial, which are larger than the preceding, is hatchet-formed. The knob of the antennæ is but little longer than broad, and is composed of close articulations; the third is longer than the second. The jaws are terminated by a prominent and fringed lobe. The corslet is depressed in front.

These insects are found on flowers; their larvæ devour those of some apiary insects.

Their stomach is wider in front and without wrinkles. Their intestine is short, with two swellings behind. According to M. Dufour their crop is so short, that it is almost entirely concealed in the head.

The generative organ of the male is much more complicated than that of the Melyrides, Lampyrides, and other Malacodermi. The last ring of the abdomen is broadly emarginate. These, with the *Peltis* of Fabricius, are the only coleoptera which have six biliary vessels. Their insertion is cœcal.

Attelabus apiarius, Lin. *Trichodes apiarius*, Fab. Oliv. Col. IV. 76, 1, 4, is blue, with the cases red. They are traversed by three bands of a deep blue, the last of which occupies the extremity. The larva devours that of the domestic bee, and is very injurious to the hives.

That of another species, (*Trichodes alvearius*, Fab. Oliv. *ibid*. I. 5, a. b. Reaum, Insect. VI. viii. 8, 10,) almost similar to the preceding, but having a spot of blueish black on the scutellum, lives in the nests of the *mason-bees* (*G. osmie*) of Reaumur, and lives at the expense of their posterity.

NECROBIA, Lat. *Corynetes*, Fab.

Have the four palpi terminated by an articulation of the same size, in the form of an elongated and compressed triangle; the second and third articulations of the antennæ are almost equal, and the terminal knob is elongated, and with loose articulations. The front of the corslet presents no depression.

Necrobia violacea, Oliv. Col. *ibid.* 76, *bis.* I, 1; *Dermestes violaceus*, Lin., is small, of a violet or greenish blue, with feet of the same colour. Its cases have points disposed in longitudinal series. It is very common in spring, in houses. It is also found in carrion. (*See* Olivier, genus *Necrobie*, and Schœnh. Synon. Insect. I. 2, page 50.)

We shall terminate this tribe by a sub-genus, in which the two last but one articulations of the antennæ, more or less dilated at the internal side, in the manner of teeth, compose, with the last, which is ovaliform, a serrated or semi-pectinated knob. The palpi are terminated by a larger articulation, either in the form of an elongated and compressed triangle, or hatchet-formed. Such are

ENOPLIUM, Lat. *Tillus*, Oliv. Fab. *Corynetes*, Fab.*

The fifth tribe of the Malacodermi, that of PTINIORES, has for type the genus *Ptinus* of Linnæus, and some others which are derived from it, or approximate to it the most. The body of these insects is of a consistence tolerably solid, sometimes almost ovoid, or ovaliform, sometimes almost cylindrical, but generally short, and rounded at the two ends. The head is almost globular, or orbicular, and received for a considerable part in a corslet, very much arched, or vaulted, in the form of a hood. The antennæ of some are filiform, or proceed narrowing towards the end, whether simple, flabellate, pectinate, or serrated; and those of the others are terminated abruptly by three larger and much longer articulations. The mandibles are short, thick, and denticulated under the point. The palpi are very short and terminated

* *Tillus serraticornis*, Oliv. Col. II. 1. 2.; *T. Weberi*, Fab.; ejusd. *T. Damicornis*; *Dermestoides*, Scheff. Elem. Entom. 138; *Corynetes sanguinicollis*, Fab. See Schœnh. Synon. Insect. I. 2. pag. 46.

by a larger articulation, almost ovoid, or in a reversed triangle. The legs are without denticulations, and the spurs of their extremities are very small. The colours are always obscure, and but little varied. All these insects are of small size. When they are touched, they counterfeit death, lowering the head, inclining the antennæ, and contracting the feet. They remain some time in this apparent lethargy. Their motions are, in general, rather slow. The winged individuals seldom take flight for the purpose of escaping. Their larvæ are very injurious to us, and have a great resemblance to those of the Scarabæi. Their body, often curved in an arch, is soft, whitish, with the head and feet brown and scaly. Their mandibles are strong. They construct for themselves, with the fragments of the substances which they have gnawed, a shell, or they are transformed into nymphs. Other species establish their domicile in the country, in old wood, stakes, and under stones; otherwise, their habits are similar.

Such are the general characters of the genus.

Ptinus, Lin.

Some have the head and corslet, or the anterior moiety of the body, more narrow than the abdomen, antennæ always terminated in an uniform manner, simple, or very little serrated, and, at least, almost as long as the body.

Ptinus (Proper), Lin. Fab. *Bruchus*, Geoff.

Have the antennæ inserted between the eyes, which are prominent or convex. Their body is oblong.

They remain for the most part in the interior of houses, principally in the garrets and inhabited parts. Their larvæ gnaw herbals, and the prepared and dried skins of animals. The antennæ of the males are longer than those of the females, and in many species, these last individuals are deprived of wings.

P. far. Lin. Fab. *P. latro*, *striatus*, F. Oliv. Col. II. 17, i. 1, 3. ii. 9, Var. of the male; a line and a half in length, of a clear brown; antennæ of the length of the body; corslet having on each side a pointed eminence, and two others rounded and covered with a yellowish down in the interval; two transverse bands, greyish, and formed by hairs on the cases.

According to Degeer, it lives on flies, and other dead insects with which it meets. Its larva makes a terrible devastation in herbals, and collections of natural history.

P. imperialis, Fab. Oliv. *ibid*. I. 4, remarkable for two spots of the cases, representing, by their union, the rude figure of an eagle with two heads. Lives on old wood.

This species, it appears to us, ought to be placed in the genus *Hedobia*, of the catalogue of the collection of M. le Comte Dejean. It differs from that of Ptinus, by having more separated antennæ, a little serrated, and particularly by the tarsi, which are short, and composed of articulations almost in the form of a heart, more broad, especially the last. The crotchets of this last are even concealed. In the Ptini these tarsi are narrow, with the last articulation in the form of an inverted cone. The antennæ are approximated at their base.

I have frequently found on excrements the *P. Germanus*, (Lat. Crust. et Insect. 1. p. 279), which has several relations with *P. latro*. (*See*, for the synonimy of the species of this genus, Schœnhen, Syn. Insect. II. p. 106.)

Gibbium, Scop. *Ptinus*, Fab. Oliv.,

In which the antennæ are inserted in front of the eyes, which are flatted and very small, in which the scutellum is wanting, or is not very distinct, and whose body is short, with the abdomen very large, swelled, almost globular, and semi-transparent. The antennæ are more slender towards their extremity, and the cases are synostosized. These insects also make their domicile in herbals and collections.*

The others have the body either oval, or ovoid, or almost cylindrical; the corslet of the breadth of the abdomen, at least at its base; the antennæ sometimes uniform and serrated, or pectinate, sometimes terminated by three articulations much larger than the preceding. They are shorter than the body.

PTILINUS, Geoff. Oliv. *Ptinus*, Lin.,

Whose antennæ, from the third articulation, are strongly pectinate, or plumed in the males, and serrated in the females.

These insects live in dry wood, and pierce it with small holes. It is there also that they couple; one of the sexes is outside, and suspended in the air.†

XYTELINUS, Lat. *Ptilinus*, Fab.,

To which we unite OCHINA of MM. Ziegler and Dejean, have the antennæ simply serrated in the two sexes.‡

DORCATOMA, Herbst. Fab.,

In which the antennæ finish abruptly by three larger articulations, and of which the penultimate two are in the form of the teeth of a saw. They are composed of nine articula-

* *Ptinus Scotias*, Fab. Oliv. Col. *ibid.* 1. 2. Panz. Faun. Insect. Germ. V. 8; *P. sulcatus*, Fab.

† *Ptilinus pectinicornis*, Fab. Oliv. Col. II. 17. *bis.* 1. 1; *P. pectinatus*, Fab.; *ejusd. P. serratus*; *Ptinus denticornis*, Var. Panz. *ibid.* VI. 9. XXXV. 9.

‡ *Ptilinus pollens*, Germ.; *Ptinus serricornis*, Fab In the *Ochina hederæ* the antennæ are a little longer than those of the Xytelinæ, a little less serrated, with the second and third articulations almost of equal length. I have not examined the other species of Ochina, mentioned by M. Dejean in his catalogue (p. 40).

tions only. (*Dorcatoma dresdensis*, Herbst. Col. IV. xxxix. 8.)

Anobium, Fab. Oliv. *Ptinus*, Lin. *Byrrhus*, Geoff.,

In which the antennæ are equally terminated by three articulations, larger or longer, but of which the penultimate two are in an inverted or elongated cone, and that of the end oval, or almost cylindrical; they have eleven articulations.

Many species of this genus inhabit the interior of our houses, where they do much mischief in their first state, that of the larva, by gnawing the planks, joists, wood, furniture and books, which they pierce with small round holes, similar to those made with a very fine gimlet. Their excrements form those little dusty heaps of worm-eaten wood, which we often see on floors. Other larvæ of anobium attack flour, wafers which are kept in drawers, collections of birds, insects, &c.

The two sexes, to call each other at the time of their amours, and come together, strike several times rapidly and in succession, with their mandibles, the wainscoting in which they are placed, and answer each other mutually. Such is the cause of the noise, similar to that of the accelerated ticking of a watch, which we often hear, and which superstition has named the *death-watch*.

A. tesselatum, Fab. Oliv. Col. II. 16. 1. 1. Three lines in length, of an obscure and dead brown, with yellowish spots, formed by hairs; corslet smooth; cases without striæ.

Ptinus pertinax, Lin. *A. striatum*, Fab. Oliv. *ibid*, I. 4, blackish; corslet having at each posterior angle a yellowish spot, and near the middle of its base a compressed elevation, divided into two in front, by a depression; cases with punctuated striæ. It prefers, according to the observations of Degeer, suffering itself to be burned at a slow fire to giving the least sign of life when it is held.

Anobium pertinax, of Fabricius; *Striatum*, Olivier, Panz. *ibid.* LXVI. 5, very much resembles the preceding, but it is smaller, and has no yellow spots at the posterior angles of the corslet. It is very common in houses. M. Dufour has observed that some appendages form around its pylorus a sort of ruff.

A. paniceum, Fab. *A. minutum*, ejusd. Oliv. *ibid.* II. 9, is very small, fawn-coloured, with the corslet smooth, and the cases striate. It reddens farinaceous substances, and ravages collections of insects when suffered to multiply there. It establishes itself also in cork. (*See* Schœnh. Synon. Insect. I. 2, p. 101.) Some species of Fabricius are referable to the genus *Cis*.

The third and last section of the Serricornes, forming also a final tribe, that of Xylotrogi, and being distinguished, as we have already said, from the preceding two, by having the head entirely disengaged, is composed of the genus

Lymexylon, Fab.

We divide it thus:

Some have the maxillary palpi much larger than the labial, pendant, in the form of a comb or tuft in the males, and terminated by a large ovoid articulation in the females. The antennæ are short, a little widened towards their middle, and slender towards the end. The tarsi are filiform, with all the articulations entire. The posterior four are long and very slender. Those whose elytra are very short, in the form of a small scale, compose the genus

Atractocerus, Palis. de Beauv. *Necydalis,* Lin. *Lymexylon*, Fab.

The antennæ are compressed, and almost fusiform. The corslet is square, and the abdomen depressed.

A. necydaloides, Palis. de Beauv. Magaz. Encyc. *Necy-*

dalis brevicornis, Lin. *Lymexylon abbreviatum*, Fab. *Macrogaster abbreviatus*, Thunb. is found in Guinea, and appears to differ but little from another species received from Brazil. The Museum of Natural History possesses a second, much smaller, and perfectly distinct, enclosed in amber. Another is found in Java.

Those in which the elytra are of the length of the abdomen, or but little shorter, form two sub-genera.

In *these* the antennæ are compressed, serrated, and with transverse articulations. The corslet is almost squared. Such are,

HYLECOETUS, Lat. *Meloë, cantharis*, Lin. *Lymexylon*, Fab.

B. dermestoides, *Meloë Marci*, Lin., the male; *Lymexylon Morio*, Fab., and *L. proboscideum*, item; *Cantharis dermestoides*, Lin., the female; *L. dermestoides*, Fab. item; Oliv. Col. II. 25, I. i. 2, item. The female is six lines in length, of a pale fawn-colour, with the eyes and breast black. The male is black, with cases sometimes blackish, sometimes reddish, with the extremity black. In Germany, England, and to the North of Europe.

In *those* the antennæ are simple, little or not at all compressed, and almost moniliform. The corslet is almost cylindrical.

LYMEXYLON (Proper), Fab. *Cantharis*, Lin. *Elateroides*, Schœff.

L. flavipes, Fab. male; ejusd. *L. navalis*, fem.; Oliv. *ibid.* I. 4, of the length of the preceding, but more narrow, of a pale fawn-colour, with the head, the external edge, and the end of the cases black. This last colour predominates a little more in the male. This insect is very common in the oak forests of the North of Europe, but rare enough in the environs of Paris. Its larva is very long, and very slender, almost similar to a wire. It so multiplied some time back, at Toulon, in the naval timber-yards, that it occasioned very great destruction.*

The others have the palpi very short, and similar in the two sexes. The antennæ are always simple, and of the same thickness throughout. The tarsi are short, and the penult articulation is bilobate in some of them.

The body is of a solid consistence, with the top of the head unequal or furrowed, and the corslet almost square or suborbicular.

CUPES, Fab.,

In which the antennæ are composed of articulations almost cylindrical, and in which the penult of the tarsi is bifid.

The mandibles are unindenticulated under the point. The palpi, the jaws, and the tongue, are uncovered. The tongue is bilobate, and the chin is almost semi-orbicular. Two species are known, and both proper to North America.†

RHYSODES, Latr. Dalm.,

In which the antennæ are grained, and all the articulations of the tarsi entire.

The mandibles, as far as it appears to me, are narrowed and almost tricuspidate at their extremity. The chin is corneous, very large, buckler-formed, and termi-

* The *lymexylon proboscideum* of Olivier, the individual of which has served as a type to his description, and which now forms a part of the collection of M. le Comte de Jousselin, at Versailles, ought to form a peculiar genus. See also the *Lymexylon flabellicorne* of Panzer, Faun. Insect. Germ. XI. 10.

† *Cupes capitata*, Fab. Lat. Gen. Crust. et Insect. I. viii. 2. Coqueb. Illust. Icon. Insect III. XXX. 1.

nated above by three teeth or points. The palpi are very short.

Notwithstanding the number of articulations in the tarsi, this genus appears to approach cucujus, and even some of the brenta with short proboscis, in the two sexes. The habits are the same as those of the Xylophagi. *Rhysodes exaratus*, Dalm. Analect. Entom. p. 93. M. Leon Dufour has just discovered this species in the Pyrenees.

SUPPLEMENT

ON THE

SERRICORNES.

In the first section of this family, the first genus to be noticed is that of Buprestis, upon which very little can be added to the details of the text. This genus, which is very distinct, and very easily to be recognized, resembles that of *elater*, but differs from it chiefly in the corslet or thorax, which is destitute of a spring.

The larvæ of these insects live in woods. The perfect insect usually remains on trees, bushes, plants, and flowers. It is very frequently to be found in timber-yards.

The buprestes walk very slowly, but are remarkable for their agile flight when the weather is hot and dry. Some of them suffer themselves to fall among the brambles, when one approaches to take hold of them.

This genus furnishes the most beautiful coleopterous insects which can adorn either the domain of nature, or the cabinet of the naturalist. Most part of the species are clothed with such brilliant colours, that Geoffroy has thought proper to designate them all under the generic appellation of *Richard*. The origin of this name is as singular as its application is fantastical. It was originally given to the jay, in consequence of the facility with which that bird was taught to pronounce the word. In the buprestes we find the brilliancy of polished gold on an emerald ground, or the purest azure playing over a golden surface. Very often the same individual presents a mixture of many brilliant metallic colours. These insects are not much varied, and they are but little numerous in the north of Europe; more abundant towards the southern countries of France, and very much varied, and very common in the warmest climates of the two hemispheres. It is accordingly from such climates that the largest and finest species are imported into Europe.

The genus Buprestis of Linnæus is not, however, distinguished by the splendid tints with which the insects are adorned, since it comprises one of the most extensive families of the coleoptera. Indeed, we are informed that Professor Klug, the conservator of the Berlin Museum, has stated that the cabinet of that repository is enriched with not less than five hundred species.

The classification of these beautiful creatures, singularly enough, scarcely attracted the attention of naturalists, with the exception of Schonherr in his Synonima Insectorum, and the Russian naturalist, Eschscholtz, who has divided these insects into the following genera:

Sternocera, the types of which are, B. chrysis, castanea, interrupta, sternicorius of Fabricius.—*Jalodis*. B. hirsuta, fascicularis, variolaris, pilosa, hirta.—*Acmaeodera*. B. gibbosa, tæniata, adspersa, ornata, cylindrica.—*Trachys*.—*Apanisticus*.—*Agrilus*. B. novem maculata, exasperata, *Schonh*. bicornis. Dej. penicillata. *Klug*. lucogaster. *Wilden*. Rubi, elata, æreicollis. Dej. leucostica. *Kirby*.—*Capnodes*. B. cariosa, tenebrionis, carbonaria, *Klug*. tenebricosa, tarterica. —*Buprestis*. B. florentina *Dahl*. mariana, lugubris, cuprea, &c.—*Pœcilonotus*. B. conspersa, rutilans.—*Dicerea*. B. Berolinensis ænea, alni, acuminate.—*Ancylochira*. B. flavomaculata, strigosa, 8 guttata, punctata, cupressis, hœmorrhoidalis, rustica.—*Stigmodera*. B. grandis, variabilis, macularia, eximia, and principialis *Klug.*—*Conognathus.* B. equestris and amæna Kirby.—*Anthaxia.* B. cyanicornis, manca, nitida, quadrimaculata.—*Melanophila.* B. appendiculata, tarda.—*Chrysobotus.* B. impressa, chrysostigma affinis.—*Belionota.* B. scutellaris, stigma, canaliculata F. and B. écunoné, Latr. Regne An. V. 4. 13. 2.

We are enabled to add four species, three of which we consider as new. The first we call *Whitehillii. Hope's MSS.* It is dark green, varied on the thorax with yellowish white; also with five rows of downy spots of the latter colour, on each elytron, which are likewise covered with smaller spots of the like colour; legs blackish green, tarsi dilated, beneath, purplish copper covered with yellowish hairs; eighteen lines in length, and is an inhabitant of Poonah. India. From Mr. Hope's cabinet.

Buprestis flavomarginata. Hirsute, the thorax æneous black, punctate, with the margin yellow. The elytra æneous black, with deeply punctuated striæ, near the tip, with two angulated transverse fasciæ red, margin yellow, tip acute; the length is six lines, and the species is from Brazil.

Buprestis bifasciata. Linear, golden green, with a few elevated longitudinal lines of violet; with two transverse fasciæ of the same, near the tip, which is red, and dentated; this species is six lines long, and is from Mexico.

The last species we have to notice is *Buprestis penicillata. Klug.* The head with two horns, bronze, the thorax tuberculate, with the sides dilate, crenate, blackish bronze, the scutellum with the middle dilate, angulate, obscure bronze; the elytra depressed, carinate, punctate, bronzed, sprinkled with white and black hairy spots, with the tip attenuate, denticulate, black, with two transverse angulate striæ, the tip truncate with black tuft of hair at the angles, and likewise a similar tuft on each elytron beyond the middle.

This species, according to Eschscholtz, enters into the genus agrilus, which, however, seems distinct from that so described by Mr. Curtis.

The sub-genus MELASIS was established by Olivier on a single species, in which he thought that he recognized the *Elater buprestoides*. Illiger, however, and some other naturalists consider the insect of Linnæus as different from that of Olivier, and accordingly designate this last species under the name of *Elateroides*. This opinion seems yet to remain in doubt; but the description given by M. Paykall of the *Elater buprestoides*, of Sweden, or *Melasis flabellicornis*, of Fabricius, and the figure of Herbst, which is cited as synonimous, agree perfectly well to the insect named by Olivier, *Melasis buprestoides*. This species is found in Germany and France, under the trunk of old trees, which it appears to perforate after the manner of some Anobia. These insects couple at the entrance of some of the cylindrical holes, which the external surface of these trees present. One of the sexes remains without.

The name of ELATER, or *notopeda*, in Latin, has been given to the insects composing this genus, in consequence of the leap which they make when thrown upon their back, for the purpose of recovering their feet. They very much resemble buprestes, but are distinguished from them by some observable characters, but more especially by that remarkable spring, which is observed under their corslet or thorax.

These insects are found every where in the country, on flowers, on plants, on the trunk, and under the bark of rotten trees. They are caught easily enough; but they often suffer themselves to fall, for the purpose of escaping the hand which is ready to seize them. They are all provided with wings, and sometimes make use of them to transport themselves from one place to another. They always walk with great slowness, and seem to employ their feet with difficulty. There is another character peculiar to these insects, and which has not been accorded to them without a useful purpose. We have already mentioned that their Latin name designates the faculty possessed by the insect of leaping or raising itself into the air, as it were by a sort of spring, when it is placed upon its back, and never otherwise. It leaps perpendicularly to a certain height, so that it falls pretty nearly in the same spot in which it had been placed. Its object in making this leap is to replace itself on its feet, which would otherwise prove a painful and difficult operation, in consequence of the shortness of its limbs. If it falls again upon the back, which not seldom happens, it reiterates its leaps until it finds itself upon its feet. Having explained the object of these leaps, we shall enter a little into the mechanism by which they are performed, as it has been but slightly noticed in the text.

The two external angles of the corslet are terminated in advanced, and more or less sharp points. Underneath, and exactly at the middle of the posterior edge, a third point, long, stiff, and very hard, placed on the same line with the body, advances towards the breast considerably below the edge of the corslet. This part, broad at its origin, diminishes in thickness by little and little to terminate in a blunt, or but triflingly rounded point. Underneath, and near its extremity, is remarked a small eminence in the form of a denticulation. On the breast underneath, between its anterior edge and the two intermediate feet, is found an oval hole, tolerably deep, furnished with a smooth and corneous substance. It is in this cavity that the long point of the corslet is sunk, when the insect lowers its head, which is its natural position, either in walking or in a state of repose. Such are the principal instruments which the elater employs to raise itself into the air. The following is the manner in which this little manœuvre is performed; the insect placed upon its back, lowers the head and corslet towards the plain of position ; by this movement the long spine issues entirely out of the oval hole in which it is lodged. At the same time it always applies its feet against the under part of the body, and holds them fast there, and crowded together, which secures them from the dangers of the fall. Afterwards, approximating the corslet to the breast, it pushes strongly and rapidly against the edge of the hole, the long point, which falls back like a spring in re-entering its cavity. The corslet with its lateral points, the head, and even the upper part of the elytra, all coming with force against the plain of position, also concur by their elasticity, to raise the body into the air. By holding the insect by the belly between the fingers, one may easily convince himself of the exactness of these observations. The more the plan of position is firm and hard, the greater will be the elevation of the leap. It has been also remarked that the insect rests the two lateral spines of the corslet against the inferior edge of the elytra, and that it relaxes them at the same time that it causes the long point to re-enter the whole, which operation is very capable of augmenting the elastic force which carries it upwards. It is necessary that the corslet should have a movement, and free play upon the breast ; accordingly it is very mobile, and the insect can lower it considerably.

By pressing the belly of the male between the fingers, three parts are made to come out, of which those of the sides serve as a case to that of the middle, and are concave anteriorly. In their natural situation these three parts are enclosed in a second case, which opens towards the sides, and which is supported by two concave scaly laminæ, forming a third case. The elongated portion of the middle is doubtless that which characterizes the sex of the male. From the belly of the female, by the same pressure, parts totally different are made to come out. This is a long, cylindrical piece, at the end of which are found two other elongated pieces, conical and pointed, between which a third is seen which appears to be hollow. All these parts re-enter into a sort of case, which in its turn is enclosed within the belly. The hollow part must be the conduit of the eggs, of which a tolerably great number are found in these females. They are very small, oval, whitish, and shining.

The larvæ of these insects in general are but little known, but they are supposed, for the most part, to live in wood, Degeer has described the larvæ of one species, and though his description is more full than that given in the text, we deem it unnecessary to enter upon it here.

We must, however, mention two species of elater, which, like the lampyris, possess the faculty of shining, but which by this faculty should not be excluded from the genus to which they evidently appertain in all their other characters. The first (*Elater noctilucus*) is found in America, and particularly in Jamaica and St. Domingo. According to the testimony of various writers, this insect emits light, either in obscure places or during the night, as long as it lives, from two yellow spots on the corslet, which are then transparent. The light which it sheds is so strong and brilliant, that it enables one to read the finest writing, especially when eight or ten of these insects are placed in a vessel of glass. Brown says that all the internal parts of the insect are luminous, but that the light can only escape by the two yellow spots of the corslet. To convince himself of this, he separated a little the rings of the belly one from the other, and the light was then seen through the membrane which connects the rings together. During the day these insects remain in a state of repose, and are, as it were, in a state of lethargy. They are but seldom to be found at this time. It is during the night that they put themselves in motion. They are easily to be caught by the light of a flambeau, which they will follow exactly like the phalenæ. The Indians make use of them in their nocturnal journeys, and attach them to their shoes. The women also carry on their works by the light which they emit. Brown adds, that they have the power of shining or not at will, which is another affinity which they possess with the lampyris.

The other species (*Elater phosphorus*) is smaller than the preceding, and is found at Cayenne and in Surinam. To judge from analogy, we may suppose that the two spots which this insect has upon the corslet, send forth the same luminous brilliancy as in the first species. They both resemble each other very much.

We add two new species of *Elater*, viz., *lusciosus*, *Hope's MSS.* This insect is somewhat like *Elater occulatus*, of *Fabr.*, but it is much broader, more convex, and more robust; it is black, with two black velvety eyes on the thorax, of which the irides are white. The body cretaceous, irregularly spotted, beneath black, spotted with white. This insect is from Mexico, and is nineteen lines in length. The other species is *E. aureolus.* This is like *auratus*, of *Drury*. It has the antennæ black, filiform, compressed, and subserrate; the thorax broad, depressed, of a rose green colour, with the sides elevated, and green. The elytra of a rich bronzed green, shining with fine punctated striæ; the tip acute; this is from the Tanesserim coast, and is seventeen lines in length.

After *Galba*, of Latreille, M. Guerin, has added a new genus named *Pterotarsus*. The antennæ have the first joint large, oval, the others very small, and with the last eight joints strongly pectinated; the pectinations are much longer than those of Galba. The head and antennæ, when in repose, are perfectly hidden in the thorax; the thorax is oblong globose; the elytra convex, rounded.

There are four species of the genus. The one *histrio*, of

Latr. It is red, with three black bands, one on the thorax, another at the base, and the third narrower on middle of the elytra; from the latter to the tip yellow. The species are all from Brazil, where they inhabit the trunks of dead trees, and are occasionally found creeping upon the barks.

M. Guerin has also added a genus, near *Adelocera*, which he calls *Pachyderes*. It has the antennæ pectinated, first joint large, second small, maxillary palpi, last joint very large, subsecuriform, the joints of the tarsi simple, the penultimates bilobed; the head small, thorax large, dilate, with two spines at the posterior angles; the elytra long, narrow, dentate at the tip; the only species is named *Ruficollis*.

M. Eschcholtz has formed a new genus on a single species, which is allied to *Eucnemis*, under the name of *Epiphasis*, with the following characters. The antennæ, with the four apical joints very long; the six preceding very short. The species is named *Cornutus*. It is reddish ferrugineous, the head black, the front with a compressed horn; the thorax black, with the hinder angles ferrugineous. The length is two lines and a half.

The singular genus *Chelonarium*, *Fabr.* which has but recently been regarded by Latreille as belonging to the Elateridæ, has all the aspect of the Clavicornes, in which group it was previously arranged by that author; according to M. Lacordaire, the Brazilian species of this genus are found upon the leaves and flowers of various plants, from which they drop on the approach of danger. When held between the fingers, they conceal their antennæ in the channels of the thorax, fold their legs along the sides of the body, and thus counterfeiting death, they remain in this state a considerable time.

The next genus we notice is also one of M. Guerin, which is named *Lobœderus*. The antennæ are moniliform, the first joint is very large, and the last very small, ovoid; labrum very small, mandibles strong, hooked, with a small tooth under

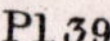

1

3

2

4

6

5

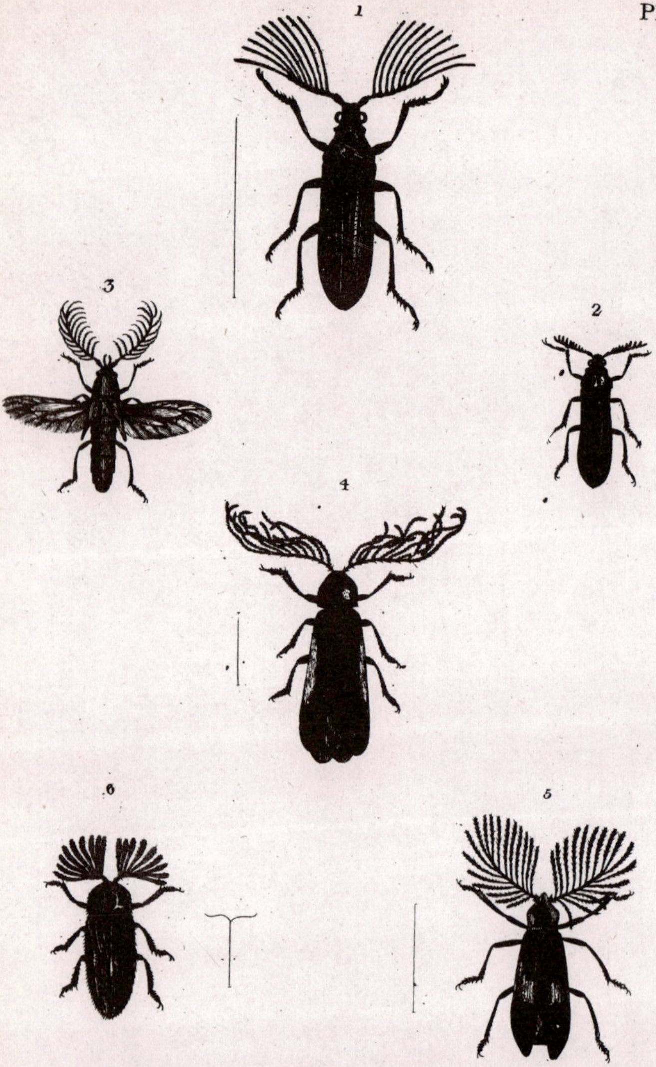

1. Callirhipis *Childreni.* G.R.Gray. 2. Callirhipis. *ruficornis.* G.R.Gray. 3. Phengodes *plumicornis.* Hoffm. 4. Cladophorus *marginatus.* G.R.Gray. 5. Calyptocephalus *fasciatus.* G.R.Gray. 6. Megolophthalmus *Bennettii.* Gray.

the points, forming a point interiorly, very hairy ; maxillary palpi short, the first very small, the second much larger, the third also as long as the first, much straighter, and the last of the same length, enlarging at the tip and truncate obliquely or securiform ; the lower lip transverse ; labial palpi short, the first joint small, the second longer, the third large, enlarging and truncate, tarsi cylindrical, thorax with a horny lobe, curved behind, and rounded at the tip under each posterior angle. The species is named by M. Guerin *Moniliformis.* It is oblong, the thorax reddish chesnut, covered with short yellowish down ; the elytra are the same colour, with longitudinal punctated striæ, and their sides furnished with yellowish down, much deeper than the thorax. This insect is from Brazil.

The Cebriones are particularly found in the most southern countries of Europe and of the north of Africa. They fly with impetuosity often in the evening or during night, almost always after storms, and often in very great abundance. They enter into houses, attracted by the light, upon which they precipitate themselves, making at the same time a humming noise. Their larvæ are unknown, and, with the exception of what we have now mentioned, the habits of the entire family, as may be seen by the text, are quite unknown.

We have figured two species of *Callirhipis ;* the first under the name of *Childreni.* This species is brown, covered with yellowish down, the eyes reddish, the legs blackish, the elytra slightly striated, and punctated. It is from Brazil. We have dedicated this species to J. G. Children, Esq., in whose cabinet is is contained. And the *Callirhipis ruficornis,* which is dark reddish castaneous ; the elytra striated, with broad punctures ; antennæ reddish : this is from New South Wales, and may probably form a new sub-genus between

Callirhipis and Rhipicera, which is also from New Holland.

We now come to the genus Lampyris. The Greeks gave indiscriminately the names of *lampyris,* and the Latins those of *cicindela, noctiluca, lucio, luciola, lucernuta, incendula,* to all those insects, which have the property of shedding during the night a phosphoric light. This same property has caused them to be vulgarly termed *glow-worms.* The modern entomologists ought, without doubt, to have applied themselves to arrange insects under the same denomination, only in proportion as they present the same generic characters; but as it is only by long observations and continued labours that this final end of science can be attained, the lampyris proper were for a long time confounded with *telephorus* and *malachius,* under the name of *cantharis.* Geoffroy, in separating them from *telephori,* has nevertheless associated them with *lycus,* and Linnæus has again associated them with *lycus* and *pyrochrus.* Fabricius, enlightened by the errors of his predecessors, was the first who properly distinguished this genus, and assigned to it its particular characters.

All the insects which emit light, could not have failed to arrest the attention of the observers of nature; accordingly, the lampyrides have been for a long time very well known to naturalists. They received the name of glow-worms, because the females, which are most frequently met with, are deprived of wings, and all these females emit light during the night. Some males are deprived of this faculty of shining, or possess it only in a feeble degree. The luminous faculty of the glow-worms is situated underneath the two or three last rings of the abdomen. These are yellow spots, from which proceeds, during darkness or obscurity, a very lively light, of a greenish or bluish white, as is the case with all phosphoric lights. This light, according to some authors, does not depend on the influence of any external cause, but solely on the will of the insect.

The glow-worms are found in summer, after the setting of the sun, in meadows, on the sides of roads, and near bushes. In the countries where these insects are very common, during the quiet nights in the fine season, the males flutter in the air, which they seem to fill with sparkles of fire; and the females, which during the day remain concealed under the grass, discover themselves during the evening and night, by the brilliant light which they shed. As long as these insects are at liberty, their light is very regular; but once in our power, they shine very irregularly, or not at all. When they are disturbed, they emit a frequent light. Being placed upon their back, they shine almost without interruption, in making continual efforts to turn themselves round.

The luminous matter of these insects has excited the curiosity of many philosophers. It has been the object of frequent experiments, which have furnished some very interesting observations, which we shall here relate. Forster having announced that the light of the glow-worms was so strong, and so continuous in the oxygen gas, that one might read by it with facility, M. Beckerheim, in verifying this fact, discovered that these insects live a long time, *in vacuo*, and in different gases, the nitrous acid, muriatic and sulphurous excepted, in which they die in less than eleven minutes.

He has also ascertained the following particulars:—

That they have never diminished the goodness of the gases in which they have lived, whatever length of time they may have remained there; that, on the contrary, hydrogen gas has become detonating, in consequence of the sojourn of these animals in it; and that many gases, tried before and after, have appeared to be ameliorated from the same cause.

That in whatever gas these glow-worms may have been, the light has never appeared to be augmented.

That this light is produced by small luminous bodies, which the insect can cover with a membrane.

That after having removed these luminous points from the body of the insect, without endangering it, it has continued to live, but without any re-appearance of light.

That these luminous points, removed from the living insect, and exposed to the action of many gases, have produced light there for different spaces of time, from which this author seems to conclude that the duration is greater in the oxygen gas than in the others.

The experiments made by Dr. Carradori on the *lampyris Italica*, have furnished him with the following results:—

These insects shine at will, in each point of their abdomen, which proves that they possess the faculty of moving all the parts of this viscus, independently one of the other.

They can render their phosphorescence more or less lively, and prolong it for as long a time as they think proper.

The faculty of shining does not cease in consequence of the laceration, or incision of the abdomen.

M. Carradori has seen a part of the abdomen separated from the rest of the body, which had been almost extinct, become all at once luminous for some seconds, and afterwards be extinguished insensibly. Sometimes he has beheld a similar portion cut off, pass suddenly from the finest brilliancy into a total extinction, and afterwards resume its former light. M. Carradori attributes this phenomenon to a remnant of irritability, or a stimulus produced by the air.

A slight compression is sufficient to deprive the lampyrides of the faculty of shining.

The phosphoric matter once expressed loses in a few hours all its splendour, and is converted into a white and dry substance. A piece of the phosphoric abdomen, put into oil, shone but feebly, and very soon became extinct in water. Another shone with the same vivacity in the air, and for a longer time; the phosphorus of these insects shines equally in the barometric vacuum. The author has recognised that the phosphorescence is a property, independent of the life of these insects, and that it is attached rather to the degree of softness of the phosphoric substance. Dryness suspends the light; softening in water regenerates it, but only after a certain given time of dessication. Reaumur and Spallanzani have made the same observation respecting the pholades and medusæ.

By plunging the lampyrides alternately into lukewarm and cold water, they shine with vivacity, in the first, but are extinguished in the last; in hot water the light disappears by little and little. M. Carradori has tried on the lampyrides and their phosphorus, the action of the different saline and spirituous fluids, in which they have exhibited the same sort of phenomena as the other phosphoric animals. These experiments have proved that the phosphoric matter of the lampyris undergoes a dissolving action only through the medium of water.

The larva of lampyris has much resemblance to the female; it is provided with six scaly feet placed on the first three rings. This larva, though furnished with strong jaws, which might lead one to suppose that it was carnivorous, yet lives on herbs and the leaves of different plants. It walks very slowly and by the assistance of the hinder part of its body; as soon as it is touched, it draws back its head and remains immoveable; when it is suffered to want humid earth it becomes feeble and languishing.

When insects are about to be transformed into the nymph state, the skin is usually cleft or broken in the middle of the top of the head, and thus leaves a sufficient aperture to give a passage to the rest of the body. The larva of the lampyris appears to adopt another method of getting rid of its skin, which is cleft on each side of the body, through the full extent of the first three rings. The upper part of these rings is detached altogether from the under, and the larva draws its head out of the skin which covers it, pretty nearly in the same manner as one would draw his hand out of a purse. The two lateral clefts give a very spacious aperture to the insect, to issue from the old skin, and it brings this easily about in the course of a few minutes by contracting and elongating the rings of the abdomen alternately.

As soon as the larva is disengaged from its skin, it bends the body into an arch or semi-circle, and is then in its true state of nymph; but it is still observed to move and elongate the head as well as the antennæ and feet, but slowly. It also gives some motion to its body.

The observations of Degeer prove that the female lampyris shines in the larva state, and in that of the nymph, as well as in that of the perfect insect, which proves that Nature has not endowed it with this faculty, principally for the purpose of attracting the male, as some authors have supposed. Nevertheless, it certainly appears that the male takes advantage of this light to lead him to the female.

The females of the European lampyrides observed by Degeer lay a great number of eggs on the turf or grass on which they live. These eggs are endued with a viscous matter, which serves to fix them upon the plants. Their shell is nothing but a soft and flexible skin, so that the slightest touch will crush them.

We may insert here three new genera, viz.:—

Calyptocephalus. The antennæ long, bipectinated, with the laminæ long, slender; hairy on the sides. The mandibles slender, simple, rather curved, and hairy. The labrum long and rounded. The maxillary palpi, with the last joint, large and acute, sub-triangular. The thorax prolonged over the head, in the manner of a hood. The species is named *fasciatus.*

It has the thorax yellow, with a medial spot of black. The elytra black, with a yellow band across the middle, rather curved. The antennæ blackish; the legs pale. The length is seven lines and a half; it is from Demerara.

Cladophorus. The antennæ strongly pectinated, with laminæ long, slender, and hairy; the last joint of the maxillary palpi sub-securiform; the second joint very long, cylindrical. The anterior tarsi with membraneous palette beneath. The species we have named *Marginatus.*

This insect is black, with the sides of the elytra yellow. The length is five lines; it is from Brazil.

Megalopthalmus. Antennæ short, with first joint very large, the second very small, the other strongly pectinated; the labrum triangulate; the mandibles ample, long, curved; the eyes very large, round, placed behind the trophi. The maxillary palpi long, the first and third joints very short; the second and fourth longer, the latter large and acute; the labial palpi shorter, but of the same form. The thorax rounded, covering the head. This species we have named *Bennetti.*

It is yellowish brown. The elytra with elevated striæ. The antennæ and legs blackish. The eyes black. The length is five lines; it is from Columbia.

The TELEPHORI are found in tolerable quantities towards the end of spring in the meadows, on the flowers and plants which grow there; the nature of the telephorus does not appear to be equally mild with that of other insects in general. If we are to believe Degeer it is carnivorous, and lives on prey, nor does it spare its own consimilars any more than any other insects. This accurate observer has seen the female lay the male prostrate, hold him reverted between her feet, open the belly, and gnaw him without compassion. The telephorus flies with facility and promptness, especially when the sun is shining. Coupling takes place in a very little time after the final metamorphosis; but it is necessary that the male should choose his time and act with precaution, if he does not wish to run the risk of being devoured.

Under the head of the larva of telephorus is found the labrum, to which are attached the four barbles. It is mobile. The larva can push it very forward, and afterwards draw it back into its cavity, which it does continually when it walks, for the purpose of feeling at the same time, with the barbles, all the objects that it meets. Underneath the last ring is the anus, which has a sort of edge, of a clear brown, a little raised, so that it forms a sort of nipple, with a sinking in the middle. When the larva walks, it applies, at each step, this nipple, against the plane of position, and makes use of it as a seventh foot. These larvæ live in the earth, the humidity of which appears to be necessary to them; and if it is desired to preserve them in existence, it is necessary to furnish them frequently with fresh and humid earth. When they are placed on the surface of the ground they very quickly bury themselves beneath it. M. Latreille was inclined to think that these larvæ live on roots, but from the observations of Degeer, it appears that they are also carnivorous, and in case of necessity will feed on individuals of their own species, as well as on earth-worms.

Towards the end of the month of May, the larvæ of *T. fuscus* change into nymphs in the ground, without presenting the slightest appearance of a shell. These nymphs are nearly six lines in length, and their body is a little arched. Their colour is of a very pale red, or reddish white. One may clearly distinguish all the parts of the perfect insect, the arrangement of which is the same as in so many other nymphs. The belly, which the nymph moves from time to time from one side to the other, is divided into rings, and terminated by two small points. In the month of June these insects quit the nymph state, and appear under the form of the perfect telephorus.

The MELYRIDES, proper, which may be considered as the type of the tribe so called, are usually, as it would appear, found on flowers, and are in general peculiar to Africa, more especially to the Cape of Good Hope. Their other habits are utterly unknown.

The MALACHII are very common insects, which live generally on flowers. Some authors have, however, observed, that they do not content themselves with the honeyed juice of plants, and that they attack other insects for the purpose of feeding on them. Their habits are conformable to those of the telephori, but they present a singularity (as also does the last mentioned sub-genus) too remarkable not to deserve mention. When they are taken in the hand, we observe issuing from the sides of the corslet and belly, two vesicles, very red, inflated, soft and irregular, composed of three lobes. These four vesicles contract, re-enter the body of the insect as soon as it has ceased to be touched, and leave in the same place only marks of a red spot. Some French writers have given to these appendages the appellation of "*cocardes*." It is difficult to satisfy curiosity respecting the function or utility of this very singular part. These insects have sometimes been deprived of one or all of these vesicles, without having appeared less agile, or less lively. Some species have, at the extremity of their elytra, an emargination with a projecting point or crook in the middle. This character appears peculiar to only one of the sexes.

The larvæ of the malachii are not exactly known. Nevertheless it is presumable that they live in wood, for the perfect insect, newly issued from the nymph, is frequently found in timber-yards.

We now come to the division of CLERII, which contains the

genus Clerus of Geoffroy. This naturalist separated, under this name, certain species from the *Attelabus* of Linnæus, who having merely established it on the forms of the antennæ and the head, united together many insects of the most distinct kind. It has since been still further diminished in extent, and forms at present only a part of the tribe Cleri.

With respect to the insects in general comprehended under this latter title, it appears that their females deposit their eggs in places which serve as a retreat to the larvæ of other insects, so that these last may answer the purpose of food to their progeny. After having undergone their metamorphoses, the clerii betake themselves to flowers to suck the honey therefrom, or remain on the trees, or in the old wood in which they had lived before. They incline their antennæ and feet, folding back their limbs when they are taken.

As for the *Cleri* proper, there are some slight differences in their habits from those of other insects of this division. For instance, the *opili* and *thanasimi* operate their metamorphoses in the carious trunks of trees, or in old wood, and it is on these bodies that it is necessary to seek them, when they are in their perfect state. It is not the same with the cleri, as we shall presently see.

These coleoptera are often bristling with hairs, and generally adorned with lively and brilliant colours disposed in transverse bands over the elytra. They are found on flowers. When they are seized they incline their head, fold back their feet, and counterfeit death. They fly, however, with a tolerable degree of lightness. What however should most engage our attention, is their larvæ, which are much less remarkable in themselves than for the places they inhabit. That of *C. apiarius* (*Trichodes apiarius*, Fab.) introduces itself into the nest of *Megachile muraria*, finds the means of piercing their cells, and of sheltering itself against their sting, for the purpose of feeding, without fear, on their larvæ

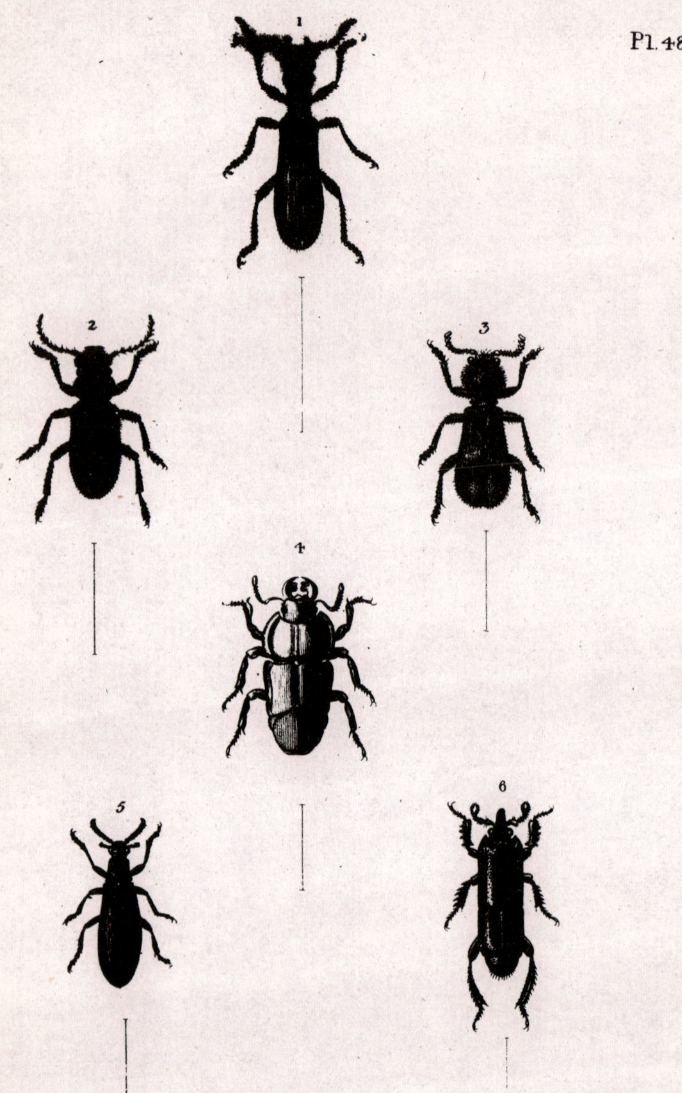

1. Cymatodera *Hopei*. *G. R. Gray*. 2. Stygmatium *cicindeloides*. *G.R. Gray*. 3. Lasiodera *Kirbii*. *G.R. Gray*. 4. Hololepta *glabra*. *Gray*. 5. Stilpnonotus *eurypiformis*. *G.R. Gray*. 6. Trypanœus *denticollis*. *G.R. Gray*.

and nymphs. It is of a fine red, and furnished with six short scaly feet, and two small crooks equally scaly, placed near the hinder part of the body. About a year is necessary for its transformation into the perfect insect. It is in the same place where it has lived and grown that its metamorphosis takes place. Although provided with wings and elytra, the hardness of which may secure it, it is not easy to conceive how the perfect insect can go to deposit its eggs in the domicile of these *Apiariæ*, without being at least repulsed by the number of these insects, if not wounded by their stings.

It has been supposed that the *Megachile apiaria* itself collected the eggs of this clerus among the stamina of flowers, and that it transported them into its habitation, with the wax which it brings there. This conjecture is not altogether improbable, but yet it cannot be adopted, until confirmed by observation in a more positive manner.

The larva of another species, *clerus alvearius*, destroys those of our domestic bees.

After Opilo of Latreille, we add two new genera, with the following characters.

Cymatodera. This has the labrum transverse, short, bilobate; the maxillary palpi with the last joint short, broad, subsecuriform; the labial palpi with last joint truncate, oval; the tarsi cup-shaped and lobed; the antennæ filiform, the last joint the largest. The species we name *Hopei.* It has the head, thorax, the base of the elytra, and a subtriangular spot on each side of the elytra, black; the rest of the elytra, antennæ, palpi and legs, pale yellowish red; the head and thorax punctured, hairy; the elytra striated with two rows of impressed dots. This species is from Mexico, and the height is ten lines and a half.

The next genus we have named *Stigmatium*, with the maxillary palpi elongate, triangular; the labial with the joints cylindrical, the last long; the labrum very small, formed of two lobes; the eyes very large; the antennæ filiform, with the first joint large, the last rather long, and placed on the margin of the eyes; the anterior tarsi with the last joint bilobed; the posterior tarsi with basal joint obliterate; the species we have named *Cicindeloides*. It is hirsute, brown, the elytra with a few longitudinal striæ of elevated dots, also with two transverse fasciæ, and the tip yellow; the legs pale, with the tarsi black. The length is eight lines.

We also add a new species of *Clerus*, with the specific name of *Kirbyi*. It has the head with the front yellow, the antennæ and thorax black, hairy; the elytra yellow, with two black fascia near the base; the apex black, with two large orange spots. The species is from Brazil. The length seven lines.

The Tilli frequent plants and flowers, and live on the honeyed juices, which they find in abundance there. But their larvæ are never met with in similar places, and there is reason to believe that they live in the substance of wood, or in the ground, which distinguishes this genus from that of *chrysomela*, whose larvæ live on plants, and gnaw the leaves.

Before we speak of the habits of Necrobia, it may be as well to say a word on its classification

Almost all the insects which attack animal substances, as well as the majority of those which live in dead wood, or which destroy our furniture and provisions, have been for a long time designated under the generic name of Dermestes. Linnæus, the creator, as we may call him, of Entomological science, having but a small number of insects to present, thought proper to unite them under but few divisions, and those easy to be distinguished. The genera which that celebrated naturalist established, being clear and precise, sufficed, during a certain period, for the researches which were made; but since this science has become more generally cultivated;

since the manners and mode of life of insects have presented to us an infinity of wonders, not before suspected to exist; since philosophers have had the good sense to discover that the study of these little animals had its application in the arts and in medicine, and that it was connected with the animal and vegetable economy; above all, since their number, in our collections, has exceeded that of plants, naturalists have been obliged to form new sub-divisions from time to time, and to multiply genera in proportion to the progress of discovery.

Geoffroy very well appreciated the natural relations of the insects of which we are now treating, when he placed one of the most common species, (*Dermestes violaceus*, Lin.) in his new genus *clerus*, which is composed for the most part of many *attelabi* of the Swedish naturalist; but this species, and some others analogous, presenting certain peculiar characters, have been united into another generic section by M. Latreille, called *Necrobia*, a name compounded of the Greek word, *necros*, (carcase,) and *bios*, life, because these insects are usually found in dead and decomposed animal bodies.

The necrobiæ are adorned with very fine colours. Their walk is slow, and their flight by no means rapid.

They are sometimes found on flowers, and on the leaves of plants; but they more particularly frequent carcasses and the dried spoils of animals. The larva, which feeds on these last-mentioned substances, has the body elongated, soft, and composed of many rings. It has six scaly feet, and two scaly crooks near the anus. Its growth is tolerably quick, and it undergoes its metamorphosis in the same places in which it has lived.

As for the tribe of Ptiniores, in general, it consists of small insects, which, like some of the last, counterfeit death when they are seized. Most of them are fond of obscure places, pierce with little holes all the substances which they gnaw, whether in their perfect state, or under the larva form. Some of them live in old wood, others attack collections of animals, books, substances prepared with flour, &c.

The genus, or sub-genus Ptinus, is very distinct, and very easy to be recognized. The insects which compose it cannot be confounded either with the capricornes, among which Linnæus first placed them, nor with anabium, where he finally left them. The number of articulations in the tarsi distinguish them sufficiently from the first, and the form of the corslet, and especially the filiform antennæ, from the second.

The Ptini are very small insects. They are commonly found on walls and in houses, principally in garrets and inhabited places. It is more rare to meet with them in the country. Similar to many other insects, when they are taken, they draw back the head, apply the antennæ and feet against the body, and by the semblance of death, appear to wish to escape the danger by which they are menaced.

The larvæ of the ptini (not described in the text) are hexapod. The body composed of several rings, not very distinct, in consequence of the wrinkles and rugosities by which it is covered, is soft, cylindrical, and slightly hairy. Its posterior part is curved underneath, which makes it appear raised, as it were, into an arch. The feet are short, and terminated by a single crook. The head is hard, scaly, and furnished with two small jaws, tolerably strong. These larvæ feed on dry plants, and dried animals, which are not in a state of putrefaction. In consequence of this, they are very destructive to herbals, hay, collections of animals, furs, and other objects of considerable value, which thus require great care in their preservation. Linnæus, after Cramer, informs us, in his *Systema Naturæ*, that these mischievous larvæ may be destroyed by means of arsenic and alum.

The larvæ of PTILINUS, like those of Anobium, live in dead wood, and form there small, round, and deep holes. They have a scaly head, provided with two corneous mandibles, hard and trenchant, and six small scaly feet. They undergo their metamorphosis in the wood, and do not issue from it, but under the form of the perfect insect.

The insects forming the genus ANOBIUM, were originally classed by Linnæus under *Dermestes*, with which they have certainly some relations; but they differ from them in having longer antennæ, terminated in a less thick and more elongated knob, and by their denticulated mandibles. They have also many relations with the ptini.

Their French name (*Vrillettes*, wimbles) designates the instinct which they have received from nature, which impels them, in their larva state, to gnaw wood, making small round holes in it, such as might be made with a wimble. Those insects are commonly seen to escape in the spring, from the wood in which their nymphs were enclosed, and, attracted by the rays of the sun, creep along the windows, or scaffolding, and other wood-work. Their colours without brilliancy, their habits without industry, and their form without grace or beauty, are not likely to render them very interesting in our eyes. Like the *dermestes*, as soon as they are touched, they sink their head into the corslet, apply the legs and tarsi exactly against the thighs, conceal the antennæ entirely between the head and the lower edges of the corslet, and then resemble an inanimate body. But what distinguishes them from the Dermestes is, the invincible obstinacy with which they remain in this species of lethargy. If we may believe Degeer, neither water nor fire can induce them to throw it off. They will suffer themselves to be completely burned, rather than give any sign of life. When they are no longer touched, and suffered to remain tranquil, they recover from this state by little and little; but it is not until after a long repose, that they recommence to move. They walk slowly, and in an indolent sort of style. They very seldom make any use of their wings, although they are tolerably strong, and much longer than the elytra.

The larva of these insects, well known by its ravages, is more calculated to fix our attention. Old wood-furniture, worm-eaten, and sifted with round and cylindrical holes, indicate at once its work, and its habitation. It is a small white, soft, and elongated worm, which has six small and short feet. Its head is scaly, and terminated by two jaws in the form of strong and trenchant pincers, which answer the purpose of gnawing the wood on which it is to live, and which it gives back in small and very fine grains, connected together, but which may easily be reduced into an almost impalpable powder, and which fill the little cavities which the larva has just made and abandoned. In proportion as it acquires an increase of growth, it enlarges its dwelling; and when it is come to its full size, and feels the necessity of transformation, it carpets with some threads of silk the bottom of the hole, or canal, which it has hollowed, becomes changed into a nymph, and issues forth in the form of the perfect insect. It is not only in houses that this larva is to be found, but also in fields, gardens, and every where, where there is dry wood, proper to serve it as an asylum, and furnish it with aliment. There is one species which works on substances of less hardness than wood. It attacks bread, flour, paste, wafers a long time kept in drawers. It forms furrows and canals in these matters, as the other species do in wood.

This is the proper place to notice a singular little phenomenon, and one which has given rise to a number of conjectures. There is often heard in a chamber, and when one is alone, and a profound silence prevails, a small continued noise, like the ticking of a watch. It ceases the moment any motion is made, and does not recommence until after the return of silence. Some have attributed this noise to a small species of spider, others to a very little insect designated by Linnæus under the name of *Termes pulsatorius*, and under that of *Hemerobius pulsatorius* by Fabricius. Rolander pretends that this sound is produced by the female of this same *termes*, by giving with its head reiterated blows on the wood: Geoffroy believed, that it was occasioned by a species of Anobium, which strikes with redoubled blows the old wood for the purpose of piercing it, and thus procuring itself a lodging. The spider, of which mention has been made, possesses no instrument of sufficient hardness and strength, for the production of such a noise. The *termes*, equally unprovided with similar means, is, moreover, too small to produce a sound of so much strength. Geoffroy was undoubtedly right, when he attributed this phenomenon to a species of the genus now under consideration. This fact has been confirmed by the observations of M. Latreille, who, moreover, justly claims the exclusive merit of having discovered the cause, or rather the design of this ticking. The two sexes, striking reiteratedly the wood with their mandibles, call each other reciprocally, approximate, and finish by an union. The female deposits her eggs in the clefts and crevices of the wood; but her mandibles, much less strong than those of the larva, do not seem adapted to enable her to gnaw the same substance. The metamorphosis of the anobia takes place towards the surface of the wood: if it occurred at too great a distance from this, the perfect insect could not issue from its prison, but must certainly perish there. We know the precautions which the larvæ of *bruchus* and *tinea*, which feed on the farinaceous substance of grains, take, for the purpose of facilitating the issue of the perfect insect; Wherefore, should not the larvæ of anobium take similar precautions? The larva approaches by little and little the surface of the wood, so that at the moment of its metamorphosis there remains but a feeble barrier through which the perfect insect can pierce with facility. M. Olivier conjectures, that the noise which we hear may possibly be occasioned by the blows of the larva against the wood, for the purpose of ascertaining its thickness. Analogy, however, would lead us to incline to the opinion of M. Latreille, that this noise is intended to facilitate the approximation of the sexes, and the re-production of the species. At the same time it is probable, that further observations are necessary for the complete elucidation of this interesting phenomenon.

Superstition, which has availed itself of every thing for the purpose of painting its terrors, and strengthening its tyranny over the human mind, has not suffered this little phenomenon to escape it. The vulgar name of *death-watch* given to this insect, sufficiently announces the popular prejudice respecting it; and there can be little doubt that the fate of many a nervous and superstitious patient has been accelerated, by listening in the silence and solitude of night, to this imagined knell of his approaching dissolution.

The larva of *Anobium paniceum* feeds on farinaceous substances, and on bread which has been preserved for a long time. In such substances it forms itself a shell, is changed into a nymph, and issues forth at the end of some time, in the form of the perfect insect.

Of the insects comprehended in the third and last section of the serricornes, we can say but very little. The habits of most of them are altogether unknown. We can only observe of LYMEXYLON, that the larvæ of this genus live in the interior of wood, which they pierce or furrow, assisted by the *capricornes*, the *lepturi*, and *lucani*; the lymexylons often contrive to render a tree carious, to such an extent, that it will perish after having been in a decaying state for a considerable time. It is on the trunk of those same trees that the perfect insect is to be found, either at the time when it has just arrived at that state on issuing from the nymph, or when the female returns thither for the purpose of depositing her eggs.

THE

FOURTH FAMILY

OF

PENTAMEROUS COLEOPTERA,

THAT OF

Clavicornes,

Having, like the preceding, four palpi, and cases covering the upper part of the abdomen, or its greater portion, differ from it by having the antennæ almost always thicker towards their extremity, often even in a perfoliate or solid knob. They are longer than the maxillary palpi, with the base naked, or hardly covered. The feet are not adapted for swimming, and the articulations of the tarsi, or at least those of the hinder ones, are usually entire.

They subsist, at all events in the first state, on animal substances.

We shall divide this family into two sections, the first of which shall have for common characters:—antennæ always composed of eleven articulations, longer than the head, not forming from the third any fusiform or almost cylindrical knob; their second articulation not dilated in the manner of an earlet. The last articulation of the tarsi, as well as its crotchets, of moderate length, or small. These clavicornes live out of the water, while those of the second section are aquatic, or fluvial, and thus conduct us to the palpicornes, coleoptera for the most part aquatic, and whose antennæ have never more than nine articulations.

The first section will comprehend many small tribes.

The first, that of PALPATORES, it appears to us, ought to come, in a natural series, near the pselaphi, and the coleoptera of the family of brachelytra, from their organs of manducation. Their antennæ, at least as long as the head and corslet, grow a little thick towards the end, or are almost filiform, with the first two articulations longer than the following. The head is ovoid, and distinguished from the corslet by a strangulation. The maxillary palpi are long, advanced, and swelled towards their extremity. The abdomen is large, ovaliform or ovoid, and embraced laterally by the elytra. The feet are elongated, the thighs in a knob, and the articulations of the tarsi entire.

These insects remain on the ground, under stones, or other bodies. Some of them (*Scydmænus*) frequent humid places. We shall unite them into a single genus, that of

MASTIGUS.

MASTIGUS, Hoff. *Ptinus*, Fab.

Have the antennæ composed of articulations having almost the form of an inverted cone, the first of which is very long, and the final ones but little thicker than the others. The last two of the maxillary palpi compose an ovaliform knob. The corslet is almost of an ovoid figure. The abdomen is ovaliforn. (*Mastigus palpalis*, Lat. Gen. Crust. et Insect. I. 281, viii. 5. Schœn. Sym. Ins. I. ii. p. 59, and Klüg. Entom. Mon. p. 163.)

SCYDMŒNUS, Latr. Gill. *Pselaphus*, Illig. Payk. *Anthicus*, Fab.

Have the antennæ grained, sensibly swelled towards their

extremity, and but little bent. The maxillary palpi are terminated by a very small and pointed articulation. The corslet is almost always globular, and the abdomen, almost ovoid, is proportionally shorter than that of Mastigus.*

In all the following clavicornes, the head is generally sunk in the corslet, and the maxillary palpi are never at once as much advanced, and club-formed; the whole of their physiognomy, moreover, presents differences in other respects.

The genus HISTER will form our second tribe, which we shall name with the Baron Paykull, who has studied it so well, HISTEROIDES.

In these, the four hinder feet are more separated from each other at their origin than the anterior two, a character, which of itself alone distinguishes this tribe from all the others of the same family. The feet are contractile, and the external side of the legs is denticulated, or spinous. The antennæ are almost always bent, and terminated in a solid knob, or composed of very close articulations. The body is of a very solid consistence, most frequently squared or parallelipiped, with the præsternum often dilated in front, and the elytra truncated. The mandibles are strong, advanced, and often of unequal bulk. The palpi are almost always filiform, or in a slight degree thicker towards their extremity, and terminated by an ovaliform or ovoid articulation.

* *Scydmænus Helwigii*, Lat.; *Anthicus Helwigii*, Fab.; *Notoxus minutus*, Faun. Insect. Germ. XXIII. 5; – *S. Godarti*, Lat. I. viii. 6; *S. hirticollis*, Gyll.;—*S. minutus*, ejus.; *Anthicus minutus*, Fab. See Schœnh. Insect. I. ii. p. 57. M. Duros, garde-du-corps of the King, who has a peculiar talent for discovering the smaller species of our environs, has found in an ant-hill, the S. Clavatus of M. Gyllenhall. This fact, and some others, confirm me in the opinion that these insects come with the Pselaphi, at the end of the Brachelytra.

In the relation of habits, and in consequence of the denticulations of their limbs, and some other characters, these insects seem to approach the lamellicorn coprophagi; but, from other considerations founded on anatomy, they come naturally near *Silpha*. Such is also the opinion of M. Dufour (Annal. des Sciences Nat., Oct. 1824). The digestive canal of the species which he has dissected *(Sinuatus,)* is from four to five times the length of the body. The œsophagus is very short, the oblong swelling coming immediately after, presents, through its parietes, some brownish traits which would seem to announce the existence of interior pieces proper for trituration, and if so, this swelling would deserve the appellation of gizzard; the chylific ventricle is very long, folded back upon itself, and bristling with pointed and very prominent papillæ. The hepatic vessels have six distinct insertions around the chylific ventricle *(ibid*, July, 1825). Their number, according to Ramdohr, should be but three, and each of them would have two insertions; but such a disposition of these vessels is doubtful.

These animals feed on cadaverous or stercoraceous substances; also on corrupted vegetable matter, such as dung, old mushrooms, &c. Some others make their domicile under the barks of trees. Their walk is slow; they are of a very brilliant black, or a bronze colour. Such of their larva as have been observed, (*merdarius, cadaverinus,*) feed on the same substances as the perfect insect; their body is almost of a linear form, depressed, almost smooth, soft, and of a yellowish white, with the exception of the head and the first segment, the dermis of which is scaly, and brown or reddish; it is provided with six short feet, and terminates posteriorly by two articulated appendages, and an oval and tubular prolongation; the scaly plate of the first segment is channeled longitudinally.

This tribe will comprehend exclusively, as we have said above, the genus

Hister.

M. le Baron Paykull has confined himself to detach from it some species with a very flatted form, and with which he has composed that of *hololepta*. But Dr. Leach (Zool. Misc. III. p. 76.) has established four others.

Some have the limbs, at least the anterior ones, triangular, and denticulated externally, the antennæ always uncovered and free; the body generally squared, and but little, if at all, inflated.

They may be divided into two sub-genera. In the first, that of

Hololepta, Payk.

The body is very much flatted, the præsternum does not advance upon the mouth, and the four posterior limbs have but a single rank of spines. The terminal lobe of the jaws is prolonged; the chin is deeply emarginate, and the palpi, proportionally more advanced, are formed of articulations almost cylindrical.

They remain under the bark of trees. The animal figured by M. Paykull, as the larva of a species of this sub-genus, is that of a species of syrphe or fly.

The other histeroides, whose præsternum is advanced upon the mouth, whose jaws are terminated by a short lobe, with palpi very little advanced, and composed of articulations, which, with the exception of the last, are rather in the form of an inverted cone than cylindrical, and whose chin, in fine, is slightly emarginate, will enter into the sub-genus of

Hister (Proper).

Some species, whose fore posterior limbs have, as well as

hololepta, but a single range of small spines, and live also under the bark of trees, compose the genera PLATYSOMA and DENDROPHILUS of Dr. Leach. The first* does not differ from the second, (*H. punctatus*, ejusd. VII. 5,) but that the body is flatted above, and the corslet is shorter and narrowed in front. A species of the same division (*H. proboscideus*, Payk. Monog. VIII. 4,) has a peculiar form. Its body is long and narrow, with the corslet more than half as long as broad.

The others have two ranges of spines to the four posterior limbs. These are the only ones that Dr. Leach leaves in the genus Hister.

H. unicolor. Lin. Payk. *ibid.* II., four lines in length; entirely black, shining; three denticulations to the external side of the first two legs; two striæ on each side of the corslet, and four on the external side of each case, of their length, and that which is nearest to the edge is interrupted. Very common.

The number of denticulations of the legs, that of the striæ of the corslet and the elytra, their punctuation, and the form of the body, have supplied M. Paykull with excellent characters, by means of which he has very well distinguished the species.

A last division of this tribe comprehends the very small histeroides, with thick and almost globular body, whose præsternum but little, or not at all compressed laterally, and not advanced on the mouth, is straight in front. In some (*Abræus*, Leach,) it is prolonged as far as the anterior angles of the corslet, and covers the antennæ entirely in their state of contraction. It is more narrow in others, (ONTHOPHILUS ejusdem); but here the knob of the antennæ is lodged in an orbicular and very distinct cavity, situated

* *Hister picipes*, Fab. Payk., *ibid.* VIII. 5; *H. flavicornis*, ejusd. VIII., 6; *H. oblongus*, ejusd. X. 3.

under the anterior angle of the corslet. The anterior legs are often narrow, almost linear, and without denticulations. The last upper semi-segment of the abdomen is curved inferiorly, and appears to terminate it.*

The other clavicornes have the feet inserted at an equal distance from each other. Those of these coleoptera, in which these organs are not contractile, or whose tarsi for the most are refolded against the leg, which have the mandibles most frequently projecting and flatted, or but little thick, and whose præsternum is never dilated anteriorly, will compose five other tribes.

The third tribe of the family, that of Silphales, presents five very distinct articulations to all the tarsi, and the mandibles terminated in an entire point, or without emargination or fissure,† the antennæ are terminated in a knob, most frequently perfoliate, and of four or five articulations. The jaws, in the majority, have one corneous tooth in the internal side. The anterior tarsi are often dilated, at least in the male. The elytra of the greater number have, at the external edge, a gutter with a very strong edge.

This tribe is composed of the genus

Silpha, Lin. *Peltis*, Geoff.

In these the antennæ terminate abruptly in a knob, short,

* Dr. Leach refers to *G. Abræus*, the *H. globosus*, Payk. VIII. 2;—*H. minutus*, ejusd. VIII. 1, and to his genus *Onthophilus*, the following: *H. striatus*, Payk. ibid. XI. 1; *H. sulcatus*, X. 8; *H. hispidus* of the same, XI. 2, appears to be congenerous. The genus *Ceutocerus* of M. Germar, Insect. Spec. Nov. I. p. 85, 1, 2, appears to come naturally after the histeroïdes, from the form of the antennæ, feet, &c., but the elytra cover the abdomen, and the mandibles are not projecting. I have seen no individual of this genus.

† The internal side, however, sometimes presents denticulations, and such are those of the *spherites*.

and solid, and formed by the last four articulations. The second is larger than the following. The body is almost squared, with the elytra truncated, the legs denticulated, the tarsi simple, the mandibles bidenticulated at the internal side, and the last articulation of the maxillary palpi as long as the preceding two united. The jaws have a corneous tooth at the internal edge. These insects very much resemble Hister, so much so that Fabricius has confounded them with that genus. Such are,

SPHÆRITES, Dufst. *Sarapus*, Fisch. *Hister*, Fab. *Nitidula*, Gyll.*

In those the antennæ are terminated in a perfoliate knob.

Sometimes the body is oblong, with the head strangulate posteriorly, as broad, or but little narrower than the anterior edge of the corslet. This part is in the form of a square rounded at the angles. The elytra are in a long square, abruptly and strongly truncated at their posterior extremity. The hinder thighs, at least in the males, are usually inflated. The last articulation of the maxillary palpi is a little more slender than the preceding, almost cylindrical, a little attenuated towards the end, and obtuse. The anterior tarsi are dilated in the males.

NECROPHORUS, Fab. *Silpha*, Lin. *Dermestes*, Geoff.

The antennæ, but little longer than the head, are terminated abruptly, in an almost globular knob of four articulations; the first is long, and the second much shorter than the following. The body is almost parallelipiped, with the corslet broader in front, all the legs strong, widened at their

* Dufst. Faun. Aust. I. p. 206; *Hister glabratus*, Fab. Sturm. I. xx. *Sarapus*, Fisch. Mem. de la Soc. des Natur. Moscow.

extremity, and terminated by strong spurs, and the elytra truncated at a right angle.

The jaws are destitute of a corneous onglet. The instinct which they have of burying the carcasses of moles, mice, and other small quadrupeds, have gained them the appellation of *burying beetles*. They insinuate themselves underneath, hollow the earth until the foss is sufficiently deep to contain the body, and they cause it to enter there by degrees, drawing it towards themselves. They deposit their eggs in this place, and thus their larvæ find nourishment. They are long, of a greyish white, with the upper part of their anterior rings clothed with a small scaly plate of a fawn-coloured brown, and with little points raised on the final rings. They are provided with six feet, and tolerably strong mandibles. To pass into the nymph state they sink themselves very deep in the earth, and construct a sort of lodge there, which they line with a gluey substance. These insects, as well as many others which live on cadaverous matters, have a very strong odour of musk. Their habits have latterly fixed the attention of those persons who make a trade of killing moles, and the work entitled, "L'Art du Taupier," presents us, in this respect, some facts which had escaped the attention of naturalists. These insects must have a very fine sense of smell, since in a very short time after a mole has been killed many necrophori are seen flying about it, which one would have sought for before to no purpose in the neighbourhood.

The digestive canal of Necrophorus and Silpha is at least three times longer than the body. The œsophagus is very short, and followed by an ellipsoid gizzard, the internal tunic of which, a little membranous, is bristling, at least in many species, with pointed hairs, planted in various directions, but arranged in eight longitudinal bands, separated by smooth intervals. The intestinal tube is very long, especially in the necrophori and necrodes. The surface of the intestine in the last, as well as in silpha, is all covered with prominent and granular points. It opens, either laterally or directly, into a smooth inflation, which, according to M. Dufour, may be compared to a cœcum. It receives on the side a sort of oval or oblong purse, making part of the excrementitious apparatus. The number of biliary vessels, which are slender, very long, and greatly folded, and have each a proper insertion around the extremity of the chylific ventricle, is four. It appears from the figure of the digestive canals of *necrophorus vespillo*, given by Ramdohr, that its large intestine, instead of being covered with granular papillæ, should have muscular ribands transverse, and forming annular folds.

Silpha vespillo, Lin. Oliv. Col. II. 10, i. 1, is from seven to nine lines in length, black, with the last three articulations of the antennæ red, and two orange bands, transverse and denticulated on the cases, and the haunches of the two hinder feet armed with a strong tooth. Their legs are curved.

N. Mortuorum, Fab. Panz. Faun. Insect. Germ. XLI. 3, is smaller, with the antennæ entirely black. The second transverse orange band of the elytra of the preceding species, forms here usually only a large spot of a crescent form.

It is particularly found in woods, and often on mushrooms.

N. germanicus, Fab. Oliv. *ibid.* 1, 2, a. b. often more than an inch in length. It is altogether black, with the external edge of the elytra fawn-colour, and a spot of ferrugineous yellow on the front.

N. humator, Fab. Oliv. *ibid.* 1, 2, c. differs from the preceding by the orange-colour of the knob of the antennæ. It is also constantly smaller.

North America furnishes many species, one of which especially (*grandis*, Fab.) surpasses all the others in size. This genus appears, hitherto at least, to be confined to the northern countries of this continent, and of Europe. (See for the other species, Fabricius, Olivier, and Schœnh, I. ii, p. 117.)

NECRODES, Wilk. *Silpha*, Lin. Fab.

Have the antennæ manifestly longer than the head, terminated in an elongated knob, of five articulations. The second is larger than the third. The body is an oblong-oval, with the corslet almost orbicular, broader in its middle, the legs narrow, elongated, a little widened at the end, and terminated by two spurs of ordinary size, and the cases truncated obliquely.

Species of this sub-genus are found in Europe, in the equatorial countries of the New World, in the East Indies, and New Holland.*

Sometimes the body is ovaliform, or ovoid, with the head but little or not at all strangulated posteriorly, and more narrow than the corslet. The corslet is either almost semi-circular, and truncated in front, or trapezoid, and broader behind. The elytra are rounded, or simply emarginated at their posterior extremity. The hinder feet do not differ at all, or at least very little, sexually.

The jaws are armed internally with a tooth or scaly crotchet.

Silpha (Proper), Lin. Fab. *Peltis*, Geoff.,

Whose body is almost in the form of a buckler depressed, or but little elevated, with the corslet semi-circular, truncated, or very obtuse in front, the elytra strongly edged, and hollowed into a gutter externally, the palpi filiform, and the last articulation is almost cylindrical and terminated in a point in

* *Silpha littoralis*, Fab. Oliv. Col. II. ii. 1. 8. a. b. c.;—*S. Surinamensis*, Fab. Oliv. *ibid.* II. ii;—*S. lachrymosa*, Schreib. Lin. Trans. VI. xx. 5:—*S. indica*, Fab.

many of them. The majority of them live in carcases, and thus diminish the quantity of miasma which would otherwise proceed from them. Some others climb on plants, and particularly the stalks of corn, where the smaller helices are found to eat the animal. Others remain on elevated trees, and devour the caterpillars. Their larvæ are equally agile, live in the same manner, and often assemble in great numbers. They have much resemblance to the perfect insect. Their body is flatted, composed of twelve segments, whose posterior angles are acute, with the posterior extremity more narrow, and terminated by two conical appendages.

In the majority of the species, the two anterior tarsi of the males are alone more dilated than the others. The antennæ are thickened insensibly, or terminate abruptly in a knob of four articulations at the most. The second and third articulations are but little different; the last of the maxillary, is at most, of the length of the preceding, and often a little shorter, and a little more slender.

The species in which the extremity of the antennæ is distinctly perfoliate, or composed of articulations which, with the exception of the last, are transverse, and more broad than long, in which this knob is abrupt, and whose elytra are emarginate at their extremity, at least in the males, form the genus THANATOPHILUS of Dr. Leach. (*Silpha sinuata*, Fab. Oliv. *ibid.* II. 12; *S. dispar*, Illig. Gyll. &c.)

Those in which the elytra are entire, but which otherwise have similar antennæ, compose that which he terms OICEPTOMA.

S. Thoracica, Lin. Fab. Oliv. Col. II. ii. 1, 3, a, b, whose body is black, with the corslet red, silky, and three raised and flexuous lines, the external one of which is shorter, forming a keel, and terminating near a transverse tubercle on each elytrum. In the male the posterior extremity of these elytra finishes in a point at the suture. This species particularly inhabits the woods.

Another species proper to the forests, but which commonly sojourns on the young oaks to live on the caterpillars there, is *S. quadripunctata*, Lin. Fab. Oliv. ibid. I. 7, a, b. Its body is black, with the border of the corslet and the elytra yellowish. Each of them have two black points, one at the base, and the other in the middle.*

The silphæ, whose antennæ are equally perfoliate at their extremity, but whose nob is gradually formed, alone preserve, in the method of the same naturalist, the generic denomination of silpha. These species remain habitually in the fields, on the sides of paths, &c.

S. lævigata, Fab. Oliv. *ibid.* I. 1, a, b, which is of a shining black, very much punctuated, with the corslet much more narrow in front, and the elytra without raised lines.

S. obscura, Lin. Fab. Oliv. *ibid.* II. 18, of an obscure black, with the corslet truncated in front, the elytra more deeply punctuated, and three raised lines, but not very prominent, short, and the intermediate one longer on each elytrum.

S. reticulata, Lin. Panz. Faun. Insect, Germ. V. 9, of an opaque black, with the corslet truncated in front, three raised lines on each elytrum, the exterior of which is stronger, forming a keel, terminated by a tubercle, and there are transverse wrinkles in the intervals.†

In some, the antennæ are not exactly perfoliate at their extremity, the last articulations being almost globular. These are the Phosphuga of the same. (*S. atrata*, Fab. *ejusd. Pedemontana*, var.; Oliv. *ibid.* I. 6.)

A species of silpha belonging to Germany, and which

* Add, *S. rugosa*, Fab. Oliv. II. *ibid.* 17;—*S. laponica*, Fab.

† Add, *S. opaca*, Fab. Herbst. Col. LI. 16;—*S. tristis*, Illig., &c.

might form a separate sub-genus, (*Necrophilus*, Latr.) is removed from the preceding by many characters. The four anterior tarsi are similar, and dilated at their base, the first two articulations being sensibly broader, at least in the males, than the following two. The third articulation of the antennæ is longer than the preceding, and the last five form abruptly a perfoliated knob. The last of the maxillary is as long as the preceding two united. This species is the *Silpha subterranea* of Illiger, and several other entomologists.

Agyrtes, Frœh. *Mycetophagus*, Fab.

Have the body tolerably thick, convex or arched above, not in the form of a shield, with the corslet almost squared, a little broader than long, and a little more narrow in front, the external margin of the elytra inclined and without canal, and the last articulation of the maxillary palpi thicker and ovoid.*

Some clavicornes, which appear to us, from many characters, and from their habits, to approach the agyrtes, but whose mandibles are cleft or bidenticulated at their extremity, will compose a fourth tribe, that of Scaphidites. Their tarsi have five very distinct and entire articulations. Their body is ovaliform, narrowed at the two ends, arched or convex above, thick in the middle, with the head low, and received posteriorly in a trapezoid corslet, not at all, or feebly edged, and broader posteriorly. The antennæ are in general as long at least as the head and corslet, and terminated in an elongated knob of five articulations. The last articulation of the palpi is conical. The feet are elongated and slender. If we except some species (*choleva*) the tarsi are almost identical in the two sexes.

* *Agyrtes castaneus*, Gyllenh. Insect. Suec. I. iii p. 682; *Mycetophagus castaneus*, Fab.; *M. spinipes*, Panz. Faun. Insect. Germ. XXIV. 20. I suspect that the *A. subniger* of M. Dejean is only the female.

This tribe compose the genus

SCAPHIDIUM.

SCAPHIDIUM (Proper), Oliv. Fab. *Silpha*, Lin.

The last five articulations of their antennæ are almost globular, and compose the knob. The maxillary palpi are but little prominent, and terminate gradually in a point, the penultimate articulation being but little thicker than the last, at their junction. The body has a navicular form, with the corslet a little edged, and the cases truncated. They live in mushrooms. But a small number of species are known, one of which belongs to Cayenne, and the others to Europe.

CHOLEVA, Latr. Spence. *Catops*, Fab., *Peltis*, Geoff. Have the club of their antennæ composed of articulations, for the most part almost in the form of a whirligig, and more or less perfoliate; the maxillary palpi very prominent, and terminated abruptly in the manner of an oval; the body ovoid, with the corslet plane, without edges. The first four articulations of the anterior tarsi, and the first of the intermediate, are dilated in the males of some species. (*Catops blapoides*, Germ.)

In the *choleva*, properly so called, the antennæ are about the length of the head and corslet; their eighth articulation, or the second of the knob, is sensibly shorter than the preceding and the following, and even sometimes but little distinct. The last is semi-ovoid and pointed. In *Mylæchus*, Latr. Oliv. *Catops*, Payk. Gyll., the antennæ are shorter; the eighth articulation is larger than the preceding, and almost equal to the following; the last is rounded and obtuse at the top.

The fifth tribe, that of NITIDULARIA, approaches that of Silphales, in the buckler-formed and edged body, but the mandibles are bifid or emarginated at their extremity. Their tarsi seem to be composed but of four articulations, the first and the following one, in some, not being apparent except underneath, and forming there but a small projection, the penult in the others being very small, and in the form of a knot, enclosed between the lobes of the preceding. The knob of the antennæ is always perfoliate, of three or two articulations, and usually short, or but little elongated.

The palpi are short, filiform, or a little thicker at their extremity. The elytra are short or truncated in many. The feet are but little elongated, with the legs often widened at their extremity, and the tarsi furnished with hairs or cushions. The habitation of these insects varies according to the species: they are found on flowers, in mushrooms, corrupted meat, and under the bark of trees. They form the genus

NITIDULA.

In some the knob of the antennæ has but two articulations, and the fore part of the head advances in the manner of a semi-circular hood, flatted, covering the mandibles, and the other parts of the mouth.

COLOBICUS, Latr.

In this sub-genus and the following, the tarsi, proceeding from the point in which they are mobile, seem to have but four articulations, of which the first three are much shorter than the last, entire, and simply furnished underneath with hairs, in greater or less abundance, as well as in many of the Cleri of Olivier; the first, properly so called, appears only below, and makes there a slight projection. It is also furnished with hairs. The palpi of Colobicus, and those of the following sub-genus, are terminated by an articulation a little thicker than the preceding.

In the other Nitidularia, the knob of the antennæ is of

three articulations, and the head does not advance above the mouth.

Sometimes the first articulation of the tarsi, as well as in Colobicus, is very short, the following three are elongated, entire, equal, and simply furnished with hairs underneath. The palpi are thicker at their extremity.

THYMALUS, Latr., *Peltes*, Fab. *Silpha,* Lin.

In the species whose body is almost hemispherical (*limbatus*), the knob of the antennæ is proportionally shorter ; the third articulation and the following are more attenuated than the second ; the spurs of the legs are extremely small.

Sometimes the first three articulations of the tarsi, at least those of the males, are short, broad, emarginate, or bilobate ; the fourth is very small, little or not at all apparent ; the maxillary palpi, at least, are filiform.

In these, the limbs, or at least the anterior ones, are widened at their extremity in the form of a reversed angle ; the first articulation of the antennæ is usually larger than the second, the elytra are generally truncated, or very obtuse at the end.

In the two following sub-genera the third articulation of the antennæ is sensibly longer than the following ; the knob is formed abruptly, almost orbicular or ovaliform.

IPS, Fab. *Nitidula*, Oliv. Latr. *Silpha*, Lin.,

Whose body is always oblong-oval, depressed, with the posterior extremity of the abdomen uncovered. One of the mandibles (the left) is as it were truncated and tridenticulated at its extremity, and the other widened, and broadly emarginated or concave at the same end. The terminal lobe of the jaws is elongated.

Some species of Fabricius seem proper to be referred to his genus *Engis*.

NITIDULA (Proper), Fab. *Nitidula. Strongylus*, Herbst. *Silpha*, Lin.,

In which the two mandibles are narrowed towards the end, and terminate in an emarginate or bifid point.

Some are flatted, oblong, or ovoid, the others are orbicular and gibbous, or proportionally more convex than the preceding. Accordingly some authors have placed certain species of them in genera of an analogous form, but very different, such as Spheridium and Tritoma.

We find in great abundance on flowers, *N. ænea*, Fab. ejusd. *N. viridescens rufipes*, Var. Oliv. Col. II. ii. 12: III. 20, a, b. V. 33, a, b. It is small, oblong-ovoid, of a brilliant bronzed green, very much punctuated, with blackish antennæ, terminated by a large obtuse knob. The corslet is transverse, slightly emarginated in front, edged laterally, and the feet sometimes of a blackish brown, sometimes fawn-coloured.

In these, the second and third articulations of the antennæ are almost of the same size, and the knob is elongated in the form of a reversed cone or pear.

CERCUS, Lat. *Catheretes*, Herbst. Illig. *Dermestes*, Lin. Fab. *Sphæridium*, Fab. Gyll. *Nitidula*, Oliv.

The body is depressed, with the elytra truncated. The first two articulations of the antennæ are much larger in the males of some species than in their females, and perhaps this sub-genus ought only to comprehend such species. The others should be thrown back into the preceding.

In those the legs are long, narrow, almost linear; the elytra cover the abdomen, and are not truncated.

The body is oval, with the corslet trapezoid; the knob of the antennæ is oblong, the first two articulations are almost equal, and the third is but little longer than the following.

Byturus, Lat. Schœn. *Dermestes*, Geoff. Fab. Oliv. *Ips*, Oliv.

A sixth tribe, that of *Engidites*, analogous to the last, as to the emargination of the extremity of the mandibles, is distinguished from them by their not jutting out, or but very little, and merely at the sides, on the labrum. The body is ovaliform or elliptical, with the anterior extremity of the head a little advanced, in an obtuse or truncated point. The tarsi have five distinct articulations, entire, and for the most part a little hairy underneath. The penult is simply a little shorter than the preceding. The antennæ are terminated in a perfoliate knob, of three articulations. The elytra entirely cover the abdomen. The palpi are a little thicker at their extremity. Some species, very small, live in the interior of houses, and they are often found behind the panes of cross-bar windows.

These clavicornes will be united into a single genus, that of

Dacne.

Dacne (Proper), Lat. *Engis*, Fab. Dej. *Erotylus*, Oliv.

Their antennæ are terminated abruptly, in a tolerably large knob, orbicular or ovoid, compressed, composed of close articulations, of which the middle one at least is much more broad than long. The third articulation is longer than the preceding and following.

The middle of the posterior edge of the corslet is dilated behind, or lobate, and the superior extremity of the chin is advanced and terminated in a truncated or bidenticulated point.

Cryptophagus, Herbst. Schœn. *Dermestes*, Lin. Fab. *Ips*. Oliv. Latr. *Antherophagus*, Knoch.,

Whose moniliform antennæ, with the second articulation as

large or larger than the preceding, terminate in a knob less abrupt, and more narrow than in Dacne.

The antennæ of *Antherophagus* are thicker in proportion, composed of more transverse articulations, and terminated almost gradually in a knob; proceeding from the second to the eighth, they are almost equal. The *Cryptophagus silaceus* of M. Gyllenhall, has on each side of the under part of the head, a projection in the form of a tooth or horn. The *Tryphyllus* of MM. Megerle and Dejean do not differ from Crytophagus, but in the number of articulations of the tarsi.

We shall now pass on to some tribes in which the presternum is often dilated in the manner of a chin-cloth, and which differ from the preceding by having their feet altogether, or in part contractile; the tarsi may be free, but the legs at least are folded against the thighs. The mandibles are short, generally thick, and denticulated. The body is ovoid, thick, furnished with scales or slight hairs, which colour it diversely. The antennæ are usually shorter than the head and corslet, and straight. The head is sunk in the corslet as far as the eyes. The corslet is little edged or not at all, trapezoid, wider posteriorly; the middle of its posterior edge is often a little prolonged or lobate. The larvæ are hairy, and feed, for the most part, on the spoils or carcasses of animals. Many of them are extremely hurtful to entomological collections.

Those then whose feet are not completely contractile, the tarsi remaining always free, with the legs narrow and elongated, will form our seventh tribe (Dermestini), and the genus

Dermestes.

Aspidiphorus, Zieg. Dej.

Are the only ones of this tribe whose antennæ present but

ten distinct articulations, and whose palpi are very short, and, inflated underneath, afterwards proceed into a point. The body is orbicular. (*Nitidula orbiculata*, Gyll.)

Among those whose antennæ have eleven distinct articulations, and whose palpi are filiform, or proceed thickening, we shall first separate those whose antennæ are not received in any special fossets of the under part of the corslet. The presternum rarely advances on the mouth, with the sole exception of *Dermestes undatus* of Fabricius, and *Limnichus*.

In some the antennæ are terminated abruptly in a perfoliate knob, large, and formed by the last three articulations.

DERMESTES, Lin. Geoff. Fab.,

In which the antennæ are similar, or but little different in the two sexes. The length of the last articulation does not remarkably exceed that of the preceding.

Some species commit great ravages in furrieries and cabinets of natural history. Accordingly, Degeer designates them under the title of *dissectors*. In fact, *D. Lardarius* cuts and reduces to pieces the insects in the collections into which it penetrates. The others devour carcasses.

D. Lardarius, Lin. Oliv. Col. II. 9. i. 1. is black, with the base of the elytra ashen, and punctuated with black. Its larva is elongated, diminishing insensibly in thickness from front to posteriors, of a marrow-brown above, white underneath, furnished with long hairs, with two kinds of scaly horns, on the last ring. It casts its excrements in the form of long threads.*

MEGATOMA, Herbst. Lin. Geoff. Fab.

Differ from Dermestes only by the knob of their antennæ,

* Add, *D. Vulpinus, murinus, affinis, laniarius, tesselatus, trifasciatus* of Gyllenhall. (Ins. Suec. I. p. 145. et seq.)

which is much more elongated in the males than in the females. The last articulation is in the form of an elongated or lanceolate triangle.

Dermestes Pellio, Lin. Oliv. *ibid.* II. ii. is but two lines and a half in length. Its body is black, with three white points on the corslet, and on each elytrum. They are formed by a sort of down. The larva is greatly elongated, of a reddish, shining brown, furnished with red hairs, of which those of the posterior extremity form a tail. It walks in a sliding manner, and as it were by jolts, which the perfect insect does as well as the Dermestes.*

In the others, such as

LIMNICHUS, Zieg. Dej.

The antennæ thicken insensibly, and terminate by a larger and ovoid articulation. They are grained, and lodge under the anterior angles of the corslet. The jaws terminate by two lobes, the exterior one of which is narrow, and in the form of a palpus. The labial palpi are very small, and the last articulation of the maxillary is larger than the preceding, and ovoid. (*Byrrhus sericeus*, Duft.; *B. Pygmæus*, Sturm.)

In all the following sub-genera, the antennæ, or at least their knob, lodges in particular and lateral cavities of the under part of the corslet. The presternum is always dilated or advanced in front.

In these the knob of the antennæ is perfoliate, and not solid.

ATTAGENUS, Latr. *Megatoma*, Ejusd. *Dermestes*, Fab.,

In which the knob of the antennæ is very large, almost in a

* Add, *Dermestes megatoma*, of Fab., of which his *Macellarius* appears to be the female; *D. emarginatus*, of Gyllenhall; *D. undatus*, of Fab. The presternum in this last species advances on the mouth.

saw-form, and composed only of three articulations, of which the first and last especially, are larger.

The body is ovoid, short, and but little convex. The last articulation of the maxillary palpi is larger and ovoid.*

TROGODERMA, Latr. Dej. *Anthrenus,* Fab.

In which the knob of the antennæ has four articulations at the least.

The body is ovoid, oblong, and the palpi are filiform.†

Now we find the knob of the antennæ solid, or composed of very close articulations. The body is ovoid, short, all covered with small perishable scales. The corslet is lobate posteriorly.

ANTHRENUS, Geoff. Fab. *Byrrhus,* Lin.,

Whose antennæ terminating in a knob, in the form of an inverted cone, lodge in short cavities, made under the anterior angles of the corslet.

These Coleoptera are very small, live on flowers, in their perfect state, and gnaw, in the larva form, dried animal substances, especially insects in collections. These larvæ are oval, and furnished with hairs, many of which are denticulated. They form aigrettes, and the final ones are prolonged behind, in the manner of a tail. Their last skin serves as a shell for the nymph.

Byrrhus verbasci, Lin. Oliv. Col. II. 10. 1. 2, grey above, of a reddish yellow underneath, with the posterior angles of

* *Dermestes serra,* Fab.; *Attagenus serra,* Latr. Hist. Nat. des Crust. et Ins. IX. p. 244. *ejusd. Megatoma serra,* Gener. Crust et Ins. I. viij. 10; *Anthrenus viennensis,* Herbst. Col. VII. cxv. 10. K.

† *Anthrenus elongatus,* Fab.; *A. ruficornis,* Latr. Gen. Crust. et Insect. II. p. 59;—*A. versicolor,* Creutz. Ent. vers. I. ii. 21. a.;—*Dermestes subfasciatus,* Gyll. Insect. Suec. I. p. 155.

the corslet; two transverse bands on the cases, and a spot near their extremity grey.

GLOBICORNIS, Latr.,

In which the antennæ terminated in a globular knob, lodge in fossets prolonged, as far as close to the posterior angles of the corslet.*

The eighth tribe, that of BIRHII, differs from the preceding by having the feet perfectly contractile, the legs folding back upon the thighs, and the tarsi on the legs, so that the animal appears, when these organs are contracted, and applied on the under part of the body, to be absolutely without feet, and inanimate. The legs are usually broad and compressed. The body is short and gibbous.

This tribe is principally composed of the genus

BYRRHUS, of Linnæus.

NOSODENDRON, Latr.

Which are removed from the other Byrrhi, by having their chin entirely uncovered, very large, and in the form of a buckler. Their antennæ are terminated abruptly in a short knob, perfoliate, and with three articulations.

They are found in the fissures of trees, particularly of the elm.

BYRRHUS (Proper), Lin. *Cistela*, Geoff.

Differ from nosodendron, by their chin of the ordinary size, and enclosed, at least partially, by their presternum, the anterior extremity of which is dilated.

In some the antennæ thicken insensibly, or terminate in an elongated knob, formed of five or six articulations.

* *Megatoma rufitarsis*, Lat. Gen. Crust. et Insect. II. p. 35; *Dermestes rufitarsis*, Panz. Faun. Insect. Germ. xxxv. 6.

B. pilula, Lin. Oliv. Col. II. 13. i. 1. Three or four lines in length, black underneath, of a blackish brown, or soot-colour, and silky above, with small black spots, interrupted by others more clear, disposed in lines.

M. Waudouer has discovered the larva of a variety of this species. It is narrow, elongated, with the head thick, the plate of the first segment large, and the last two longer than the others. It remains under moss.

Another species (*Striato-punctatus*, Dej.) having the antennæ conformed in the same manner, forms, by reason of its tarsi, the first articulation of which is very small, and concealed between the lobes of the preceding, a particular division.

Another byrrhus, very small, and bristling with hair, has antennæ, terminated in a knob, with three articulations. This species forms the genus Prinodes, of MM. Megerle and Dejean. (*Anthrenus hirtus*, Fab. Panz. Faun. Insect. Germ. XI. 16).

From this consideration we may also detach from the Byrrhi some other analogous species (*B. crinaceus*, Zieg.; and *B. setiger*, Illig.) in which the knob of the antennæ is composed of but two articulations, the last of which is much thicker, and almost globular.

All the Byrrhi generally remain on the ground, in sandy places.

N. B. The genus *Murmidius* of Dr. Leach belongs, according to him, to this tribe. The antennæ have but ten articulations, the last of which forms an ovoido-globular knob.

We cannot mark the Clavicornes of our second section, though very natural, but by the union of many characters. Some of these insects are removed from all the other clavicornes, by reason of their antennæ, with nine or six articulations. These are they which, in this point of view, seem to approach the most to the following family. The antennæ of the other Clavicornes of the same section, are composed of eleven or ten articulations; but sometimes they are but little longer than the head, and form from the third articulation an almost cylindrical or fusiform knob, arched, and a little serrated. Sometimes they are almost filiform, of the length of the head and the corslet. But here, as in the most part of the other sub-genera of the same division, the tarsi are terminated by a large articulation, with two strong crotchets at the end. Those of some have but four articulations.

The body of these Coleoptera is generally ovoid, with the head sunk as far as the eyes, in a trapezoidal corslet, edged laterally, and terminated posteriorly by acute angles, the presternum dilated anteriorly (the *Potamophili* excepted), and the feet imperfectly contractile. They are found in water, under stones, near shores, and often sunk in the mud. By the construction and shortness of their antennæ, some of them (*Dryops*) have an affinity with the gyrini.

I shall divide this section into two tribes. The first, that of Acanthopoda, is remarkable for the flatted limbs, tolerably broad, and armed externally with spines. The tarsi are short, with four articulations, the crotchets of which are of the ordinary size, and the body of the animal is depressed. The presternum is dilated. The antennæ are a little longer than the head, arched, with eleven articulations, the last six of which form a knob, almost cylindrical, and a little denticulated, like a saw. The second is short, and without dilatation.

This tribe is composed of a single genus, that of

Heterocerus, Bosc. Fab.

These insects remain in the sand or in the mud, near the borders of rivulets and marshes, and come out of their holes when they are disturbed by any one walking near them, or by the sound of footsteps. The form of their legs permits them to dig in the earth, and conceal themselves there. The tarsi can be folded back upon them. It is there also that the larva lives, which the late Mr. Miger was the first to observe.

H. marginatus, Fab. ejusd. *H. lævigatus*, Panz. Faun. Insect. Germ. XXIII. is a small blackish insect, silky, with small yellowish, or reddish spots, of which the number and form vary, and sometimes even disappear on the elytra.

M. Gyllenhall remarks that the tarsi have really five articulations, but that the first is small and oblique. (Insect. Suec. p. 138.)

The second tribe, that of *Macrodactyla*, embraces the clavicornes, with simple and narrow legs, long tarsi, all composed (with the exception of a single sub-genus (*Georissus*), well distinguished from all the others of this tribe by its antennæ of nine articulations, of which the three last form an almost solid knob) of five distinct articulations, the last of which is large, with two strong crotchets at the end. The body is thick or convex. The corslet is less rounded, and terminates most frequently on each side by acute angles.

This tribe has as a principal type, the genus

DRYOPS, of Olivier,

Or that of *Parnus* of Fabricius, which is divided in the following manner:

1st.—Those whose antennæ, never scarcely longer than the head, are composed of from ten to eleven articulations, which, proceeding from the third, form a knob, almost cylindrical, or a little fusiform, arched, and somewhat serrated.

POTAMOPHILUS, Germ. *Parnus*, Fab.,

Which, without knowing the establishment of this sub-genus, we have named (Regne Animal III. p. 268.) HYDERA, have their antennæ uncovered, not lodging in peculiar cavities, a little longer than the head, with the first articulation, almost as long as the following taken together, and the second short and globular. The palpi are prominent, the mouth is entirely naked, the presternum not advancing upon it, a character exclusively peculiar in this tribe to this sub-genus.*

Dryops (Proper), Oliv. *Parnus,* Fab.,

Whose antennæ, shorter than the head, are received into a cavity situated under the eyes, and covered, in a great part, by the second articulation, which is large, dilated, in the form of a palette almost triangular, and forms a projection in the manner of an auricle. From thence comes the name *Dermeste à oreilles*, given by Geoffroy to the most common species. The palpi are not projecting.†

2nd.—Those whose antennæ, composed of eleven articulations, are filiform, or a little thicker towards the end, and at least, almost as long as the head and corslet.

Elmis, Lat. *Limnius,* Illig.

Are found in the water, under stones, or on the leaves of *nenuphar.*

3rd.—Those whose antennæ, always very short, present but nine or six articulations, and which terminate in a knob, almost solid, oval, or nearly globular.

* *Parnus acuminatus,* Fab. Panz. Faun. Insect. Germ. VI. 8; *Dryops picipes,* Oliv. III. 41. 1. 2.

† Latr. Gen. Crust. et Insect. 55. Schœn. Synon. Insect. I. ii. p. 116. The *Dryops* of Dumeril presents some differences in the length of the feet, the form of the antennæ, and the corslet, and after which Dr. Leach has thought proper to form, with this species, a peculiar genus, *Dryops.* The other species enter into that of *Parnus.*

MACRONYCHUS. Müll. Germ.

Have five distinct articulations on the tarsi; the body oblong, antennæ of six articulations, the last of which, composed perhaps of three, forming an oval knob. They are susceptible of being folded under the eyes.*

GEORISSUS. Latr. Gyll. *Pimelia*, Fab.,

In which the tarsi appear composed of only four articulations. The body is short, inflated, and almost globular, with the abdomen embraced by the elytra. The antennæ present nine articulations, and terminate in a round knob formed by the last three.†

* *Macronychus quadrituberculatus*, Müll. Illig. Mag. V. Latr. Gen. Crust. et Insect. II. p. 58. *Parnus obscurus*, Fab. Germ. Insect. Spec. Nov. I. p. 89.

† *Pimelia pygmæa*, Fab.; *Georissus pygmæus*, Gyll. Insect. Suec. I. iii. p. 675. *Trox dubius*, Panz. Faun. Insect. Germ. LXII. 5.

SUPPLEMENT

ON THE

CLAVICORNES.

On the genus which commences this extensive family of Coleopterous insects, (Mastigus,) we cannot glean any information in addition to what is given in the text, that could interest our readers. We shall, therefore, proceed to the next tribe, which is composed of the genus Hister.

This genus cannot be confounded with any other in the insect class. The antennæ prevent us from comprehending its members, with *lucanus*, *scarabæus*, *copris*, *trox*, and with which, however, they have some relation in the form of the anterior feet. And though they have some slight resemblance to *Dermestes*, *Anthrenus*, *Spheridium*, and *Byrrhus*, in the form of the body and the mode of life, they are distinguished from them by generic characters of the most essential kind.

These insects are found in cow-dung, excrement of other kinds, carcases, and in slaughter-houses, on the blood which remains there dried up. Some species live under the bark of dead or carious trees. They are to be met with during spring, summer, and, in fact, for a considerable portion of the year. Sometimes they may be seen running through the ground, on the sand, and in the highways and foot-paths; when one attempts to touch them, similar to the dermestes, and byrrhi, they glue their feet and antennæ against the body, suspend all motion, as if they were dead, and remain in this position as long as their fear may endure. Their larvæ live in the earth, in dunghills, and in carcasses.

The larva of *H. Cadaverinus* was observed by M. Latreille under dried human excrements. This larva creeps, or rather drags itself along, than walks ; it can go backwards. Its skin is so slippery that it escapes the fingers.

The genus Silpha has been confounded with *Cassida*, *Elophorus*, *Spheridium*, *Necrophorus*, and *Nitidula*. But all those genera are easily distinguished from it, by the characters which have been assigned to them.

The Silphæ sufficiently indicate by their disgusting dirtiness, and the fetid odour which they exhale, what their manner of living is, and what the character of their ordinary habitation. They are sometimes found in the fields; but they habitually seek out sombre and retired places which conceal the carcases, and excrements of animals, which constitute their food. One circumstance proves that their disagreeable odour is the effect of those animal substances in a state of putrefaction, in which they are continually groping, and on which they feed. This is, that those which are but just born, and have not yet made use of this description of aliment, have no unpleasant odour whatsoever. Thus we find, that the office which may be assigned to those insects, in the general economy of Nature, is to purge the earth of those impurities, which the destruction and decomposition of living beings must be incessantly producing. The same is the case with the great majority of the larvæ of flies, of dermestes, of necrophori, and of certain staphylini, &c.

When the silphæ are taken in the hand, they cause to issue from the mouth and anus, a drop of black and muddy liquor, the odour of which is of the most abominable atrocity. This fluid is not produced by the effect of compression, since, in proportion as it is dried up, it immediately re-appears, until its source is entirely exhausted. It doubtless serves to accelerate the putrefaction of flesh, and to prepare for these insects the nutriment which Nature has intended as most suitable for their support.

The larvæ of the silphæ live in the earth, in dunghills, but more especially in carcases. Here they are frequently to be found along with the perfect insects; they are also sometimes seen to run along the ground. The body is more or less elongated in the different species, flatted, and composed of twelve rings or segments, terminated laterally by an angle tolerably acute, and the last of which is furnished with two conical appendages. They have six short feet composed of three pieces only. The last, which appears to enclose the tarsi, is terminated by a single crook. The head is small, and armed with two strong jaws. It has filiform antennæ, a little longer than the jaws, and composed of three articulations only. These larvæ run with tolerable quickness, and are not attached to their prey like many larvæ, which, when they have consumed their provision, perish. They, on the contrary, seek for fresh food, and know how to provide for their new wants. They bury themselves under ground, for the purpose of undergoing their metamorphosis.

If this genus does not furnish a very great number of foreign species, although the known species are sufficiently great, it is, doubtless, because in the warm climates, where such insects must more especially abound, the naturalist has not been tempted to proceed in search of them, to the infectious and dangerous places which they inhabit.

The Necrophori are insects of a tolerably large size. The name was bestowed upon them by Fabricius, in reference to their peculiar habits of burying the carcases which they meet with to serve as food to their young.

The strong and disagreeable odour emitted by these insects,

announces the places which they inhabit, and the substances on which they feed. They answer the purpose, like a multitude of other insects, of consuming rotten flesh, and excrementitious matters, which would otherwise infect the atmosphere. Instinct, always in accord with organization, causes them to seek with eagerness the dead bodies of little animals, for the purposes of food. It is a truly interesting spectacle to see them attracted from a considerable distance by a cadaverous odour, associating together in their enterprize, combining their efforts, and peaceably enjoying the fruits of their labours. Thus, scarcely does the corruption of the dead body of a mouse or a mole begin to be perceptible, than they flock to the place in greater or less numbers, and hollow the earth with much activity around, underneath the animal, which sinks into it insensibly, and without our seeing the operators, the work is finished before our eyes, and the whole carcase disappears. Four or five of those insects can bury in this manner a mole in the space of four-and-twenty hours. Then, sheltered from fear and danger, they enter the body which they have interred, and feast upon it at their leisure. It is also in these carcases that the females deposit their eggs, and there that larvæ which proceed from them are destined to live.

But this subject is so very interesting that we cannot thus briefly dismiss it without a more detailed notice—and we gladly avail ourselves of the account of this curious phenomenon, given by Mr. Kirby, nor shall we do the author the injury of altering his language. He says,

"These beetles, however, in point of industry, must yield the palm to one *(Necrophorus vespillo)*, whose singular history was first detailed by M. Gleditsh in *the Acts of the Berlin Society* for 1752. He begins by informing us that he had often remarked that dead moles, when laid upon the ground, especially if upon loose earth, were almost sure to disappear in the course of two or three days, often of twelve hours. To ascertain the cause, he placed a male upon one of the beds in his garden. It had vanished by the third morning; and on digging where it had been laid, he found it buried to the depth of three inches, and under it four beetles, which seemed to have been the agents in this singular inhumation. Not perceiving any thing particular in the mole, he buried it again; and on examining it at the end of six days, he found it swarming with maggots, apparently the issue of the beetles, which M. Gleditsh now naturally concluded had buried the carcase for the food of their future young. To determine those points more clearly, he put four of these insects into a glass vessel, half filled with earth, and properly secured, and upon the surface of the earth two frogs. In less than twelve hours one of the frogs was interred by two of the beetles; the other two ran about the whole day, as if busied in measuring the dimensions of the remaining corpse, which on the third day was also found buried. He then introduced a dead linnet. A pair of the beetles were soon engaged upon the bird. They began their operations by pushing out the earth from under the body, so as to form a cavity for its reception; and it was curious to see the efforts which the beetles made, by dragging at the feathers of the bird from below, to pull it into its grave. The male having driven the female away, continued the work alone for five hours. He lifted up the bird, changed its place, turned it, and arranged it in the grave, and from time to time came out of the hole, mounted upon it, and trod it under foot, and then retired below and pulled it down. At length, apparently wearied with this uninterrupted labour, it came forth, and leaned its head upon the earth beside the bird, without the smallest motion, as if to rest itself, for a full hour, when it again crept under the earth. The next day in the morning the bird was an inch and a half under ground, and the trench remained open the whole day, the corpse seeming as if laid out upon a bier, surrounded with a rampart of mould. In the evening it had sunk half an inch lower, and in another day the work was completed, and the bird covered. M. Gleditsh continued to add other small dead animals, which were all sooner or later buried; and the result of his experiment was, that in fifty days four beetles had interred, in the very small space of earth allotted to them, twelve carcases: viz., four frogs, three small birds, two fishes, one mole, and two grasshoppers, besides the entrails of a fish, and two morsels of the lungs of an ox. In another experiment a single beetle buried a mole forty times its own weight and bulk in two days. It is plain that all this labour is incurred for the sake of placing in security the future young of these industrious insects, along with a necessary provision of food. One mole would have sufficed a long time for the repast of the beetles themselves, and they could have more conveniently fed upon it above ground than below. But if they had left thus exposed the carcase in which their eggs were deposited, both would have been exposed to the imminent risk of being destroyed at a mouthful, by the first fox or kite that chanced to espy them."

The same learned author observes, that this insect must of necessity be gifted with a very acute sense of smelling, which enables it to scent out putrid carcases at a remarkable distance. What seem to him to be the organs of smell we have already noticed in our Supplement on Insects in General.

The larvæ of the necrophori are long, of a greyish white, with the head brown. Their body is composed of twelve rings, furnished anteriorly at their upper part, with a small scaly plate of a ferrugineous brown. The plates of the last rings are provided with small elevated points. The head is hard, scaly, armed with mandibles tolerably strong and trenchant. They have six scaly feet, very short, attached to the first three rings of the body. When arrived at their full growth they sink into the earth, to the depth of more than a foot, form for themselves a sort of oval box, which they invest with a kind of gluey substance to consolidate its parietes, and there they become changed into the nymph-state. The perfect insect comes out at the end of three or four weeks.

The Necrodes are closely allied to the last sub-genus, and present nothing worthy of remark here.

Passing over the intervening divisions, which present nothing worthy of remark, we come to the genus Nitidula, a name derived from *nitidus* (shining), and given to these insects by Fabricius, for no very assignable cause. It is but little suited to them in reference to their external forms, but M. Latreille has retained it in consequence of the generality of its adoption.

The Nitidulæ differ from Silpha in their bifid mandibles, and toothless jaws; from Dermestes in the same mandibles, the antennæ, the figure and proportions of the articulations of the tarsi, and the general form of the body. They approximate more to the *peltis* of Fabricius, to Colobicus, to Byturus, and to Cercus.

The nitidulæ are found in carcases, or dried animal substances, under the rotten bark of old trees, in mushrooms, and even upon flowers. The species which frequent flowers fly more frequently than those which are found on carcases, or under the bark of trees. The sombre, obscure, or at all events not very brilliant colours of the great majority of the nitidulæ, contrast a little with the generic name which they have received. Yet we cannot help thinking, notwithstanding its unsuitableness, that its retention was more judicious than the substitution of another, which, without adding to our knowledge, always tends to increase a confusion in nomenclature, which unfortunately exists already in a far too great degree.

The larvæ of Nitidulæ very much resemble those of Silpha. Their body is flatted, of an oval elongated form, composed of twelve rings, terminated laterally in a rather acute angle. The last ring, similar to that of the larvæ of Silpha, is furnished with two small conical appendages. These larvæ have six short and scaly feet. When arrived at their full growth they bury themselves in the ground, for the purpose of undergoing their metamorphoses. The nitidulæ constitute altogether a very numerous genus. When the sap flows from wounds in a tree, it is, says Mr. Kirby, attended by various beetles, and among others, the nitidulæ are very busy in this way. Their operations prevent the tree from healing as soon as it otherwise might do. If the bark be any where separated from the wood, these *subcortical* insects are sure to take their station inside it, and prevent a re-union of the bark and tree. The *nitidula grisea* is remarkable for its ravages in this kind. These insects will devour the wood of the willow and sallow, which in the course of time are hollowed to such an extent by those depredators, as to be very easily blown down.

The Byturi have very great relations with the last mentioned insects, and also with the *Cerci*, differing, however, from the latter, in the length of their elytra, which cover almost entirely the major part of the abdomen. They are very commonly found on flowers in the spring season, throughout all Europe.

The Dacne was at first confounded with his Ips, by Fabricius, a genus of Coleoptera very different from it. These insects creep upon the earth, or sand, and are not unfrequently found under stones.

The Dermestes are insects which have been for a long time known in consequence of the ravages which their larvæ commit upon articles of the most valuable kind. The perfect insect seems to live only for the purpose of fulfilling its final destination. It is often found on flowers, and if it frequents animal substances, it is more with the design of depositing its eggs there than to commit any fresh depredations. The voracity of the larvæ of Dermestes is especially destructive to cabinets of natural history, and collections of furs. They entirely destroy the birds, the quadrupeds, the insects, and all other prepared animals which are preserved in such places. They destroy furs, from which they cause the hairs to fall, and even gnaw the skin itself. They also attack the carcases of animals of every species which are to be found in the fields, and consume all the fleshy substance and tendinous parts, dissecting them even to the bone, and reducing them to complete skeletons. They are to be found in offices, cupboards, and in short, in every place which may conceal the animal food which is suitable to their constitutions. Bacon, feathers, horn which may have been left for a long time in drawers, are not spared by these robbers. It is exceedingly difficult to guard against their ravages. They escape research in consequence of their minuteness, and their perseverance bids defiance to precaution.

The majority of the Dermestes seek out retired places, and generally such as are neglected and dirty, and seem to shun the impression of the light. They are attached to repose, and seldom put themselves in motion but when they are disturbed by any noise around them, or touch those bodies which contain them. They are seldom seen on the surface of such bodies. Buried in the interior, they withdraw themselves from our regards, and seem to quit their retreat with fear and trembling. Their gait is timid and uncertain. When we are habituated to reflect on the external signs of those passions which predominate over ourselves, and consider the Dermestes, at the slightest aspect of danger, running backwards and forwards, and endeavouring to escape; at the least touch suspending its walk, or drawing back its antennæ and feet, remaining obstinately in a state of feigned death, and wishing to impose, by an appearance of firmness, or deceive by stratagem; we may fancy that we recognize all the combined movements which fear and reflection can inspire in the principle of self-preservation.

The body of the larvæ of these insects is not very hairy, and is composed of twelve very distinct rings. They have a scaly head, provided with very hard and trenchant mandibles. They have six scaly feet, terminated by a claw. The extremity of their body is remarkable for a tuft of very long hairs. They have two antennæ, and some very short barbles. They change their skin several times, and the spoils remain entire. When they are about to change into nymphs, they seek some retired place, contract their bodies, and without wearing any shell, change into the perfect insect in the course of some little time. It is towards the end of summer that these larvæ acquire their full growth, and then cause the greatest degree of destruction in collections of natural history, and in furrieries.

It would be very much to be desired to discover efficacious means of removing the Dermestes, and other destructive insects, from collections of animals exposed to their ravages. All dealers in objects of natural history imagine, or pretend, that they are in possession of secrets, the efficacy of which, according to them, is always infallible. But every one is aware of the small degree of confidence which ought to be reposed in every thing which is announced as a secret. Nevertheless, a French apothecary, M. Becœur, has invented something of this kind, which has been tried with sufficient success to merit a preference, and obtain publicity. Its preparation is this:—Take of quick lime half an ounce, of salt of tartar a drachm and a half, of camphor five drachms, of arsenic four ounces, of white soap four ounces; dissolve the camphor in a sufficient quantity of spirit of wine; pound up the soap along with it, and preserve the whole mixture in a glass vessel, properly stopped, to be used as occasion may require.

The BYRRHI are oval, almost globular insects, which have very great relations with Dermestes, Anthrenus, and Spheridium, being only distinguished from them by the form of their antennæ. When these insects are touched, they withdraw their head into the corslet, apply their antennæ and feet against the body, and counterfeit death. They remain for some time in this position, after which they resume their walk. They are to be met with in the fields, on the edges of roads, and other such like places. They very rarely employ their wings.

The most common species of this genus is *Byrrhus pilula*, which is found throughout Europe, in fields and pathways, and in sandy places.

The HETEROCERI frequent the borders of marshes, and dig holes there with their feet for the purpose of concealing themselves. But they very speedily come out of them if the earth about them be disturbed.

The GEORISSI have some relations with Byrrhus. They are very small insects, and seem to frequent, by preference, humid and aquatic situations.

We have now noticed all that is of any popular interest in this extensive family of Coleoptera; but as our readers will see, we have been under the necessity of passing over many of its sub-genera, as affording no materials for this portion of our work.

THE FIFTH FAMILY

OF THE

PENTAMEROUS COLEOPTERA.

PALPICORNES

Present us, like the preceding, with antennæ terminating in a knob, and usually perfoliate, but with nine articulations at the most in all, inserted on the lateral and advanced edges of the head, but little longer than it and the maxillary palpi, and often even shorter than these latter organs. The chin is large and buckler-formed.

The body is, in general, ovoid, or hemispherical, gibbous, or arched. The feet are in many adapted for swimming, and have then but four or five very distinct articulations, but the first of which is much shorter than the following; all the articulations are entire.

Those whose feet are proper for swimming, with the first articulation of the tarsi much shorter than the following, and whose joints are entirely corneous, will compose a first tribe, that of HYDROPHILI, which embrace the genus

HYDROPHILUS of Geoffroy.

Linnæus has formed of it but a single division (the first) of his genus *Dytiscus;* but the anatomy of these insects differs essentially. The digestive canal of the hydrophili has much analogy from its length, exceeding four or five times that of the body, and from its contexture, to that of the lamellicornes, and does not approach to that of the carnivorous coleoptera, but in the relation of the biliary vessels. They have neither the natatory bladder, nor the excrementitious apparatus, which characterizes the hydrocanthari. In the females alone, this apparatus is replaced by the organs secreting the matter, which is proper to form the cocoon which encloses the eggs, and the anus presents, for this effect, two filaments. In fine, the masculine genital organs have the greatest relations with those of the coleoptera of the preceding family.*

Some, whose body is sometimes oval, oblong and depressed, or elongated and narrow, with the corslet uneven, and contracted posteriorly, the legs slender, furnished with small spurs, and the tarsi filiform, little or feebly ciliate, and terminated by two strong crooks, have antennæ (always composed of nine articulations,) finishing in a knob, almost in the form of an inverted cone, slightly perfoliate, or almost solid, and the extremity of the mandibles entire or terminated by a single tooth. These palpicornes are all very small, swim little or badly, inhabit stagnant waters, and sometimes remove from them, to conceal themselves in the earth, or under stones. They compose the family of the HELOPHORIDEA of Dr. Leach, a denomination which reminds us of the genus ELOPHORUS of Fabricius.

In some, the length of the maxillary palpi does not exceed that of the antennæ, or is even inferior to it. The chaperon is entire, or without any remarkable emargination.

Sometimes the maxillary palpi are terminated by an articulation which is thicker and ovaliform.

* "The conformation and structure of the male genital organs of the palpicornes fully justify the place which M. Latreille has assigned them in the entomological scale." (Leon. Dufour, Ann. des Sc. Nat. VI. p. 172.)

ELOPHORUS, Fab. *Silpha*, Lin. *Dermestes*, Geoff. *Hydrophilus*, D. G.

Have the body oval, the corslet transverse, and the eyes but little raised. These are the Elophori of Fabricius, with the exception of the following sub-genera.

HYDROCHUS, Germ. *Elophorus*, Fab.,

Which are distinguished from the preceding, only by their narrow and elongated form, the corslet in a long square, and the prominence of their eyes.*

Sometimes the maxillary palpi terminate in the manner of an awl, or by a more slender, short, and conical articulation.

OCHTHEBIUS, Leach. Germ. *Elaphorus*, Fab. *Hydræna*, Illig. Latr.

The corslet is almost semi-orbicular.†

In others, the maxillary palpi, terminated by an articulation larger than the preceding, in the form of a spindle, and pointed at the end, are much longer than the antennæ and the head. The chaperon is strongly emarginate. They have otherwise the appearance of the ochthebii.

HYDRÆNA, Kugel, Leach.‡

The other Hydrophilii have the body ovoid, or almost

* *Elophorus elongatus*, Fab.;—*E. crenatus*, ejusd.;—*E. brevis*, Gyll. See Germ. Insect. Spec. Nov. I. p. 90.

† *E. Pygmæus*, Fab.; *Hydræna riparia*, Latr.;—*Hydræna margipallens*, Latr.; *Elophorus marinus*, Gyll. See Germ. *ibid.* p. 90.

‡ *E. minimus*, Fab. Gyll.; *Hydræna riparia*, Kugel; *H. longipalpis*, Schœnh. Germ. Faun. Insect. Europ. VIII. 6. See for the other species, Germ. Insect. Nov. Spec. I. p. 93.

hemispherical, and generally convex or gibbous, with the corslet always more broad than long, and smooth, the legs terminated by strong spurs, and the tarsi very frequently ciliated. The extremity of their mandibles presents two teeth. They embrace the family of the HYDROPHILIDEA of Dr. Leach, or the genus hydrophilus of Fabricius.

Some have but six articulations in the antennæ, and their chaperon is emarginated. Such are

SPERCHEUS, Fab.

Spercheus emarginatus, Fab. Panz. Faun. Insect. Germ. XCI. 4. M. Bourdon, a French naturalist, who is at present exploring the States of the Republic of Columbia, first discovered this species in the neighbourhood of Paris.

In the following, the antennæ are always composed of eight or nine articulations, and the chaperon is entire, or slightly concave at the anterior edge.

One species which has been communicated to us by our friend Dr. Leach, presents some singular characters, which have determined me to consider this insect as the type of a new sub-genus,* that of

GLOBARIA,

Which I shall name thus, because its body is almost spherical, compressed laterally, and it appears capable of putting itself into a ball, after the manner of the agathidii. Its

* It seems to come more naturally near the Berosus of Dr. Leach; but in consequence of the number of articulations in the antennæ, I have deemed it proper to place it immediately after the Sperchei. One might, however, reverse this order, commencing with this sub-genera, which have nine articulations in the antennæ, and terminating by those in which they have one and three less, or by Globaria and Spercheus.

antennæ appear to me to be composed of but eight articulations, the fifth of which is dilated in the manner of a spine at the internal side, the following in an inverted cone, elongated, the seventh cylindrical, and the last or eight, conical. These last articulations form a very elongated knob, almost cylindrical, and terminating in a point. The maxillary palpi are a little shorter than the antennæ. The eyes are bulky and projecting. The corslet is almost semi-lunar. The elytra embrace the abdomen entirely. The breast is without a sternal spine. The four posterior legs, have at their extremity a bundle of silky hairs, almost as long as the tarsus. The scutellum is small, and in an elongated and narrow triangle.

The only known species (*G. Leachii*) is small and exotic. I believe it comes from South America.

All the other hydrophillii have nine articulations to the antennæ, and the knob is ovaliform or ovoid. The body is not susceptible of being contracted into a ball.

The largest species have the two intermediate articulations of the antennary knob, or the seventh and eighth, in the form of a kidney, or irregular crescent, obtuse at one of their ends, elongated, arched, and pointed at the other, with a vacancy or remarkable separation between them; the first of this knob is cupular, more elongated on the anterior side. The middle of the sternum is raised into a keel, and terminates posteriorly in a point, more or less long and very sharp. The maxillary palpi are longer than the antennæ, with the last articulation shorter than the preceding. The tarsi, especially the last, are compressed, garnished with a fringe of hairs on the internal side, and terminated by two generally small crooks, unequal and unindented below. The scutellum is tolerably large.

These species will compose the genus,

Hydrophilus (properly so called), Geoff. Fab. Leach. *Dytiscus*, Lin.

Here the sternal spine is strongly prolonged behind. The last articulation of the anterior tarsi of the males is dilated in the manner of a triangular palette. The scutellum is large. These are the *Hydrous* of Dr. Leach. Zool. Misc. III. p. 94.

The larvæ resemble a sort of worms, soft, with a conical and elongated form, provided with six feet; with the head tolerably large, scaly, more convex underneath than above, and armed with strong and crooked mandibles. They respire by the posterior extremity of the body. They are very voracious and injurious to ponds, by devouring the spawn.

H. piceus, Fab. Oliv. Col. III. 39. 1. 1., is an inch and a half long, oval, of a dark brown, polished, as it were, or incrusted with a varnish, with the knob of the antennæ partly reddish, and some striæ, not greatly marked, on the elytra, the posterior extremity of which is rounded, externally, and prolonged into a little tooth at the external angle.

It swims and flies remarkably well, but walks badly. Its sternal point may sometimes wound when held in the hand and allowed the liberty of moving.

The anus of the female has two filaments, with which she forms an ovoid shell, surmounted with a point, in the form of an arched horn, and of a brown colour. Its external tissue is a gummy paste, liquid at first, and subsequently hardening and becoming impenetrable to water. The eggs which it envelopes are disposed there with symmetry, and held together by a sort of white down. These eggs float on the water.

The larva is depressed, blackish, wrinkled, with the head of a reddish brown, smooth, round, and capable of being turned backwards. This faculty affords it the means of seizing the small shell-fish, which swim on the surface of the water. Its back serves it as a sort of resting, and in this sort of table it breaks the shells, and devours the animals which they enclose. The body of these larvæ becomes flaccid when they are taken up. They swim with facility, and have, underneath the anus, two fleshy appendages, which serve to sustain them on the surface of the water, the head being under, when they come there to respire. According to M. Miger, who has furnished us with these observations, (Ann. du Mus. de l'Hist. Nat. XIV. 441.) other larvæ of the hydrophili are deprived of these appendages, and do not swim, and suspend themselves like the preceding. The females of these species swim with difficulty, and carry their eggs under the abdomen in a silken tissue; but these species belong to the last sub-genera of this tribe.

That of the *hydrophilus* proper, of Dr. Leach, is composed of species whose tarsi are identical in both sexes, and not dilated, whose pectoral spine terminates with the posternum, and whose scutellum is proportionally smaller.*

In all the following hydrophilii, the two intermediate articulations of the knob of the antennæ are perfectly transverse, of a regular form, not prolonged in the manner of a tooth, at one of their ends, and without a vacancy between them. The last is obtuse or rounded at the end. The breast presents neither keel nor spine. The tarsi are much less adapted for swimming, little or not at all ciliate, and terminated by large crooks equal and simple.

Those whose maxillary palpi are much longer than the antennæ, with the last articulation shorter than the pre-

* Refer to the Hydroüs of Dr. Leach, beside *piceus*, the following species of Fabricius: *ater*, *olivaceus*, *rufipes*, &c. Those which the latter names *caraboïdes*, *ellipticus*, &c. are the hydrophilii proper of the English naturalist.

ceding, ovoid, cylindrical, whose body is but little raised, with the end of the elytra truncated, or very obtuse, compose the genus,

LIMNEBIUS of Dr. Leach.

Those whose maxillary palpi are but little longer than the antennæ, with the last articulation as long, or longer than the preceding, almost ovaliform, and whose body is gibbous, are comprised by the same English naturalist in two other genera. The one, that of

HYDROBIUS,

Has the eyes depressed, or but little convex. The anterior extremity of the head is not narrowed abruptly, and the base of the corslet is of the breadth of that of the elytra.†

BEROSUS

Have, on the contrary, very prominent eyes, the anterior extremity of the head abruptly narrowed, and the corslet more narrow at its base than the elytra. The body is very gibbous. (*H. luridus*, Fab.)

The second tribe, SPHŒRIDIOTA, is formed of terrestrial palpicornes, with tarsi composed of five very distinct articulations, and the first at least as long as the following. The maxillary palpi are a little shorter than the antennæ, with the third articulation larger, inflated, and in the form of a reversed cone. The maxillary lobes are membranous.

The body is almost hemispherical, with the presternum prolonged into a point at its posterior extremity, and the legs spinous. The anterior ones are palmate, or digitate in the large species. The antennæ are always composed of nine articulations, or simply of eight, if we consider the last as

* *H. griseus, truncatellus*, Fab.

† *H. scarabœoides, melanocephalus*, &c.

an appendage of the preceding. (See *Elater* and many other genera of coleoptera.)

These insects are small, and inhabit dung and other excrementitious substances. Some species remain near the edge of waters.

They compose the genus,

SPHÆRIDIUM, of Fabricius,

But from which it is necessary to separate many species, which Olivier has already done. Dr. Leach even retains only those in which the anterior tarsi are dilated in the males. Such is,

Dermestes scarabœoides, Lin.; Oliv. Col. II. 15, 1 and 3, II. ii. It is of a shining black, smooth, with the scutellum elongated, the feet very spinous, a spot of blood-red on the base of each elytrum, and their extremity reddish. These spots diminish, or are obliterated in many individuals.

The species whose tarsi are similar in the two sexes, and in which the knob of the antennæ is loosely imbricated, compose the genus CERCYDION of this naturalist. We might, from the consideration of the form of the legs, the disposition of their spines or denticulations, divide the sphœridia into many other sections, which would facilitate the study of the species, the number of which appears to have been too much multiplied.

SUPPLEMENT

ON THE

PALPICORNES.

THE first tribe of this family, the *Hydrophilii*, is composed of the great genus HYDROPHILUS, which Geoffroy separated under this name from the *Dytiscus* of Linnæus. They form a genus composed of a good many species, the major part of which inhabit Europe. Some of them are of large size; others under the middle size. The colours of all are obscure, or of no great brilliancy.

These insects are found in fresh waters, in lakes, rarely in rivers, but most frequently in marshes and ponds. They swim tolerably quick, but with less celerity than the Dytisci. It is usually at the approach of night that they issue from the water to fly and transport themselves from one marsh or pond to another. Accordingly these insects are found, as well as all those of the tribe of *Hydrocanthari*, in the smallest collections of water, even in those which are formed by rain in the inequalities of the soil. They make a humming sound as they fly along, similar to that made by the Scarabœi.

Degeer has informed us, that the Hydrophili, as well as the *Dytisci*, are carnivorous, and very voracious, and that they live on such other aquatic and terrestrial insects as they can catch. This fact, which the examination of the internal parts of these insects would seem to contradict, has, nevertheless, been confirmed by M. Felix Miger, in a Memoir on the Metamorphoses of the Hydrophili, inserted in the 14th Vol. of the Annals of the Museum of Natural History of Paris. He has fed, for more than a month, the *H. caraboides* with water-snails. He has also seen *H. piceus* devour, with avidity, these same mollusca and aquatic larvæ. But he also remarks, according to his own actual observations, that its principal food consists of aquatic plants, which might lead us to believe that it is carnivorous only from circumstances, or rather that it is omnivorous. Its intestinal canal has been opened, which is very long, and it has always been found filled with the debris of vegetable substances.

In the larvæ, on the contrary, the intestines are so short, that they do not exceed in length that of the entire body. The hydrophili, in the perfect state, have an additional attribute, which characterizes the herbivorous insects. This is, that they are provided with two very long cœca, which end in the intestinal canal, near the anus, and are folded on themselves, in a spiral figure, in the cavity of the abdomen.

The nature and form of the masticatory apparatus also seem to furnish a fresh proof that these coleoptera are herbivorous. Yet some insects, though herbivorous, are found to be provided with solid jaws, and strongly denticulated.

Although the hydrophili can live a long time under water, they are, nevertheless, obliged to respire the air from time to time. They bring themselves to the surface of the water, and to arrive there they have only to hold their feet in a state of repose, and suffer themselves to float. Being lighter than the water, they swim above it at once. The hinder part of the body is then applied to the surface of the water, and seen a little above it. They then raise their elytra a little, or lower the end of the abdomen, so that it forms a vacancy between the elytra, and the hinder part of the body, which thus remains dry. The external air then penetrates easily between the elytra and the abdomen, so that the air easily enters there. It is carried to the stigmata placed underneath the elytra, along the two sides of the abdomen. When the insect is desirous of returning to the bottom of the water, it quickly approaches the abdomen to the elytra, and stops at the same time the vacancy which is between them, so that the water can never penetrate it.

Lyonnet has informed us of a very singular fact, the truth of which is no longer to be doubted, after the excellent observations of the naturalist, whom we have quoted a little above. It is, that these insects can spin, and make a sort of nest or shell of silk, of an ovoid figure, in which they lay and enclose their eggs. Degeer has found similar nests floating on the water, and filled with eggs, from which there issue forth small larvæ, which cannot be mistaken for any other than those of hydrophilus or dytiscus.

He was never able to catch the moment in which they were engaged in the construction of shells of this kind, but Lyonnet was more fortunate, and informs us that he has beheld the insect working at these shells, that it makes them with the hinder part of the body, and that it adds a sort of brown horn, a little curved, and solid. The use of this horn appears to him to be to retain the shell, when a gust of wind, or any other accident might overturn it.

These facts, as well as some others, are well developed in the memoir of M. Miger. He speaks at first of the mode of sexual intercourse in the hydrophili, which takes place in the usual manner of other insects; but the male crooks himself to the external edge of the elytra of the female, and supports himself by means of the last articulation of the anterior tarsi, which is formed like a triangular palette, or a trowel. This observer was witness of the manner in which the female lays its eggs, and the singular shell which encloses them. We shall translate his account of these phenomena, as the subject is one of very great curiosity and interest.

" I beheld her attach herself to the reverse side of a leaf, which was floating on the water, place herself across it, and elongating her first pairs of feet, rest them on the upper part of this leaf, and on each side, so as to make it assume a slight curve. The abdomen was strongly applied to the reverse of this leaf, and there were left visible at its extremity, two appendages, which advanced and retired with quickness, and from which there appeared to issue forth a white and gummy fluid. This fluid was the principle of the shell, and the appendages were the instruments of spinning in the hydrophilus. Considering these instruments more attentively, I saw that they deposited here and there, under the leaf, around the abdomen, and without passing it, some silvery threads, which applied successively on each other, formed a small semi-circular pouch, in which the extremity of the abdomen was, as it were, engaged. At the end of about ten minutes, the hydrophilus, withdrawing its feet from the upper part of the leaf, turned itself round quickly, and placed its head below, without, on that account, taking away from the shell the extremity of the abdomen.

In this new position the insect kept itself nearly motionless, the four anterior feet being extended, and the two others strongly attached underneath the leaf, and on each side of the shell. During nearly an hour and a half, I distinguished easily through the tissue, all the movements of the spinning instrument. It was a pencil or brush, with two slips, or thin tufts, which proceeded from right to left, and from top to bottom, with much agility, in the interior of the shell, and which invested the parietes and the external edges of it with this gummy fluid of which we have been speaking. This shell worked in this manner, and always inside, grew, thickened, and finally became so compact, that it was no longer possible for me to distinguish the movements of the spinneret. In the mean time some small bubbles of air commenced to escape from the interior of the shell; I thought that it was the eggs which occasioned this displacement; in fact, at the moment in which the hydrophilus removed its abdomen from the extremity of its elytra, I approached to it a strong magnifier, and I perceived distinctly small oblong and whitish bodies, which were placed side by side, and which the spinnerets gradually covered with a white and transparent fluid. In three quarters of an hour the business of laying was over. The insect withdrew its abdomen by little and little from the under part of the leaf, closed its shell rather imperfectly, and assumed a new position.

It remained for it to form the point which terminates this shell. To bring this about, the hydrophilus, always having the head underneath, brought back its hinder legs on the leaf, and placed them on each side of the shell. The elytra, the extremity of which was on a level with the water, were removed from the abdomen, and passed, by some lines, the anus, which was very much dilated. The spinnerets were no longer concealed by any thing. One might easily follow all their movements; those were continual and rapid. It required, however, more than half an hour for the hydrophilus to form this point. The insect carried here and there, above the shell, and on the edge of the leaf, a slender and yellowish thread, which, at the very instant, acquired some degree of firmness. Speedily new layers were applied on the first, and as the last always exceeded the preceding by some lines, there was insensibly formed a slender and conical appendage, of a citron-yellow colour, which was raised about an inch above the surface of the water. This labour being achieved, the hydrophilus directed its spinneret lightly, from top to bottom, along the point, and bringing back by degrees, its entire body under the water, it abandoned the shell, which from this moment was completed. The whole of these labours lasted for about three hours."

M. Miger has seen three hydrophili thus spinning their shells under his immediate inspection. But he was unable to follow them but once in their first labour, because it is difficult, at this time, to observe, without interrupting them; but this is not the case when the oviposition has commenced. The insect may be drawn out of the water, and even a portion of the shell removed, without the hydrophilus shewing any apprehensions of the observer, or discontinuing its operations.

Three females which this naturalist had put into a vessel filled with water, but without any foreign body which was proper to serve as a fixed point for their shells, did not spin, but laid, however, all the three, a sort of cartilaginous, oblong, and yellowish shell, of the bulk of a grain of barley, and which, being detached from the anus, fell to the bottom of the vessel. Having opened them at the end of some days, he found then neither eggs nor fluid. Not being certain whether these individuals spin their shells, he does not venture to decide, if these last are the production of an abortion, or a mass of superabundant fluid. But as they did not enclose any egg, and as they were much smaller than the ordinary shells, this last opinion appears the most probable.

The spinnerets are scaly, conical threads, two lines in length, and composed of two articulations, the first of which is of a clear fawn-colour, spotted with brown, and the second of this last colour, and much smaller. It is terminated by a white and transparent lash. Two other conical appendages, but fleshy and inarticulated, are placed as well as the preceding between two sorts of corneous and semi-circular lips, terminating the last ring of the abdomen. The fleshy portion of this ring, by the facility of contraction and dilatation which it possesses, and by the continued movements in all directions which the insect can communicate to it, concurs principally to the execution of the work. It is it which puts in action the two spinnerets.

The air which the hydrophili possess the faculty of keeping in reserve, under their elytra, and which is often observed in the form of a bubble, serves them for the purpose of respiration while they are forming their shells, and preserves their eggs from the dangerous influence of the water. It has been said that these shells floated isolatedly on the water, and that the horn or point which terminates them, served as a sort of mast for this little skiff. This, however, only happens when they are empty, for M. Miger has found always, that a shell filled with eggs is overturned by its own weight, and that the insect has always need of a support, such as dry or living plants, to lay the foundations of its edifice. He thinks, with reason, and as had been already suspected, that this point of the shell is destined for the introduction of air. The examination of the nature of the kinds of fluids, of which the shells are composed, seems also to strengthen this opinion. The liquid which serves to form the external tissue is a sort of liquid and gummy paste, which is melted, and amalgamates with all that surrounds it, and which becomes, by dessication, a tolerably flexible envelope, but at the same time so compact, that it hinders the water from penetrating into the interior. The second fluid, that which envelopes the eggs at the instant of their emission, is a slight down of great whiteness, which keeps them in their places, and without which they might be bruised. Finally, the fluid of which the point is composed, is a silky, porous, and lustrous tissue, very like that of the shells of the lepidoptera. It appears proper for the introduction of the air, and the water penetrates it with facility from the moment it is submerged. The shell is ovoid, whitish, with the point of a deep brown, and which, flat at first upon one side, becomes rounded in drying, and tubular in all its length. At its base is the aperture prepared for the issue of the larva. It is closed by some threads, which, by means of the air enclosed in the shell, hinder the water from introducing itself. It happens, however, that some eggs, placed near this aperture, are not always preserved, and they become decomposed. The interior of the shell, after we have removed that part opposite to that which is nearest to the surface of the water, presents from forty-five to fifty little cylinders, slightly inflated, and curved towards their summit, of the length of two lines, grouped, in crossing, at the middle of the shell, all in a vertical position, and having each a particular lodge, formed by the cottony down of which we have spoken. This down retaining the eggs, is attached to the upper part of the shell, and leaves, at the lower part and all around, a void space which extends as far as the aperture. From this disposition of the eggs, we may well conceive how the shell, a little isolated, is retained upon the water in the position which is natural to it.

The egg undergoes a sort of development. It swells, assumes a brown and shining tint, and it is possible to distinguish the form of the larva, and particularly the eyes. The pellicle of the egg breaks, and we see this larva issue forth, which is twice as large as the egg, and begins to stir about in all directions. It disengages itself with so much the more facility, because there is an empty space above the head. All the larvæ have this part of the body directed to this side. Thither they repair at their birth, and there they rest during more than twelve hours without taking any nourishment, and are stirring about one against the other. They leave no vestiges of the cottony cases which kept them separated. M. Miger has several times observed that the shells become more heavy immediately after the birth of these larvæ. He explains this fact, by supposing that there is then introduced by the points, a greater quantity of air.

The eggs disclose the young generally in the space of from twelve to fifteen days, at least in the month of May, which was the time in which M. Miger made his observations. The young larvæ have scarcely quitted their nest, than they are seen to re-enter there, issue forth anew, and sport all around it, up to the moment when the want of nourishment obliges them to separate.

These larvæ change their skin several times in the water, and in the usual manner. Those which M. Miger attempted to rear, at their issuing from the egg, perished, probably from not having had the sort of nourishment which was suitable to them. But he preserved many of those which he had taken at a more advanced age, and which were more robust. They did not make at first, when they were seized, any motion to defend themselves; but soon after they made themselves so flaccid, that he imagined that he was touching nothing but a simple spoil. Taken by the tail, they immediately contracted themselves, and lost nearly one third of their length. They became quickly agitated, and squirted forth, with a slight noise, from the posterior extremity of their body, a fetid, and blackish water. We know that many other aquatic larvæ have the faculty of drawing in, and rejecting water by their hinder part. The larvæ of the libellulinæ make use of it to carry them forwards. Those of the dytisci thus humectate their bodies when, being out of the water, they are exposed to too great a heat. It appears that those of the hydrophili employ this means for defending themselves.

M. Miger put these larvæ into fresh spring water. They swam there with great facility, agitating their feet in a lively manner. They often came to the surface to respire, applying there the posterior extremity of their body, where the organs of respiration are situated. They keep them even on a level with the water when they are in a state of repose. They then hook themselves to aquatic plants, and often one to another, extending themselves horizontally. Those which the naturalist, just mentioned, reared, never fought amongst themselves. They lived on aquatic insects, and particularly on water-snails, to which they are very partial. Lyonnet has observed that their head is a little inclined backwards, to enable them to seize the mollusca with more facility, and that their back serves as a resting-point on which to break the shell, and a table on which to eat the animal. After having seized it with their mandibles, they bend themselves backwards, raise the back a little, and rest the snail upon it. In this attitude, with their head a little turned, it bears more perpendicularly on the shell, which gives them a greater facility in breaking it, and then eating the animal.

M. Miger confirms the truth of these observations. He has given to these larvæ some small pieces of raw meat, on which they have lived during a longer space than fifteen days.

When they are on the point of passing into the nymph state, they issue from the water, gain the shore, and employing their mandibles and feet, they dig for themselves in the earth an almost spherical cavity, very smooth, about eighteen lines in diameter, and presenting no aperture. Their body, in this retreat, is placed upon the belly, and curved into an arch. They still preserve their form for about the space of ten days. Their skin afterwards opens on the back, as far as the fourth ring, commencing from the head, and the nymph thus makes itself a passage. It is thirteen or fourteen lines in length, whitish, terminated by forked appendages, and presents, at each of the anterior angles of its corslet, an aigrette of three corneous and recurved hairs. Its body, the head of which is inclined under the corslet, and its abdomen a little curved, is supported in this lodge only by means of these various appendages. No other part is in contact with the interior parietes of its dwelling, which circumstance, according to M. Miger, preserves it from an injurious degree of humidity. If this nymph be turned on the back or side, it soon re-establishes itself in its natural position, by the movements of the rings of its body. But it does so with difficulty in a cavity of a different curve.

The nymph state endures about three weeks, during which the corneous parts become coloured by little and little. Turned upon its back, and causing its feet and the rings of its body to move, the insect contrives to disengage itself from its envelope, which was at first cleft at its superior part. The elytra being applied under the belly, go and place themselves over the back; the wings are unfolded and remain extended, until they have acquired a certain degree of consistence. The insect soon draws them under the cases, which are still white and soft, and places itself upon its feet, which are as yet but infirm. At the end of twenty-four hours it has acquired the brown colour which is proper to it; but it still remains for twelve days in the earth, and without moving. When this period has elapsed, it has acquired all its strength. The flexibility of its elytra, the compressibility of its body, and the play of its other external organs, impart to it the means of breaking its prison. It escapes through a very small aperture. According to the very interesting observations of M. Miger, the total duration of these metamorphoses, beginning with the oviposition, is about ninety-eight days; of this number the insect has passed sixty days in the larva form.

According to this skilful naturalist, the larvæ of the hydrophili are of two kinds. The first, which he designates under the name of *swimmers*, (*nageuses*,) have near the respiratory organ, short and fleshy appendages, and which, when these larvæ come to the surface of the water to respire, serve to keep their head down, and the body, as it were, suspended by its posterior part. They swim with agility. The other larvæ, which he calls *rampant*, are deprived of these appendages, do not swim, and remain constantly on a level with the water. They do not suspend themselves like the first, but turned on the back, they proceed onwards with considerable swiftness, and make horizontal and vermicular movements. Both these sorts of larva undergo their metamorphoses in the earth. Their nymphs have always two small appendages at the posterior extremity of the body, and some corneous hairs or aigrettes at the anterior edge of the corslet. The lateral tubercles of the body of the larva are very much diversified in their forms and dimensions. But these modifications have not appeared to M. Miger, to be connected with any sensible changes in their habits. The hydrophili, which are essentially swimmers, may proceed from larvæ placed in these two divisions. Their females spin the shell or cocoon, of which we have given the description· But the species which swim with difficulty, appertain, in general, to the second division. Their females carry under the body a silky tissue enveloping the eggs. They have filiform tarsi, less appropriate to the purposes of natation than those of the species of the first, with the crotchets of the tarsi entire or scarcely unidenticulated.

The Elophori differ from *Silpha* in the antennæ; from *dermestes*, in the mandibles and jaws; from *hydrophilus*, in the mandibles, jaws, palpi, and tarsi.

The Elophori are small insects which live in the water, and usually swim at the surface, where they remain on duck-weed, *confervæ*, and other aquatic plants. According to Schrank, they feed on the larvæ of other insects, and the spoils of frogs. It has been remarked, that when this insect is in the water, it always conceals its antennæ under the head, and only suffers the barbles to appear, which it keeps in continual motion; but when it walks on dry ground, it advances first its antennæ. The larva is unknown. The *Elophorus aquaticus* is found through Europe in fresh and stagnant waters.

On the remaining sub-genera of this tribe there is nothing which can be added here.

The insects of the genus Sphœridium have been confounded by Linnæus and Geoffrey with *Dermestes*, and by Degeer with *Hister*. But they are distinguished from both one and the other, by the number of articulations in their antennæ, their insertion, the size of their maxillary palpi, the labium, and various other characters.

They approximate much nearer to hydrophilus, &c. But independently of their not living in the water, they are removed from them by the character of their tarsi, their jaws, and the size of the second articulation of their maxillary palpi. Some males, like certain of the hydrophili, have the last articulation of the anterior tarsi dilated. These insects feed, for the most part, on the dung of animals, especially on cow-dung, and are of small size, hemispherical form, black and shining. The *Leiodes* and *Phalœri* have, in these last relations, some resemblance to them, and were even at first placed in the same genus. But they are strongly distinguished from them by the number of articulations in their tarsi.

THE

SIXTH AND LAST FAMILY

OF THE

PENTAMEROUS COLEOPTERA.

THE LAMELLICORNES

PRESENT us with antennæ inserted in a deep fosset, under the lateral edge of the head, always short, of nine or ten articulations, for the most part, and terminating in all in a knob, usually composed of the last three, which are in the form of plates or laminæ—sometimes disposed like a fan, or in the manner of the leaves of a book, opening or shutting in the same way—sometimes turned, and emboxing concentrically; the first, or the under one of this knob, having then the form of a half-tunnel, and receiving the others, sometimes disposed perpendicularly to its axis, and forming a sort of comb.

The body is generally ovoid or ovaliform, and thick. The external side of the two anterior legs is denticulated, and the articulations of the tarsi, some males excepted, are entire and without either brush or cushion underneath. The anterior extremity of the head is advanced, or dilated, most frequently in the manner of a hood. The mentum is usually large, covers the lingua, or is incorporated with it, and carries the palpi. The mandibles of many are membranous, a character not observed in any other coleoptera. Often the males differ from the females, either by elevations in the form of horns, or tubercles on the corslet or head, or by the size of their mandibles.

This family is very considerable, and one of the handsomest of the insects of this order, in relation to the size of the body, the variety of forms of the corslet and head, considered in the two sexes, and often also in the species which live in the perfect state, on vegetable substances, by the brilliancy of the metallic colours by which it is adorned. But the most part of the other species which feed on decomposed vegetables, or excrementitious substances, are commonly of a uniform black or brown tint. Some Coprophagi, nevertheless, do not yield in this respect to the preceding. All are winged, and their walk is heavy.

The larvæ have the body long, almost semi-cylindrical, soft, often wrinkled, whitish, divided into twelve rings, with the head scaly, armed with strong mandibles, and six scaly feet. Each side of the body has nine stigmata. Its posterior extremity is thicker, rounded, and almost always curved underneath, so that these larvæ, having the back convex or arched, cannot extend themselves in a right line, walk badly on a smooth plane, and fall every moment either reversed or on the side. We may form some idea of their structure from that of the larvæ, so well known to gardeners under the name of the *white worm*, that of the common May-bug. Some do not change into the nymph, but at the end of from three to four years. They form for themselves, in their sojourn, with earth, or the debris of substances which they have gnawed, an ovoid shell, or in the form of an elongated ball, the parts of which are connected with a glutinous substance, which they cause to issue from their bodies. They feed on excrements, on dung, the soil itself, the roots of vegetables, and often even on those which are necessary to our wants, from whence very considerable damage often accrues to the agriculturist. The tracheæ of these larvæ are elastic, while those of the perfect insect are tubular. The nervous system, considered in these two ages, also presents remarkable differences. The ganglia are less numerous, and more approximating in the insect arrived at its final transformation, and the two hinder ones throw out a great number of threads, disposed in radii. According to the observations of M. Marcel de Serres, on the eyes of insects, those of the majority of the lamellicornes present peculiar characters, and which approximate their organization to that of the tenebriones, blattæ, and other lucifugous insects.

The alimentary tube is in general very long, especially in the coprophagi, folded on itself, and the chylific ventricle bristles with papillæ, which M. Dufour has recognized to be sacs destined for the repository of the alimentary fluid. The biliary vessels resemble, in their number and mode of implantation, those of the carnivorous coleoptera, but they are much longer, and more attenuated.

We shall divide this family into two tribes.*

The first, that of the *Scarabœidæ,* presents antennæ, terminated in a foliated knob, and plicatile in the majority, but in others composed of emboxed articulations, either in the form of an inverted cone, or almost globular. The mandibles are identical, or almost similar in the two sexes, but the head and corslet of the male individuals often present projections or peculiar forms; sometimes also their antennæ are more developed.

This tribe corresponds to the genus

* The anatomy is, according to M. Dufour, so different, that these two tribes ought to constitute two families. The sections would then be tribes, and would form, some of their divisions, so many principal genera.

SCARABÆUS, Lin.

The alimentary tube is in general much longer than that of the Lamellicornes of the following tribe, or that of the lucanides, and the œsophagus is proportionally much shorter. The adipose tissue, or epiploon, is generally almost nullified, while here it is much more marked. But it is especially in the genital male apparatus that the Scarabeides are distinguished not only from these last, but also from all the other pentamerous coleoptera. Their testicles, according to the observations of M. Dufour, consist in spermatic capsulæ, tolerably bulky, very distinct, pedicled, and varying in number according to the genera.

The larvæ (Cuvier, Regne Anim.) have a cylindrical stomach, surrounded with three ranges of small cœcums, a narrow intestine, very short, a colon extremely bulky, and inflated, and a moderate rectum.

We shall divide this genus into many small sections, established on the consideration of the masticatory organs, the antennæ, and the habits—sections whose distinction has been confirmed by the anatomical researches of the philosopher above cited.

The COPROPHAGI, or Scarabeides of our first section, have antennæ usually composed of nine articulations, and of eight in the others, of which the last three form the knob. The labrum and the mandibles are membranous and concealed. The lobe terminating the jaws is also of this consistence, broad, and arched at the superior edge, and curved within. The last articulation of the maxillary palpi is always the greatest of all, almost ovaliform, or almost cylindrical, but the same of the labial is almost always more slender than the preceding, or very small. Behind each of these last palpi is a membranous prominence in the form of a little tongue. The mentum is emarginated. The sternum presents no peculiar prominence, and the crooks of the tarsi are always simple. The anterior tarsi are often wanting in many, either from their birth or through decay.

The alimentary tube is always very long, and this length is even sometimes (*Copris lunaris*), ten or twelve times more considerable than that of the body. The chylific ventricle, occupying the major part, is bristling with conoïd papillæ, or in the form of nails, very much folded upon itself, and sustained in this state of agglomeration by numerous trachean bridles. The intestine is filiform, and terminated by an inflation. The testicles of the coprophagi, dissected by M. Dufour, appeared to him to be composed of six spermatic capsules, orbicular, a little depressed, usually united by some tracheæ into a packet, each carried on a tubular pedicle, tolerably long, and which leads to a deferential canal of no great length. There is but one pair of vesiculæ seminales; they are filiform, very long, and very much folded.

This first section corresponds to the third division of the genus Scarabæus of Olivier, or to that of Bousier (*Copris*), but adding some Scarabæi (*Aphodus*) of this naturalist.

Some have the two intermediate feet much more separated at their origin than the others. The labial palpi are very hairy, with the last articulation much smaller than the others, or even very indistinct. There is either no scutellum, or a very small one, and the anus is uncovered.

Some Coprophagi of this division, proper to the ancient continent, with rounded body, usually depressed above, or but little gibbous, similar, or but little different, and without horns in the two sexes; in which the antennæ with nine articulations are terminated in a foliated knob, without scutellum, or any sutural hiatus indicating its place; in which the four posterior limbs, usually furnished, as well as the tarsi, with fringes of hairs, or lashes, are slender, elongated, little or not at all dilated at their extremity, truncated obliquely, and terminating in a single spur, robust, in the form of a spine or point; in fine, those in which the hood is always more or less lobate, or denticulated, form the genus

ATEUCHUS of M. Weber and of Fabricius;

But, since, restrained to the species in which the elytra have the external edge straight, or without emargination or sinus near their base, and leaving uncovered the corresponding portion of the upper edges of the abdomen. The legs and the tarsi of the last four feet are furnished with long hairs. The first four articulations of the tarsi are generally longer than in the others. The first of the labials is almost cylindrical, or in an inverted cone. The hood is most frequently divided into three lobes or festoons, and its contour presents six teeth.

These insects, which Mr. W. Macleay, in a work full of research and ingenious views, entitled *Horæ Entomologicæ*, (1st Vol. 1st part, p. 184) designates under the generic name of *Scarabæus*, as being that which they originally received from the Latins, and of which he has given in the same work an excellent monograph, enclose their eggs in a ball of dung, and even of human excrements, similar to large pills, from which circumstance they have been named by some writers *Pilulariæ*. They roll them with their hind feet, and often in company, until they have found holes proper to receive them, or places where they can bury them.

Two species of Ateuchus formed a part of the religious worship of the ancient Egyptians, and of their hieroglyphic writing. They are to be retraced in all their monuments, and under divers positions, and often depicted of gigantic dimensions. They have also represented them separately, employing even the most precious substances for that purpose, such as gold. They formed of them seals, and amulets suspended to the neck, and which they buried with the mummies. The insect itself has been found enclosed in some of their coffins.

The *Scarabæus sacer*, of Linnæus, or *Ateuchus sacer* of Oliv., Col. I. 3. viii. 59, which is found not only throughout all Egypt, but in the southern parts of France, in Spain, in Italy, and in general to the south of Europe, had been hitherto regarded as the object of this superstition; but another species discovered in the Sennâri, by M. Caillaud de Nantes, appears, in consequence of its more brilliant colours and of the country in which it is found, which was the first sojourn of the Egyptians, to have first fixed their attention. This species, which I have named *Ateuchus of the Egyptians*, (Voyage à Meroë au Fleuve blanc, IV. p. 272. Att. d'Hist. Nat. et d'Antiq. II. lviii. 10.) is green, with a golden tint, while the first is black. The hood has on one side and the other six denticulations, but here the vertex has two small eminences, or tubercles, instead of which that of the other, or the *A.* of the Egyptians, presents but one feeble eminence, elongated, smooth, and very shining. The corslet, with the exception of the middle of the back, is entirely punctuated, and even chagreened laterally, with denticulated edges. The intervals of the striæ of the elytra, are, besides, finely chagreened, and present sunken points, tolerably numerous and tolerably broad. The internal side of the two anterior limbs presents a series of small teeth. In our *Ateuchus sacer* this same side has usually two teeth, tolerably strong.

Some of these insects (*S. Esculapius*, Oliv.) and another species, *Hippocrates*, whose corslet and abdomen are shorter, rounder, and more convex, in which the first articulation of the labial palpi is also shorter and broader, in the form of a reversed triangle, compose the genus *Pachysoma*, of Mr. Kirby.*

* Besides the above-mentioned ateuchi, refer to the same sub-genus

Pl. 40

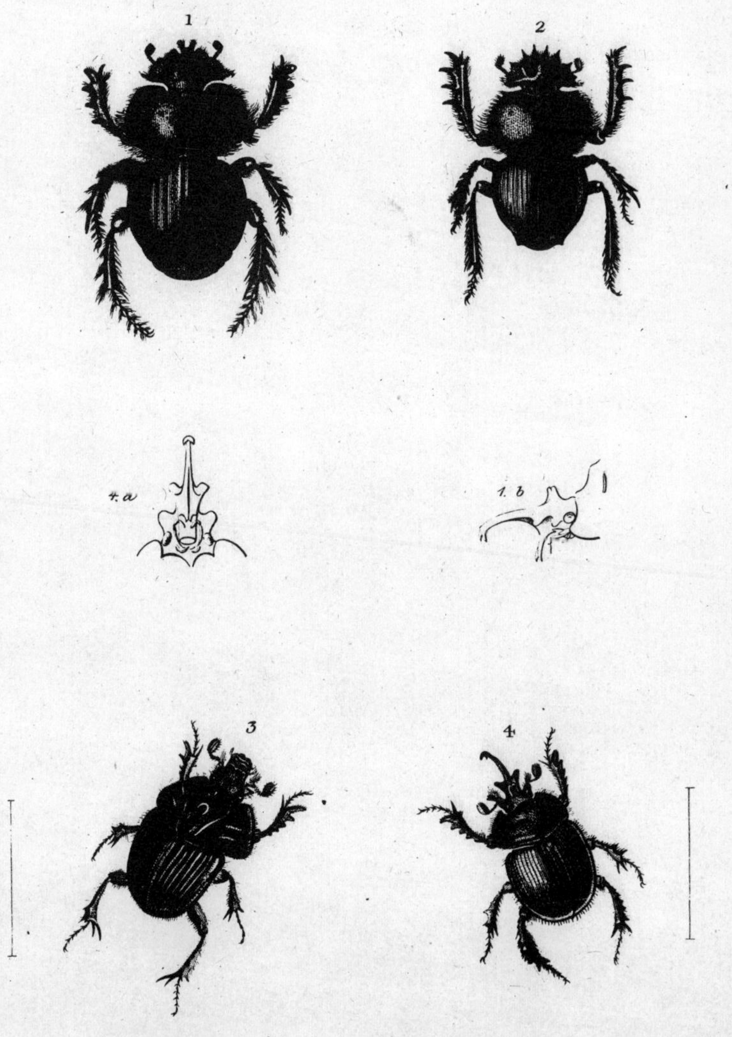

1. Pachysoma *Æsculapius*. *Kirby*. 2. Mnematium *silenus*. *Kirby*. 3. Athyreus *Bilbergii* *G.R. Gray*. 4. Elephantomus *proboscideus*. *Macleay*.

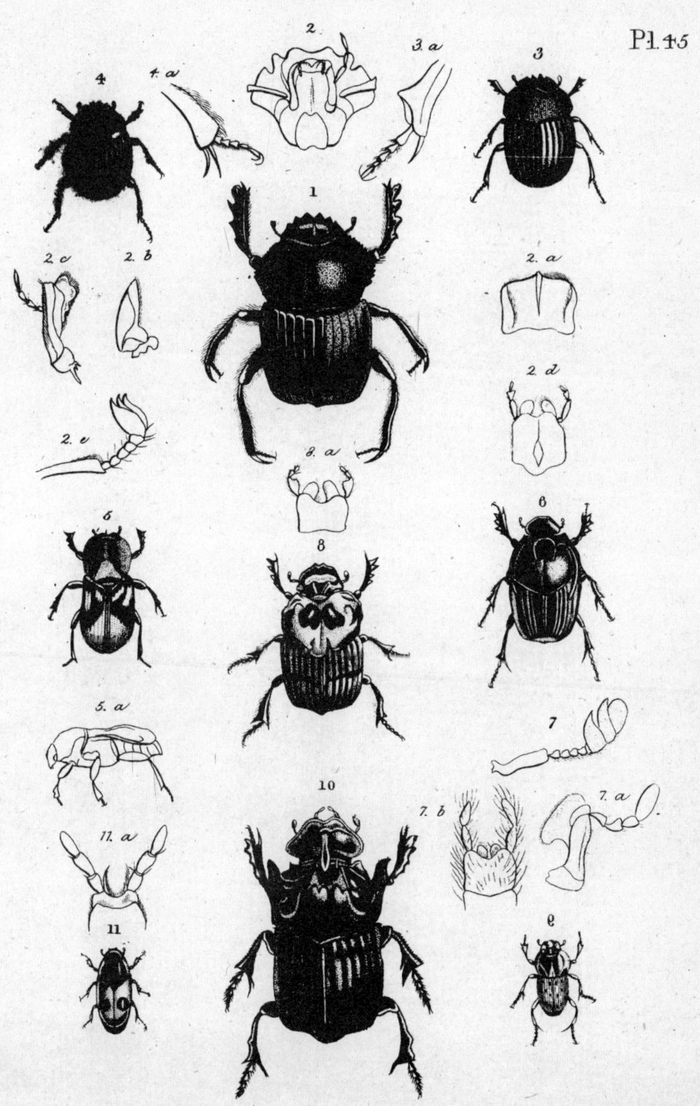

1. Ateuchus *Ægyptiorum. Latr.* 2. *Details of* Ateuchus *sacer. Latr.* 3. Circellium *hemisphericum. Latr.* 4. Coprobius *viridis. Latr.* 5. Eurysternus *fœdus. Guer.* 6. Ontophagus *rarus. Guer.* 7. *Details of* Ontop. *vacca. Latr.* 8. Phaneus *imperator. Chevrol.* 9. Oniticellus *formosus. Chevr.* 10. Copris *bellator. Chevr.* 11. Aphodius *bipunctatus. Fabr.*

The Ateuchi, whose elytra have, at the external edge, near their basis, strong emargination, are now, the

GYMNOPLEURUS, Illig.

The four posterior limbs, are, in general, simply ciliated, or furnished with small spines, and the last articulation of their tarsi is as long, or longer than the preceding taken together. The first of the labials is dilated at the internal edge, and almost triangular. The corslet has a fosset on each side.*

Other coprophagi, very analogous to the preceding, and ranged also with the ateuchi, by Fabricius, are distinguished from them by their intermediate limbs, the extremity of which, as well as that of the last two, often dilated, or in a knob, presents two spurs or spines. The hood has, in many, but four or two teeth. The first articulation of the labial palpi, is always larger than the following, and dilated at the external side. The third and last articulation is distinct. At first comes

SISYPHUS,

Which differ from the other coprophagi in their antennæ, having but eight articulations, and by reason of the triangular form of their abdomen. The last four feet are long,

the *A. laticollis, variolosus, semi punctatus, miliaris, sanctus*, &c. of Fabricius, and some others. See the fore-mentioned work of Mr. Macleay; and the Entomography of Russia, in which some species of this sub-genus and the following are perfectly figured.

* The Ateuchi, *sinuatus, pilularius, flagellatus, Leei, Kœnigii, cupreus, profanus*, &c. of Fabricius; the *Sc. fulgidus*, of Oliv., &c.; the ateuchi of Fabricius, which are proper to America, belong to the other sub-genera. Mr. Macleay (*Hor. Entom.* I. part ii. p. 510) preserves still the Gymnolopleuri with the ateuchi, or his scarabæi, but he makes a division of them, in which he indicates the species.

narrow, with the thighs in a knob. The body is short and thick. The scutel is wanting.*

CIRCELLIUM, Latr.,

In which the body is hemispherical, gibbous, with the abdomen almost semi-circular, and the lateral edges of the corslet straight, or not dilated in the middle. There is no scutellum. The hood presents four or six denticulations. (*Ateuchus Bacchus*, *Hollandiæ*, of Fab.)

COPROBIUS, Latr.

Equally without scutellum, and in which the body is ovoid, not at all, or but little gibbous, with the middle of the lateral edges of the corslet dilated, in the manner of a blunt, or rounded angle, the abdomen almost square, and the hood bidenticulated. These insects are more especially proper to the New Continent.†

The species whose four posterior limbs are proportionally shorter, dilated or widened remarkably at their extremity, with the fist articulations of the tarsi broader, compose the genus CHŒRIDIUM of M. Lepeletier de Saint Fargeau, and Serville (Encyc. Method.). We shall unite again to the Coprophili that which they name *Hyboma* (ibid).

Another sub-genus, bordering on the preceding, the species of which are also American, that which they call *Æschrotes*, but which M. Dalman had published (Ephem. Entom. 1824) before them, under another denomination, that of

EURYSTERNUS,

Differs from the preceding by the presence of a scutellum. The body is, otherwise, of an oblong-oval form, plain above,

* *Ateuchus Schæfferi*, Fab.;—*Sc. longipes*, Oliv., and some other unpublished species of the Cape of Good Hope.

† The *A. volvens*, *violaceus*, *triangularis*, *6-punctatus*, &c. of Fabricius.

with the posterior sides of the corslet cut abruptly, and in an oblique direction. The intermediate haunches are directed in the line of the length of the body, and parallel to its sides.

In all the following coprophagi, the four posterior limbs are always dilated at their extremity, and almost in the form of an elongated triangle; the intermediate, as in the last, in two strong spines, or spurs; but the head, or the corslet, or both, present in the males horns or eminences, which distinguish them from the other sex. In many, the last three articulations of the antennæ, semicupular, are emboxed concentrically. These insects are referred to the genera *Onitis* and *Copris* of Fabricius.

Two sub-genera, with antennary foliated knob, present us a character which, in this section, is exclusively peculiar to themselves. The third articulation of the labial palpi is little, or not at all distinct, and the preceding is larger than the first.

Oniticellus

Have the body oblong, depressed, with the corslet large, almost oval, and nearly as long as broad, and always smooth. The scutellum is distinct. Some simple raised lines, or tubercles of the head, distinguish the males from the females.

Onthophagus, Latr. *Copris*, Fab.

Present no scutellum. Their body is short, with the corslet tolerably thick, broader than long, whether semi-orbicular, or almost orbicular, but strongly emarginated or truncated in front. The head, and often the corslet also, is horned in the males.

S. Taurus, Lin. Oliv. Col. I. 3. viii. 63. Small, black; two arched horns in a semicircle on the head of the male; two raised and transverse lines on that of the female. Found in cow-dung.

S. Nuchicornis, Lin. Panz. Faun. Insect. Germ. I. i. and XLIX. 8. Small, black, with the cases gray, and some with small black spots. A compressed elevation, and in the form of a plate, and terminating in almost a straight point on the hinder part of the head of the male; two raised and transverse lines on that of the female; a tubercle on the anterior part of its corslet. Found with the preceding. Africa and the East Indies present many other species of them, some of which are very brilliant, but all of small size. (Dej.; see Latreille Gener. Crust. et Insect. II. p. 83.)

Two other sub-genera, presenting a scutellum or sutural hiatus, indicating its place, whose anterior feet are often destitute of tarsi, and often still more long, narrow, and arched in the males, are distinguished from all the other coprophagi by the form of the knob of their antennæ. Its first articulation, or the seventh of all, is in the form of a semi-cornet, emboxing the following, a portion of which at least is concealed, and has the form of a horse-shoe. The third, or last, is in the form of a reversed cupola. The corslet is large, and usually presents, near the middle of its posterior edge, two small fossets.

Onitis, Fab.,

In which the second articulation of the labial palpi is the largest of all, and the scutellum, though very small and sunken, is nevertheless visible.

The anterior feet are generally longer, more slender, and arched in the males. Their tarsi are most frequently wanting. The corslet, a small number excepted, is without horns. (Consult the article *Onitis* of the *Encyclopédie Méthodique*).

Phanæus, Macl. *Lonchophorus*, Germ. *Scarabæus*, Lin. *Copris*, *Onitis*, Fab.

In which the first articulation of the labial palpi is the largest of all, and dilated at the internal edge. A simple sutural vacancy indicates the place of the scutellum. The males differ much from their females, by the prominences in the form of horns, of the head and corslet; but the respective lengths of the feet are identical.

Many large and fine species of the Copris of Fabricius compose this sub-genus.*

COPRIS (proper), Geoff. Fab. *Scarabœus*, Lin.

Does not comprehend at present more than those whose antennæ are terminated by a knob with three leaves, in which the fore posterior limbs are strongly dilated, and truncated at their extremity; which have neither scutellum nor vacancy in its place; whose body is always thick, and differs above, according to the sexes; and which have the labial palpi composed of three distinct articulations, the first of which is larger, almost cylindrical, and not dilated at the internal side.

The largest species inhabit the countries of Africa and the East Indies, situated between the tropics or in their neighbourhood.

We find very commonly in Europe, *S. Lunaris*, Lin. Oliv. *ib.* v. 36., which is eight lines in length, black, very shining, with the head emarginated at the anterior edge, bearing an elevated horn, more long, and pointed in the male, short and truncated in the female. (*S. emarginatus*, Oliv. *ibid.* viii. 64.) The corslet is truncated in front, with a horn on each side. The elytra are deeply striated.†

As well as the lamellicornes of the following sections, the

* Encyc. Méthod. article *Phanée*, and above all, the work of Mr. W. Macleay, entitled Horæ Entomol. I. part I. p. 124. He refers thither the following Scarabæi of Olivier, *bellicosus*, *lancifer*, *Jasius*, *Mimas*, *Belzebut*, *festivus*, *carnifex*, &c.

† The *Copris: Antenor*, *Hamadryas*, *Midas*, *gigas*, *bucephalus*, *molossus*, *hispanus*, *nemetrinus*, *nemestrinus*, *Sabœus*, *Jachus*, &c of Fabricius. The

last coprophagi have all the feet inserted at an equal distance from each other, and a very distinct scutellum. The labial palpi are smooth, or but little hairy, with the third and last articulation larger, or at least longer than the preceding. The elytra entirely envelope the compass of the abdomen, or form to it a vault, a character which approximates them to the scarabeides of the following sections. These insects have, otherwise, very great relations, both as to the antennæ and the feet, with those of the preceding sub-genus; but the sexual differences are less marked, and often consist only in simple small eminences, in the form of tubercles. All these coprophagi, are, moreover, of small size. Many species make their appearance from the earliest days of spring. They compose two sub-genera.

Aphodius, Illig. Fab. *Scarabæus*, Lin. Geoff. *Copris*. Oliv.

The last articulation of the palpi is cylindrical, that of the labial is a little more slender than the preceding, or at all events not thicker. The jaws have not at the internal edge any appendage, or corneous and denticulated lobe. The body is rarely short, with the abdomen very gibbous, and when it presents these characters the corslet is not furrowed transversely.

S. fimetarius, Lin. Panz. Faun. Insect. Germ, XXXI. 2, three lines in length, black, with the cases and a spot of the corslet fawn-colour, three tubercles on the head, some punctuated striæ on the elytra. (*See* Schœn. Synon. Insect. I. i. p. 66. Panz. Ind. Entom. p. 7.)

Psammodius, Gyll.,

In which the last articulation of the palpi is almost ovali-

Aleuchus Tmolus, of M. Fischer (Entom. de la Russie, I. viii. 1. 2.) is a copris.

form, and the longest and thickest of all, and in which the internal lobe of the jaws is corneous, and divided into two teeth. The body is short, with the corslet furrowed transversely, and the abdomen inflated.*

This sub-genus conducts us naturally to the first of the following section, that of the Arenicoli. These scarabeides are, with the aphodii and psammodii, the only ones whose elytra entirely cover the abdomen, so that the anus is concealed; but many characters distinguish them from the latter. The labrum is coriaceous and most frequently out-edges the hood. The mandibles are corneous, usually projecting and arched. The lobe terminating the jaws is straight, and not curved internally. The third and last articulation of the labial palpi is always very distinct, and almost as long, at least, as the preceding. Some excepted, the antennæ are composed of ten or eleven articulations.

These coleoptera also live on dung, dig deep holes in the earth, fly more especially in the evening, after the setting of the sun, and counterfeit death when they are taken in the hand. M. Leon Dufour informs us that the digestive canal of Geotrupes, one of the principal sub-genera of this section, has a little less extent than that of *Copris*, and that

* I refer here, only the *Psammodius sulcicollis* of M. Gyllenhall. (Insect. Suec. I. p. 9.) The other species, the first excepted, (see *Ægialia,*) are true aphodii. See Encyc. Méthod., Article, *Psammodii.*

The genus Euparia, established in the Encyclopédie Méthodique, by M.M. Lepeletier and Serville, belong without any doubt to this section; but as they have not been marked completely, and as I have not seen the species, which serves as type, I cannot assign its place. According to them, the sides of the head are dilated, and form a triangle. The posterior angles of the corslet are emarginated, and the humeral angles of the elytra are prolonged in front, in the manner of a point. The only species pointed out is the *E. castanea.* These characters, and the colour, make me suspect that this genus is greatly approximating to that of *Eurysternus* of M. Dalman, of which we have spoken.

the chylific ventricle presents no vestige of papillæ. (Annal. des Sc. Nat. III. p. 234.)

In the *Geotrupides*, Macl., the labium is terminated by two lobes or projecting tongues. The mandibles are generally prominent and arched. The labrum is in the whole or in part uncovered. The antennæ are composed in the greater number of eleven articulations. The body is black or reddish, with the elytra smooth, or simply striated. The males have frequent projections in the form of horns, or differ externally by other characters from the individuals of the other sex. These insects live more particularly on excrementitious matters.

Some have nine articulations on the antennæ.

Ægialia, Latr. *Aphodius*, Fab.,

Have the labrum very short, transversal, scarcely apparent, and entire, the mandibles terminating in a bifid point, the internal lobe of the jaws corneous and bidenticulated; the body short, inflated, with the corslet transverse, and the abdomen gibbous. The four posterior limbs thick, incised, and the last two terminated by two compressed spurs, almost elliptical, or in the form of a spatula. The anterior two have no teeth on the internal edge. The posterior thighs are stronger.*

Chiron, Mac L. *Diosomus*, Dalm. *Sinodendron*, Fab.,

Approach, in consequence of the knob of the antennæ, rather semipectinate than foliated, to the Lamellicornes of the second tribe, and have, in fact, been placed there by Mr. W. Macleay. But they belong from the assemblage of other characters to the present section. Their labrum is

* *Psammodius arenarius*, Gyll. Insect. Suec. I. p. 6: *Scarabæus globosus*, Panz. Faun. Insect. Germ. XXXVII. 2; *Aphodius arenarius*, Fab.

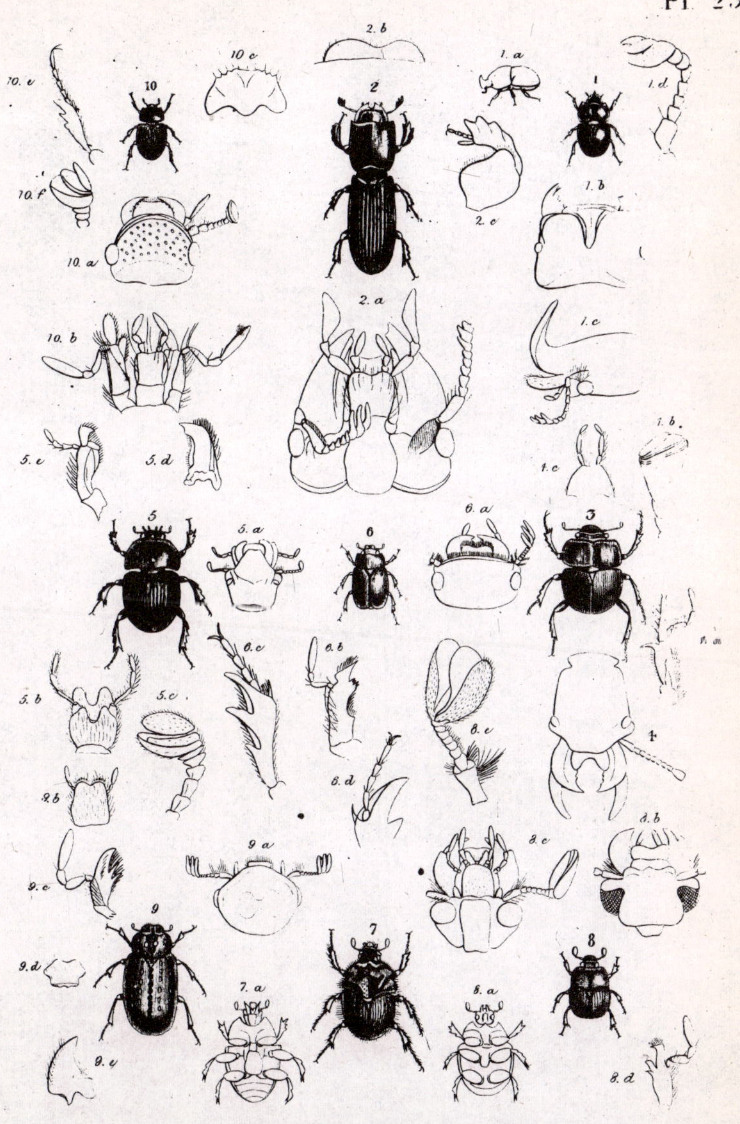

1. Ægialia *cornifrons.* Guer. 2. Chiron *grandis.* Gory. 3. Lethrus *Eversmannii.* Falderman 4. *Details of* Lethrus *cephalotes.* Fabr. 5. Geotrupes *Blackburnii.* Fab. 6. Ochodeus *rufus.* Guer. 7. Athyreus *castaneus.* Guer. 8. Bolboceras *fulvus.* Gory. 9. Trox *Æger.* Guer. 10. Hybosorus *arator.* Fabr.

entirely uncovered, large, ciliated, and quadridenticulated. Their mandibles are robust, in the form of an elongated triangle, with two teeth at the internal edge. The two maxillary lobes are coriaceous and unarmed. The body is narrow, elongated, almost cylindrical, with the corslet, longitudinal, separated from the abdomen by a deep strangulation. The abdomen is elongated, and the anterior legs broad, digitate, and provided at the end of the spur, with a silky tooth at the end. The thighs have a lenticular form, and the anterior ones are larger. The anterior extremity of the head presents a transverse range of little tubercles.*

Others have eleven articulations on the antennæ.†

Some are distinguished from all the others by the knob being in a reversed cone, and composed of articulations or leaves, turned in the manner of a funnel, and emboxed concentrically, and by their mandibles, entirely denticulated like a saw on the internal side, and presenting underneath, especially in the males, a sort of advancement or horn. The corslet is very much emarginated in front in these individuals, with the anterior angles very much prolonged forwards. The abdomen is very short, almost semi-circular, and the last feet are but very little removed from its extremity. The labial palpi are a little longer than the others, with the second articulation elongated, and the other two almost of equal length. The jaws are provided internally with hairs and lashes, in the form of small spines; their terminal lobe

* *Sinodendron digitatum*, Fab.; *Chiron digitatus*, Macl. Hor. Entom. I. Part I. p. 107; *Diasomus digitatus*, Dalm. Ephem. Entom. I. p. 4.

† This computation is sometimes doubtful, seeing that it is not always easy to distinguish the articulation which precedes the knob, and that it may in appearance be confounded with the first of this knob. The base of the second also forms a sort of knot, or rotula, which might be taken for an articulation.

is narrow and elongated. The mentum is in the form of a triangle, truncated transversely at its extremity. Such are

LETHRUS, Scop. Fab.,

Of which the species, very few in number, are proper to Hungary and the western provinces of Russia.

Lethrus cephalotes, Fab. Fisch. Entom. de la Russe, I. p. 133. XIII. i., distinguished from the other species by its colour entirely black, its corslet and elytra smooth, is, according to the celebrated Professor Gothelf Fischer, an animal very hurtful to cultivated places, because it seeks, in preference, the buds or leaves scarcely apparent, and cuts them neatly with the trenchant pincers of its mandibles. On this account it is called, in Hungary, where it does much damage to the vineyards, *Schneider* (Cutter). The breast advancing much below the abdomen, and the hind feet appearing to be inserted near the anus, it climbs very well, and returns by moving backwards. After having cut the heart of a plant, it moves backwards like a crab, carrying its prey into each hole. Each hole dug in the earth is occupied by a pair; but in the time of coupling, a male stranger often presents himself, desiring admittance. Then commences a vehement combat, during which the female closes the entrance of the hole, and always pushes the male from behind. This combat does not cease but with the death or flight of the male stranger. This philosopher has described three other species unknown before his time. (*Ibid.* p. 136, 140.)

All the other arenicoli have the knob of the antennæ composed of leaves of the ordinary form, and applied one over the other in the same direction, or like those of a book. They compose our sub-genus GEOTRUTES, or that of *Scarabæus* of Fabricius, and from which the following sub-genera have been since detached.

Those in which the knob of the antennæ is oval, or ovoid,

and in which all the leaves have, even in a state of contraction, their edges totally or partially uncovered, compose two

GEOTRUPES (proper), Lat.

Have the labrum in a transverse square, entire, or simply denticulated. The mandibles arched, very much compressed, denticulated at their extremity, and often sinuous on the external edge. The jaws furnished with a very thick fringe of hairs: the last articulation of the palpi scarcely larger than the preceding, but the same of the labials larger. The mentum profoundly emarginated; the anterior legs elongated, with a great number of teeth on the external side, and a single spur or spine at their internal extremity, and the hood in the form of a lozenge.

Sometimes the males have the corslet armed with horns. Those are the *ceratophyus* of M. Fischer, or the *Armideus* of M. Ziegler.

G. Phalangista, (*S. typhœus*, Lin. Oliv. Col. I. 3. vii. 52.) black, three advanced horns in the form of points, the intermediate one of which is shorter in front of the corslet of the male. Cases striated. In sandy and elevated places.

G. Momus, *S. Momus*, Fab. discovered in Spain, by M. le Comte Dejean, differs from the preceding by its smooth elytra, and resembles it in the rest.

G. Dispar, Male, *Ceratophyus Dispar*, Fisch. Entom. de la Russie, II. xviii. A species which is found in Italy and in Russia, has a horn upon the head and on the corslet.

Sometimes the two sexes are destitute of horns. These are geotrupes proper.

G. stercorarius, *Scarabæus stercorarius*, Lin. Oliv. *ib.* V. 39, of a shining black, or a deep green above, violet or golden green underneath, a tubercle on the vertex, punctuated stripes on the elytra, the intervals smooth, two denticulations at the base of the posterior thighs.

G. vernalis, *S. vernalis*, Lin. Oliv. Col. iv. 23, shorter than the preceding, approaching the hemispherical form, of a violet or blue black, with the antennæ black, and the elytra smooth.

OCHODÆUS, Meg. *Melolontha*, Fab.*

Have the labrum strongly emarginated and nearly in the form of a heart truncated posteriorly; the mandibles are in the form of an elongated triangle, one of which terminates in a single point, with a notch underneath, and the other in two obtuse teeth; the exterior lobe of the jaws is edged with small spines or large ciliæ, bent at the end, with two internal small corneous and equal teeth; the other, or internal lobe, formed of a pencil of silk brought to a point; the last articulation of the palpi is much larger than the preceding, and is cylindrical; the second of the labial palpi is larger than the others, and the following or last is a truncated ovoid. The fore legs have only two teeth on the outer side, and the end of the opposite side two spines, the lowest being the smallest. The body is in proportion less elevated than that of the other geotrupes, and without horns.

The Geotrupes—with a knob to the antennæ, large, orbicular, and nearly globular, the first and last laminæ of which entirely cover, when contracted, the intermediate or tenth, or form for it a sort of case—compose three other sub-genera.

ATHYREUS, Macl.

Approximate to the coprophagi by the intermediate tarsi being wider at their insertion than the others. Horæ Entomol I. i. p. 128.

* *Melolontha chrysomelina*, Fab. Panz. Faun. Insect. Germ.

ELEPHASTOMUS, Macl.

Are remarkable by the chaperon dilated on each side, and prolonged forward in their middle into a sub-quadrangular lamina, thicker and forked at the end, and moreover by the length of the maxillary palpi, which is nearly three times that of the labial. The chin is deeply emarginated, and the two mandibles are indented at their extremity. Horæ Entomol. p. 121. *Scarabæus proboscideus*, Schreib. Trans. Lin. Soc. vj. 189.

BOLBOCERAS,* Kirby. *Odontæus*, Ziegl. *Scarabæus*, Lin. Fab.,

In which, as in Ochodæus, to which they approximate closely, one of the mandibles is simple, and the other bidenticulated at the end; the maxillary palpi are scarcely larger than the labial, and the chin has no emargination.

We have one species in France, *S. mobilicornis*, Fab. Panz. Faun. Insect. Germ. xii. 2. It is small, black above, yellow underneath, with a very long horn, linear, slightly bent back, and moveable on the head. The corslet is deeply punctuated, channeled in the middle, with four tubercles in front. The elytra have punctated striæ. The body is sometimes entirely yellow. (*S. testaceus*, Fab.)

One of the sons of the celebrated traveller and ornithologist, Le Vaillant, has remarked that frogs and toads were very fond of this insect, and he has obtained a great number of them from the stomachs of these reptiles.

Our first division of Arenicolous Scarabæi terminates with those whose antennæ, as in most of the following scarabæi, have ten articulations.

* *Bolboceras Australasiæ*, Kirb. Trans. Lin. Soc. XII. 2. 3. 5; *Scarabæus quadridens*, *cyclops*, *lazarus*, of Fabricius.

The last articulation of their palpi is elongated. The maxillary lobes are membranaceous. The labrum projects less than in the preceding, and is but little advanced. The mandibles are very little indeed, or not at all indented. The hood is short, whether arched or rounded, or projecting like an angle. All these insects are very small, with the corslet hornless.

HYBOSORUS, Macl. *Scarabæus*, *Geotrupes*, Fab.

The last articulation of their antennæ is in the shape of a reversed and elongated cone, and the intermediate articulation of the knob is entirely covered by the other two, as well as in the last sub-genus. The legs are narrow and elongated. The chaperon is rounded in front. Horæ Entom. p. 120. *Geotrupes arator*, Fab.

ACANTHOCERUS, Macl.*

The first articulation of the antennæ is very large, dilated above into a laminous form, and the edges of the intermediate leaves of the knob, when it is folded, are visible. The legs, especially the four hinder, are lamelliform, and cover the tarsi folded upon them in the contraction of the feet. The chaperon goes to a point, or terminates in an angle. The corslet is nearly semilunar.

There, or in our second division of Arenicoli (*Trogides*, Macleay), the antennæ, scarcely larger than the head, are always composed of ten articulations, the first of which is large and hairy. The languette is entirely hidden by the mentum. The labrum and mandibles are scarcely seen; the latter are thick. The palpi are short. The mentum is very hairy. The jaws are armed with teeth on the inside. The

* Macl. *ibid.* p. 136; *A. æneus.* I owe the knowledge of this species to one of our most able entomologists, M. Lebefevre de Cerisy. Mr. Macleay refers to the same genus, *Trox spinicornis*, of Fab.

body, ash or dirt colour, is very rugged or tubercular above. The head is inclined, terminated by an angle, or going to a point. The corslet is short, transversal, without lateral hems, sinuous behind, with the anterior angles advanced, and their thighs cover the under part of the head. These insects produce a screaking noise, by a reiterated friction of the pedicle of the mesothorax against the inner side of the cavity of the corslet. They keep in the earth or sand, and appear to gnaw the roots of vegetables. They form the genus

TROX of Fab. and Olivier.

Mr. Macleay the younger has separated them, under the generic name of *Phoberus*, from those which have the sides of the corslet depressed, dilated, and edged with spines, and which are without wings. The posterior edge of the corslet has on each side a strong emargination, and the hood is rounded in front.*

A third section, *Xylophili*, will include the geotrupes of Fabricius, and some of his cetoniæ. Here the shield is always distinct, and the elytra do not cover the hind end of the abdomen. The hooks of the tarsi in many are unequal. The antennæ have always ten articulations, the last three of which form a foliated knob, with the intermediate leaves

* *Trox horridus*, Fab. Macl. Horæ Entom. I. 157. Trox, of Fabricius, do not change their place. See that author, Olivier, and Schœnher.

The genera *Cryptodus* and *Mœchidius*, which Mr. Macleay puts in his family of *Trogidæ*, immediately after *Phoberus*, have the hind extremity of the abdomen uncovered, and nine articulations in the antennæ, characters which seem to remove them from Trox. I suspect the Machidiæ, by reason of the form and emargination of the labrum, and of some other character, to be near Melolontha. The cryptodi are distinguished from all other scarabæi by the mentum, which almost entirely covers the mouth underneath, and even by the labial palpi being situated, as well as the tongue, behind it. These two genera have been established on Australasian insects, which I have not seen.

never entirely concealed or encased by the other two. The labrum does not project, and its anterior extremity at the most is uncovered. The mandibles are always corneous, and laterally pass the edge of the head. The jaws are corneous or solid, straight, and generally indented. The tongue is covered by the mentum, ovoid or triangular, narrowed and truncated at the end, with the angles often dilated. All the feet are inserted equidistant from each other.

A first division will include the Geotrupes of Fabricius. The males differ from their females by peculiar eminences in the form of horns, by tubercles either on the head or corslet, or on both, and sometimes also by the form of the latter. The hood is small, truncated or bidentated at the end. The labrum is almost always entirely concealed. Here the jaws terminate in a simple coriaceous lobe, crustaceous, more or less downy, and without teeth; there they are altogether scaly, go to a point, and have only a small number of teeth accompanied with hair. The mentum is ovoid, or like a truncated triangle. The chest has no projection. The hooks of the tarsi are generally equal. The shield is small or middle-sized. The colours approach black or brown.

Sometimes the jaws terminate in a coriaceous or crustaceous lobe, without teeth, and simply downy, or furnished with spinuliform lashes.

ORYCTES, Illig. *Scarabæus*, Lin.,

Whose feet differ but little in length, and whose four hind legs are thick, strongly incised or emarginated, with the extremity widened, as if starred in many.

S. nasicornis, Lin. Rees. II. 6, 7. Fifteen lines long, of a shining marron brown, with the point of the hood truncated, a conical horn, more or less long, arched behind on the head, the front of the corslet cut obliquely, with three

Pl. 24

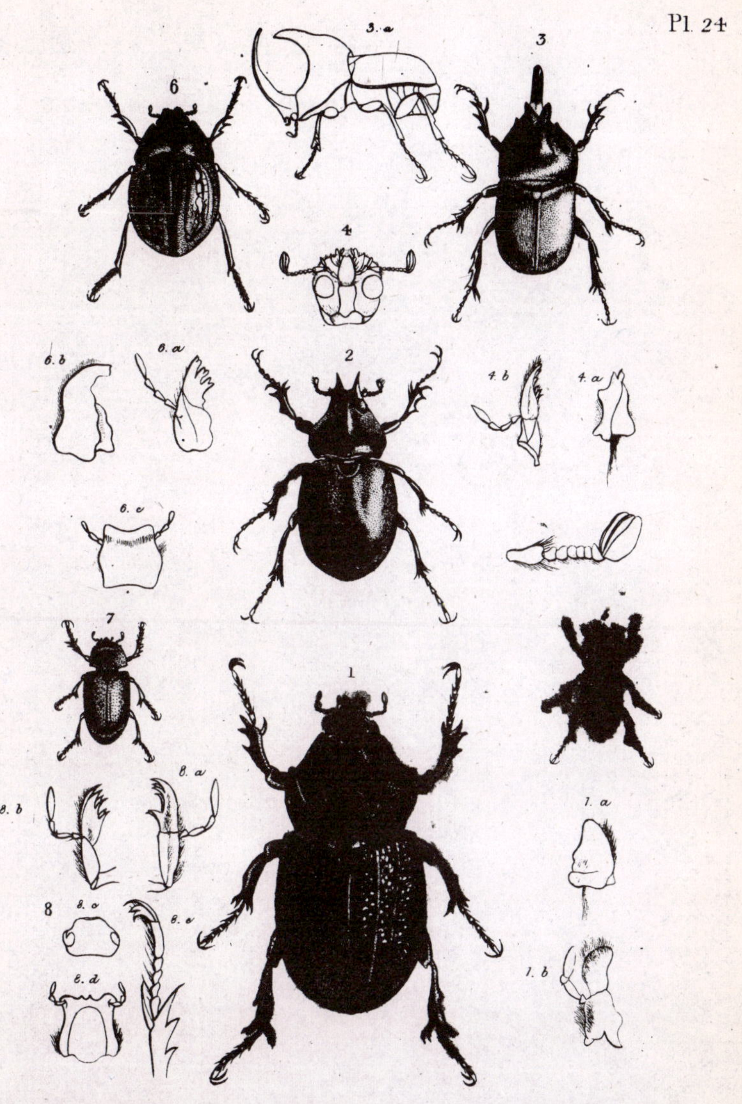

1 Oryctes *Chevrolatii.* *Guer.* 2. Ag[illegible]ala *furcata.* *Guer.* 3. Scarabæus *mentor.* *Guer.* 4. *Details of* Sc. *gedeon.* *Fabr.* 5. Phile[illegible] *cribratus.* *Chevr.* 6. Hexodon *reticulatus.* *Olivier.* 7. Cyclocephala *frontalis.* *Chevr.* 8. *Details of* Cycl *geminata.* *Fabr.*

teeth or tubercles at the elevated and posterior part of the truncation ; cases smooth. It lives as well in the perfect as in the larva state, in tan-beds.

There is found in the South of Europe another species, *G. silenus*, (Fab. Oliv. Col. I. 3, 8, 62, a—c) smaller than the last, of a brighter marron brown ; a little bent and pointed horn on the head of the male ; a deep excavation in the middle of the corslet, the last articulation of the two anterior tarsi swollen, with two very unequal hooks. The elytra are fine, and vaguely dotted.*

Agacephala, Manh.,

Whose fore feet, in the males at least, are longer than the others, and whose four hind legs are slender, or but little thick, cylindrical, and slightly dilated at their extremity, without notches or deep lateral incisions.

The labrum is entirely hidden. The lobe terminating the jaws is simply downy. The antennæ have ten articulations, and it is an error in the Encyclop. Méthod. (Art. Scarabæus) to say they have only nine.

I know two species both from Brazil. *G. Œgeon* of Fab. is probably congenerous with them.

Sometimes the jaws, commonly corneous or scaly, are more or less toothed.

* Add *Geotrupes boas, rhinoceros, stentor*, &c. of Fabricius.

G. orphnus, of Mr. Macleay, established on *G. bicolor*, of Fab., does not differ from the last. The anterior edge of the labrum is salient, or exposed. The jaws are terminated by a bundle of spinuliform lashes, arched outwardly, with a crustaceous triangular lobe. The knob of the antennæ is almost globular. His genus, *Dasygnathus*, which he places in the family of Dynastides, is unknown to us; but we suspect, from the exposition of its characters, that it approximates to the last and the following. The jaws are not toothed.

SCARABÆUS, properly so called, GEOTRUPES, Fab.

Have the body thick, convex, and the outer side of the mandibles sinuous or toothed.

The equatorial countries of both continents furnish some very remarkable species.

S. Hercules, Lin. Oliv. Col. I. 3, 1, 23, 1. Five inches long, black, with the case greenish grey, spotted with black; the male has on the head a bent indented horn, and another long, advanced, downy underneath, with a tooth on each side on the corslet.—South America. Some travellers have called it the horned fly. This species is the type of the genus *Dynastes*, of Mr. Kirby. *S. Actæon* forms another, that of *Negasoma*. See the 14th volume of the Linnæan Transactions.

S. dichotomus, Oliv. *ibid.* xvij. 156. Marron brown; a large horn forked in two branches on the head, another smaller bent and bifid at the extremity on the corslet of the male.—India.

S. longimanus, Lin. Oliv. *ibid.* iv. 27. Fulvous brown, without horns or tubercles on the head or corslet; the two fore feet are one half longer than the body, and arched.—India.

France has but one species of this sub-genus, *S. punctatus*, Oliv. *ibid.* viij. 70. The body is black, dotted, without elevation, in the form of a horn in either sex. The hood is truncated in front, with the angle thereof a little elevated, like a tooth; the middle of the head has two neighbouring tubercles.*

PHILEURUS, Lat. *Geotrupes*, Fab.

Differ from Scarabæus only in having the mandibles nar-

* Geotrupes of Fabricius, except the species before cited, forming *G. oryctes*, and of those of the following genus.

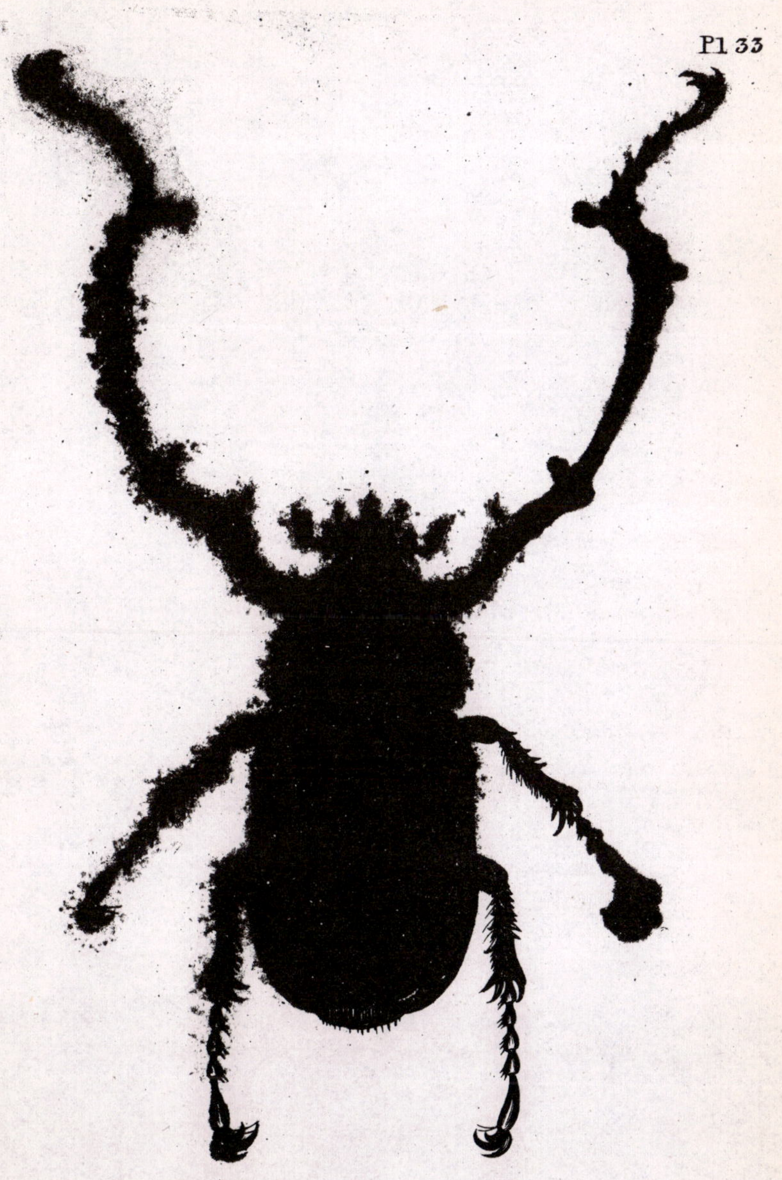

Geotrupes *longimanus.*

rowed, without sinus or tooth on the outside. The body is depressed, and the corslet dilated and rounded at the sides *

Our second division includes Scarabæi, nearly related to the last in some respects, but approaching also several Melolontha, and particularly cetoniæ, whose outward appearance they possess, but whose buccal organization is different; it is even with these that Fabricius and Olivier have placed the greater part of these insects. The body in general is shorter, rounder, and smoother than that of scarabæus, and ornamented with brilliant colours. The head and corslet are identical, and without any particular eminence in either sex. The front edge of the labrum is almost always uncovered or apparent. The jaws are entirely scaly, as if truncated at the end, with five or six strong teeth on the inner side. The mentum is in proportion shorter and longer than that of similar coleoptera, and is less narrowed above. The mesosternum is often elongated like a horn, or soft point between the second tarsi and beyond. The scutellum is commonly large. The hooks of the tarsi are generally unequal. With the exception of a small number, these Xylophili are peculiar to the equatorial countries of the new continent.

Here, as well as in all the preceding scarabæi, one does not see between the posterior angles of the corslet, and the outside of the base of the elytra, axillary pieces filling up the void included between these parts.†

* *G. didymus, valgus, depressus,* of Fab. Some inedited species of Brazil and Cayenne, having some analogy with the Sinodendrons, have the body thicker, and connect Phileurus with Scarabæus, or Geotrupes of Fab., a genus not yet sufficiently studied, with reference to its buccal organization.

† A lateral piece of the mesosternum, larger and thicker than common, and which corresponds probably with those little round scales called *Tegulæ* by some authors, to be seen at the insertion of the upper wings of Hymenoptera. See on this subject the memoir of M. Audouin on the thorax of insects.

Let us first treat of the sub-genera, in which the middle of the chest does not present any elongation like a point or horn.

HEXODON, Oliv. Fab.

Their body is almost orbicular, even underneath, with the head squared, received into a deep emargination of the corslet, the outer edge of the elytra dilated, and preceded by a furrow. The legs slender, and the crotchets of the tarsi very small and equal.

The labrum is not apparent. The knob of the antennæ is small. The jaws are strongly toothed. See Oliv. and Lat., Gener. Crust. II. 106.

CYCLOCEPHALA, Latr. *Chalepus*, Macleay. *Melolontha*, Fab.

Have the body ovoid, with the head free, the elytra slightly bordered without dilatation or lateral furrow, and the anterior tarsi terminated by a knobbed articulation; the hooks unequal, and both bifid.

The anterior edge of the labrum is apparent. The mandibles are narrow, without emargination or observable sinus on the outer side, and but little surpassing the edge.*

In the following sub-genera the sternum advances to a conical point, more or less long, pointed or round at the end, between the second legs.

The anterior edge of the labrum is always apparent. The

* *Melolontha germinata, barbata, castanea, signata, ferruginea, melanocephala, pallens,* &c. of Fab. In the first the mandibles are strong, arched, and crooked at the end. *M. signata, melanocephala,* &c. are smaller, straight, truncated, or obtuse at the end. The summits of the jaws and mentum are moreover furnished with hairs. From this we might form of these and their analogous species a peculiar sub-genus. All these insects are from South America.

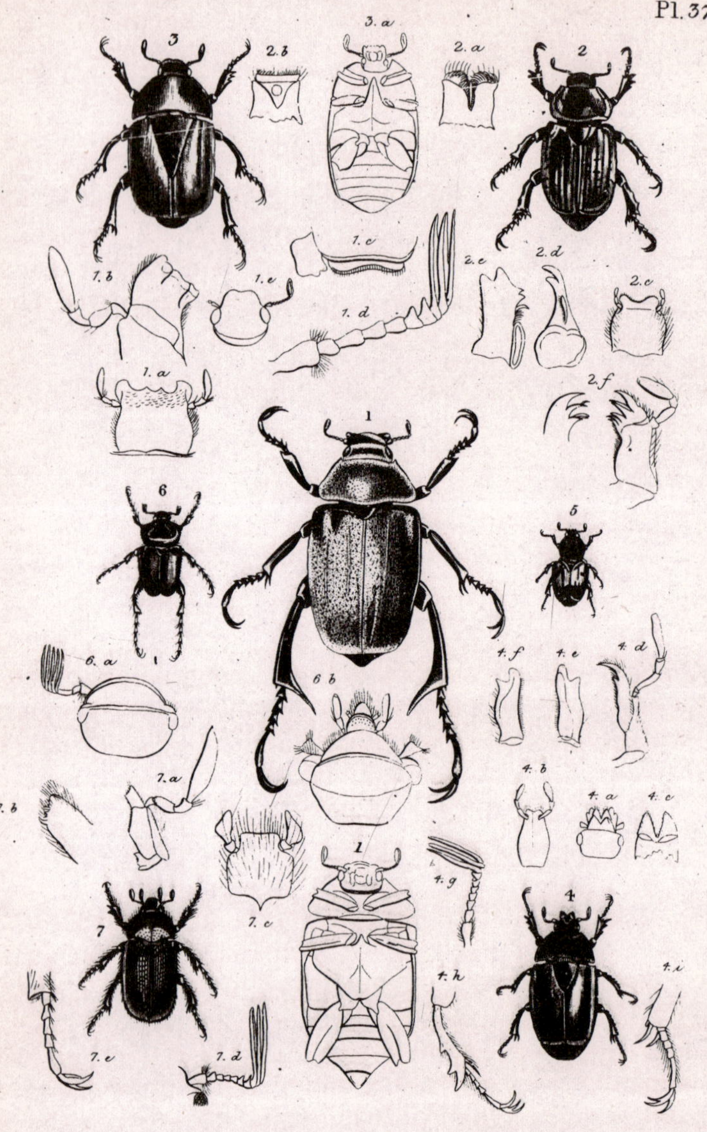

1. Chrysophora *chrysochlora*. *Lat.* 2. Rutela *nitidissima*. *Guer.* 3. Macraspis *splendida*. *Fab.* 4. Chasmodia *brunnea*. *Serv. & St Farg.* 5. Ometis *pictus*. *Guer.* 6. Pachypus *excavatus*. *Fab.* 7. Amblyteres *geminatus*. *Mac Leay.*

mandibles are commonly notched or indented on the outer edge. The hooks of the tarsi are unequal.

CHRYSOPHORA, Dej.,

The males of which have the hind legs very large, with the thighs thick, the legs bent, and terminated at the inner angle in a very strong point.*

RUTELA, Lat. *Rutela*, *pelidnota*, Macleay, Kirby. *Oplognathus*, Kirb. Macl.

Whose legs, with reference to proportions, do not differ observably in the two sexes. The mentum is almost isometrical; the shield is small, or of middle size, and the sternal point is short, not reaching to the insertion of the two fore feet. The body is ovoid or oval.†

MACRASPIS, Macleay. *Cetonia*, Fab.,

Which differ from Rutela in reference to the proportions of the mentum, which is sensibly larger than it is wide; the length of the shield, equal at least to one third of that of the elytra; and the sternal point, whose extremity reaches as far, or goes beyond the insertion of the fore feet The mandibles are nearly triangular, with the extremity pointed and emarginated. The jaws have many teeth. The mentum is in the form of an elongated square, slightly narrowed near its upper end, and without lashes on its upper edge. One of the hooks of the tarsi, or of the four anterior tarsi, at least, is bifid, and the other entire. (Horæ Entom.)

* *Melolontha crysochlora*, Lat. Voy. de Humb. c. Boupl. II. 15. 1. fem. 2. male; *Scarabæus macropus*, Shem. Mis. 380-4.

† See the Catalogue of the Collection of M. le Comte Dejean, Mr. Macleay, the younger, Horæ Entomol. I. p. 10.; the article *Rutéle*; in the Ency. Méthod. The characters of *G. pelidnota*, and *oplognathus*, do not appear to me to be sufficiently distinct.

CHASMODIA, Macleay.

Similar to Macraspis in the general form of the body, the proportions of the shield, and the sternal point, but with the mandibles narrower, and the extremity obtuse and entire. The jaws have but two teeth, with a pencil of lashes, the mentum is an elongated ovoid, obviously narrowed toward its upper end, with the edge furnished with lashes. All the hooks are moreover entire. (See the article Rutela in the Encyclop. Méthod. and Mr. Macleay's work.)

In these an axillary piece (the same as is seen in the same place in *Cetornia*, or that which M. Audouin names *epimere*,) fills the space included between the posterior angles of the corslet, and the exterior of the base of the elytra.

OMETIS, Latr.*

The genus *Melolontha*, of Fab., will compose our fourth and fifth sections

The fourth, that of *Phyllophagi*, is formed of scarabæi, nearly allied to those of the last sub-genera, but the mandibles are covered above by the hood, and hidden underneath by the jaws; their exterior side only is uncovered, but nevertheless without extending beyond; they shew nothing exteriorly of the sinus or indentations which are observable there in Rutela and other analagous sub-genera. The anterior edge of the labrum is composed, and sometimes represents the figure of a reversed and broad triangle, but more generally that of a transverse leaf emarginated in the middle. The number of the articulations of the antennæ is not constant, and varies from eight to ten; it is the same with those of the

* *Rutela cetonioides*, Ency. Méthod.; *Rutela cerata*, Germ.; *Anisoplia histrio*, Dej, but the antennæ with nine articulations. This sub-genus seems to connect these and the preceding insects with cetonia.

knob, and in many the two sexes differ in this respect. The tongue is entirely covered by the mentum, or incorporated with its anterior face, and the elytra join entirely the whole length of the suture, characters which distinguish these insects from those of the fifth section.

The family of Anophlognathi, of Mr. Macleay, and some other sub-genera, nearly allied to those of the previous section, will compose our first division. The hood is thickened before, and forms, either alone or with the labrum, a vertical facet in the form of a reversed triangle, the point of which is supported on the mentum. This last piece is sometimes nearly ovoid, very downy, with the extremity either round or truncated, and without emargination, sometimes a transversed square, with the middle of the upper edge elongated like a simple tooth or emargination. The jaws of some terminate in a coriaceous or membranaceous lobe, very downy, toothless, or with one very small, and situated near the middle of the internal edge; those of the others are entirely corneous, resembling mandibles either truncated or obtuse, and entire at the end, or else terminated by two or three teeth.

Those whose mentum is nearly ovoid, and very downy, and whose jaws terminate in a triangular lobe, equally downy, without or with very small teeth, and situated near the middle of the internal edge, form two sub-genera. The sternum has no projection.

Pachypus, Dej. *Geotrupes*, *Melolontha*, Fab.

The antennæ of the males have only eight articulations, the last five of which compose the knob. The mandibles are in the form of very thin leaflets, triangular, elongated, and entirely hidden, as well as the labrum. The terminal lobe of the jaws is very small, scarcely distinct, and toothless.

The tarsi are long, slender, downy, and terminated by two small equal and simple hooks.

In the antennæ and the form of the hood, this sub-genus approximates to Oryctes rather than to Mololontha.*

AMBLYTERES. Macleay.

Have ten articulations in the antennæ, the last three of which compose the knob. The labrum is uncovered and lobed. The mandibles are strong and scaly. The maxillary lobe is of middle size, and armed with corneous teeth on the inner side. The middle of the upper extremity of the mentum is a little elongated, truncated, with the angles rounded, and carrying palpi; their last articulation is ovoid; the same of the jaws is very much elongated and sub-cylindrical. The scutellum is large.†

In the other sub-genera, of the same division, the mentum is squared transversely, with the middle of the upper edge advanced like a tooth, entire or emarginated. The jaws are entirely corneous, resemble mandibles terminated by a strong tooth, bent, elongated, either entire and very obtuse at the end, or divided at the extremity into two or three points. The mandibles are always scaly and robust. The labrum is exposed.

Some proper to Australasia have a sternal point, and the hooks of the tarsi entire and unequal. Such are,

* *Geotrupes excavatus*, Fab. male; *Melolontha cornuta*, Oliv. Col. 1. 5. vij. 74. a. b. male: *Scarabæus candidæ*, Petag. Insect. Calab. 1. 6. a. b. male, var. black, observed also in Corsica, by M. Peyrandeau, and again in Sicily, by M. Lefèvre;—*M. atriplicis*, Fab., female of another species.

† Macleay, *Horæ Entom.* I. p. 1. 142. This naturalist does not speak of the hooks of the tarsi, nor of the sexual differences. From the description of the typical species, the corslet was without horns, the fore legs have three teeth on the outer edge; only two are to be seen on those legs in Pachypus.

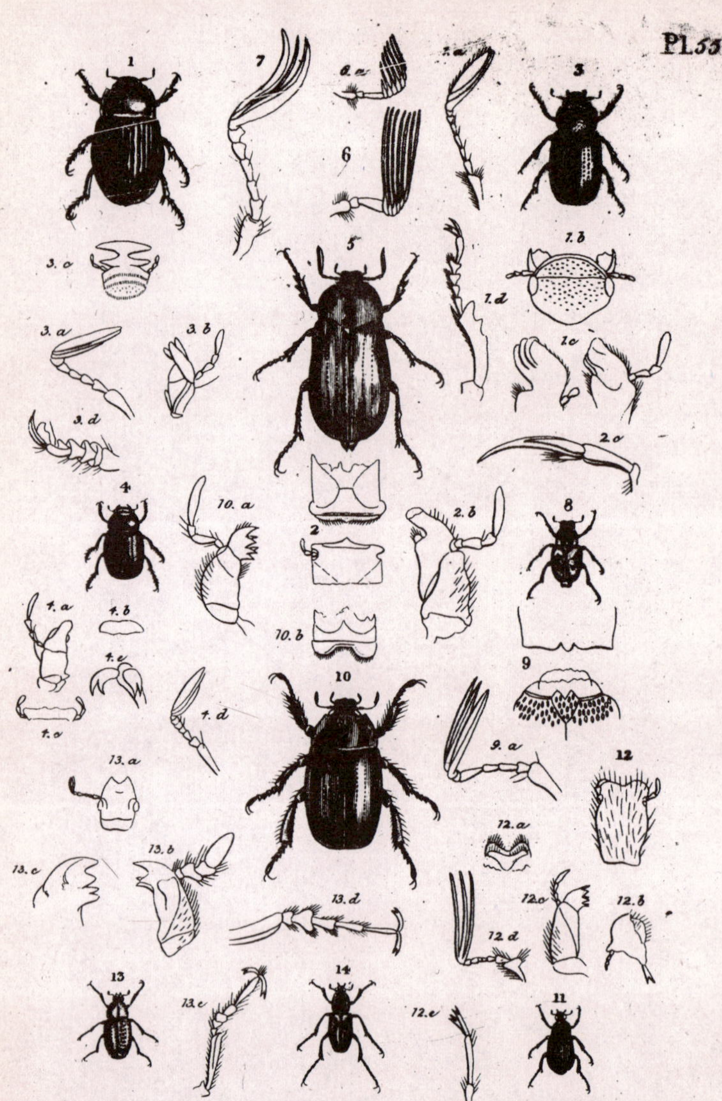

1. Leucothyreus *nitidicollis. Guer.* 2. *Details of* Anoplognathus *Latreillei. Sch.* 3. Geniates *barbatus. Kirby.* 4. Apogonia *gemellata. Kirby.* 5. Melolontha *flavida. Gory.* 6. *Antenna of* Melol. *vulgaris.* 7. *Ant. of* Rhisotrogus *pini. Latr.* 8. Ceraspis *decora. Gory.* 9. *Details of* Cer. *albida. Serv.* 10. Areoda *Kirbii. MacLeay.* 11. Serica *flavimana. Gory.* 12. *Details of* Serica *variabilis. Lat.* 13. Diphucephala *furcata. Guer.* 14. Macrodactylus *suturalis. Chevr.*

ANOPLOGNATHUS, Repsimus, Leach.

The antennæ contain ten articulations, and the extremity of the jaws is truncated, or obtuse and entire. These insects are generally rather large, and ornamented with fine colours.*

The others, which belong to the hot countries of both continents, have no sternal projection ; the hooks of the tarsi, or one of them, are bifid. Their jaws often terminate in two or three teeth.

Sometimes the antennæ have ten articulations, and the upper extremity of the jaws is entire, or at most emarginated or bidentated.

LEUCOTHYREUS, Macl.,

In which one of the hooks of the tarsi is entire and the other bifid.

The tarsi, or at least those foremost, are furnished with a brush underneath ; these are dilated in the males. The under part of the head, in the males, is more downy than in the other sex. (Macleay Hor. Entom. I. p. 1, 145.—*Melolontha sulcicollis*, Germ. Insect. Spes. No. 124.)

APOGONIA, Kirby, Macl.

In which all the hooks of the tarsi are bifid. (Lin. Trans. 401.—*A. genellata*, ejusd. *ibid.* xxi. 9.)

Sometimes the antennæ have only nine articulations, and the extremity of the jaws has three teeth.

GENIATES, Kirb.

The extremity of the mandibles is emarginated. The mentum, in the males, has underneath a sort of circular tuft, formed of serrated hairs, even, or as if cut smooth like a

* See Macleay the younger, *Horæ Entom.* I., p. 1, 143, and the twelfth volume of the Transactions of the Linnæan Society, 401—5.

brush. The first four articulations of the anterior tarsi are dilated, and furnished with tufts underneath. One of the hooks of all the tarsi is entire, and the other bifid. The anterior of the first two is accompanied at its base with a corneous leaf, emarginated beneath, round at the end, forming a sort of spur.*

A second division of Xylophilus, and which will include the family of Melolonthides of the younger Mr. Macleay, has the following characters: the labrum is formed like a transverse leaf, in general strongly emarginated underneath in the middle, so that seen in front it has nearly the shape of a reversed and semi-truncated heart. The mentum is as long or longer than it is wide, a little narrowed near the top, either subquadrate or nearly heart-shaped, the upper edge is straight, either more or less emarginated, or concave in the middle, but without a tooth-formed dilatation. The jaws are commonly scaly and armed with many (in general five or six) teeth.

This division may be separated into two groups, one of which will include *G. melolontha* of Fab., such as Illiger and myself have restricted it; and the other, *Hoplia* of the latter. The first of these subdivisions may keep the name of *Melolonthides*, and the other may receive that of *Hoplides*.

We may thus describe the first:—The number of leaves complete in the knob more than three in many; body generally thick, mandibles strong, altogether or for the most part corneous, having at most but one, membranaceous, downy, appendix situate in the concavity or emargination of the inner side, the upper extremity strongly truncated, with two or

* Kirby Lin. Trans. XII. 401; *Geniates barbatus*, ibid. XXXI. 8; *Melolontha obscura, lanata*, of Fab., the species named *Nigrifons*, by Mr. Stevens, and described in the Synon. of the Insects of M. Schœnh, and probably other species appear proper to form another sub-genus allied to Geniates, but with the tarsi not dilated.

three teeth or angular projections; all the tarsi terminate in two hooks; the first articulation of the two anterior tarsi not elongated into a bent appendage; the labrum is commonly apparent, and the maxillary teeth are robust.

The species of *Melolontha* of Fab., which will form the subgenus

MELOLONTHA, Fab., properly so called,

Have ten articulations in the antennæ, the last five or seven of which in the males, and the last five or six in the females constitute the knob. The labrum is thick and strongly emarginated underneath; all the hooks of the tarsi are equal, terminated in an entire point, and simply unindentated at the base. The posterior extremity of the abdomen finishes in general in a point or stylet, at least in the males.

Among the species whose knobs of the antennæ have seven leaves in the males, and six in the females, we will cite

Scarabæus fullo, Lin. Oliv. Col. 1, 5, III. 28, about an inch and a half long, brown or blackish, with three lines on the corslet, two ovoid spots on the shield, and many other irregular on the elytra white. The knob of the antennæ of the males is very large. It is found on the maritime coasts on the Downs.

S. melolontha, Lin. Oliv. *ibid*, I. 1, a. d.* black, downy, with the antennæ, the upper edge of the hood, the elytra, and the greatest part of the legs of a reddish bay. The

* At the very moment of sending this sheet to the press, M. Straus's work on the anatomy of this insect, was presented to the Royal Academy of Sciences, which work he has published at his own expense. We regret exceedingly not having had the opportunity of using this excellent work. M. Leon Dufour had already made known all that relates to the digestive system, and to the organs of generation. M. Chabrier had also written and figured with great exactness the muscles of the wings and thorax. M. Straus has supplied perfectly the other deficiencies.

corslet a little dilated, and marked with an impression towards the middle of the lateral edge sometimes black and sometimes red. Four elevated lines on the elytra, whose outer edge is of the ground colour. There are white triangular spots on the sides of the abdomen. The anal stylet narrows insensibly to a point.

M. Hippocastani, Fab. Oliv. *ibid*, I. 3, a, b, c, which was at first confounded with the last, is a little smaller, shorter, more convex with the elytra, bordered with black, the anal stylet shorter in proportion, and more contracted towards the end, which appears thence to be larger and obtuse.

The alimentary tube of S. melolontha is, according to Dufour (Annal. des Sciences Nat. III. 234), less extended than that of copris, but with the parietes more robust. The chylific ventricle is entirely deprived of papillæ, and is elegantly fringed at the surface by the hepatic vessels. The slender intestine is followed by a sort of colon having interior valves formed like small triangular pockets imbricated, disposed on six longitudinal series, separated by as many muscular chords. This naturalist has often found these pockets filled with a green vegetable pulp. The biliary vessels are of an extremely delicate structure, forming multiplied folds, many of them having, right and left, little barbets like a fringe. The male organ is large, very hard, terminated by two strong hooks, and has toward its hinder third part an articulation. Each testicle is an aglomeration of six spermatic capsules, orbicular as if umbilical, and each furnished with its own tubular conduit, so as to resemble those leaves designated by botanists by the term *pelta*.

These insects appear in some years in such great abundance, that they despoil of their leaves in a short time large extents of woods. The larvæ are not less destructive to our gardens, and is commonly called the *white worm*.

A fourth species, *M. villosa*, Oliv. *ibid* I. 4, is distinguishable from the foregoing by the knob of the antennæ which has five leaves in the males and four in the females. The body is a more or less deep brown, sometimes reddish on the upper side, with three grey lines of down on the corslet; the shield and under part of the body are furnished with a similar down, forming spots on the sides of the abdomen.*

Hereafter the antennæ knob has, in both sexes, only three leaflets.

Rhisotrogus, Lat.

Perfectly like the last as to the general form of the body, labrum, and tarsi; but the antennæ with nine or ten articulations have only three leaflets to each knob.†

Ceraspis, Lepel., Serv.

Have, at the middle of the posterior edge of the corslet, two small longitudinal incisions, and the space comprised forms a tooth, the extremity of which is received in a corresponding emargination of the scutellum. The antennæ have ten articulations. All the hooks of the tarsi, with the exception of the anterior ones, are unequal; the strongest of the intermediate is entire in the male; the others, and the six in the female, are bifid. The body is either covered or sown here and there with small scales.

* Add *M. hololeuca*, Fisch. Entom. of Russia, II. 28. 3; *M. Ankæteri*, ejusd. 4; *M. pilosa*, Fab. Fisch. *ibid.* 9; *M. occidentalis*, Fab., &c. See Schœnh. Synon. Insect. I. 162.

† As it is not always easy to distinguish well the number of articulations which immediately precede the knob of the antennæ, I re-unite the genus which I had named *Amphimallus*, and in which these organs have but nine articulations, to that of rhisotrogus. The *M. Solstitialis*, *pini*, *serrata*, *fervida*, *atra*, *æquioxialis*, *ruficornis*, &c. of Fabricius. The third articulation seems to be decomposed.

But few species are known, and all belonging to Brazil.*

AREODES, Leach, Macl.

Have ten articulations to the antennæ, the sternum horny, and all the hooks of the tarsi equal in those individuals which are presumed to be females (Lepel. and Serv.), and unequal in the males ; the thickest of the anterior two of the latter is bifid, and all the others are entire.

These insects have brilliant colours.

All the preceding phyllophagi, some excepted, have presented us with antennæ of ten articulations. In all the following, and of the same division, or that of Melolonthidæ, we reckon but nine.

In some the crooks of the tarsi are equal ; one of the two anterior at most is occasionally thicker.

DASYUS, Lepel. and Serv.,

In which the crotchets of the two anterior tarsi, at least in the males, are bifid, and the others entire.

SERICA, Macl. *Omalopia*, Dej.,

Which have all the hooks of the tarsi bifid, the body ovoid, gibbous (silky, and often with a changing reflexion), with the corslet broader than long.†

DIPHUCEPHALA, Dej.

Have also all the hooks of the tarsi bifid ; but the body is

* The *ceraspis pruinosa* of MM. Lepel and Serv. (Encyc. Method.) is the *M. bivulnerata* of M. Germar. The *M. variegata* of the latter appears to me to be also a Ceraspis.

† Macl. Hor. Ent. I. part I. p. 146. The *M. brunnea*, *variabilis*, *ruricola*, &c. of Fabricius. Mr. Macleay says that the antennæ have ten articulations, but I have counted but nine. The length and form of those of the tarsi vary.

Pl. 44

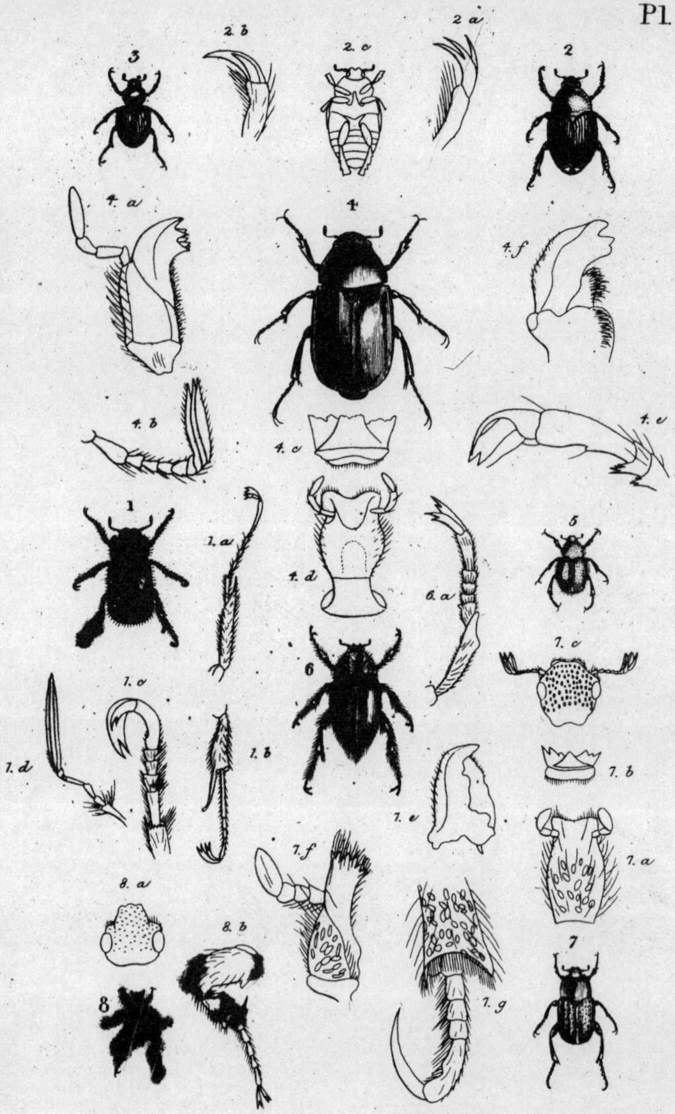

1. Plectris *tomentosa.* *Lap. & Serv.* 2. Popilia *nitidicollis.* *Gory.* 3. Anisoplia *saturalis.* *Guer.* 4. Euchlora *viridana.* *Guer.* 5. Lepisia *rupicola* *Serv.* 6. Dicrania *velutina.* *Gory.* 7. Hoplia *farinosa.* *Fabr.* 8. Dichelus *dentipes.* *Serv.*

narrow, elongated, with the corslet almost square. The first articulations of the four (male) or two (female) anterior tarsi are short, and furnished underneath with brushes. These same articulations are dilated, or broader in the first four tarsi of the males. The hood is strongly and regularly emarginate.

These insects are proper to Australasia.*

MACRODACTYLUS, Lat.

Resemble the Diphucephala, as regards the hooks of the tarsi, and the elongation of the body; but here the corslet is longer, almost hexagonal, and all the articulations of the tarsi are similar in the two sexes, elongated, and simply hairy. These insects are peculiar to the New Continent.†

In others, the hooks of the intermediate tarsi alone are unequal.

PLECTRIS, Lepel. et Serv.

The thickest of these hooks and the two of the other tarsi are bifid; the first articulation of the posterior tarsi is very long.‡

In the others, all the hooks of the tarsi are unequal. Those of the posterior two, at least, are always entire. One at least of the two or four anterior tarsi of the males, and sometimes of the females, is bifid.

POPILIA, Leach,

In which the sternum is advanced between the first feet, in the manner of a compressed and truncated, or very obtuse lamina.||

* *Melolontha colaspidoides*, Schœnh. Synon. Insect. I. 3, App. p. 101. See the Cat. of the Collect. of M. le Comte Dej.

† *M. Subspinosa*, Fab., and many other unedited species.

‡ Encyc. Method. Article, *Scarabeides*.

|| *Trichius*, 2—*punctatus*, Fab.

Euchlora, Macl. *Anomala*, Meg. Dej.

Having no sternal projection, in which one of the hooks of the four anterior tarsi is bifid in the males, and in which the body is gibbous, with the hood round and transverse.*

Lepisia, Lepel. and Serv.

Presenting no longer a sternal horn, and distinct from the preceding by their four anterior tarsi, the two hooks of which are bifid.

The *Haplides*, or phyllophagi of our third and last division, have the mandibles small, depressed, and, as it were, divided longitudinally into two parts, the internal one of which is membranous, and the other corneous; the superior extremity presents no sensible denticulations. The body is short, depressed, and broad, with the elytra narrowed posteriorly at the external side. The labrum is concealed, or but little seen;† the jaws are often but small denticulations. The two last tarsi have usually but a single hook; in those in which all have two *(Dicrania)*, the first articulation of the anterior tarsi is prolonged inferiorly, and presents at the internal side a strong hooked tooth.

M. Leon Dufour remarks, that the digestive canal of Hoplia is much less long than that of Melolontha, and that it approaches more to that of Cetonia. The chylific ventricle is smooth and flexuous. The narrow intestine is less short

* The *M. viridis, bicolor errans, marginata, cyanocephala, vitis, Julii, Frischii, holosericea, aurata*, &c. of Fabricius. See Macleay, Hor. Entom. I. part I. p. 147. The genus *Mimela*, of Mr. Kirby, appears to me to approach much to that of Euchlora, but not having seen any individual, I confine myself to this simple indication.

† In the last preceding sub-genera this piece, seen in front, only presents an edge, linear, transverse, entire, or but slightly emarginate in its middle.

than in Melolontha, and often presents at its origin an ovoid inflation. It is followed by a bulky elongated intestine, destitute of valvulous anfractuosities. The rectum is distinct from it by a pad or swelling, and well marked. The organs of generation scarcely differ from those of Melolontha.

DICRANIA, Lepel. and Serv.

Have two hooks, all equal and bifid, to all the tarsi, the anterior two of which have their first articulation prolonged inferiorly into a crooked tooth. The body is very smooth, without scales, with the scutellum tolerably large, and two strong spines at the extremity of the four hinder limbs. The lower end of the last two limbs is dilated. These insects inhabit Brazil.

HOPLIA, Illig.

Have but a single hook to the two posterior tarsi; the two of the others are unequal and bifid. The extremity of the last four limbs is crowned with small spines, none of which are manifestly longer than the others. The body is generally furnished with scales. The hood is almost square, or almost semi-circular. The thighs of the two posterior feet are moderately inflated, and their legs are long, straight, and without a hooked tooth at their extremity.

There is very commonly found in the South of France, near the borders of streams and rivers, the handsomest known species of this sub-genus *(H. formosa*, Illig., *Melolontha farinosa*, Fab. Oliv., Col. I. 5, ii. 14, a. c.). Its antennæ have nine articulations. All its body is covered with brilliant silvery scales, the upper ones of which have a reflection of violet blue, and the lower ones are a little greenish or golden.

The antennæ of some others have ten articulations.

MONOCHELES, Illig.

Do not differ from Hoplia but in their hood, which is in the form of a truncated triangle at its anterior extremity, and in the two hinder feet, the thighs of which are very bulky, and the legs are short, with a strong hooked tooth at their extremity.

Some Scarabœidæ, very near the last of the preceding section, and which had been at first united with them, in the genus Melolontha, but whose paraglossæ, or the two divisions of the tongue, form a projection beyond the superior extremity of the chin, and whose elytra are a little removed from the side of the suture, at their posterior extremity—this end being narrowed into a point, or rounded—compose a fifth section, that of ANTHOBII.

The antennæ have from nine to ten articulations, of which the last three alone form the knob in the two sexes. The lobe terminating the jaws is often almost membranaceous, silky, in the form of a pencil, coriaceous, and denticulated at the internal edge in the others. The labrum and mandibles are more or less solid, according as these parts are naked, or concealed. These insects live on flowers, or leaves.

Some have the mandibles and the labrum projecting, and two entire and equal hooks on all the tarsi.

The antennæ have ten articulations; the maxillary palpi are a little thicker towards the end, with the last articulation but little elongated, and truncated. The mandibles are corneous.

Some of these insects inhabit the North of Africa, and other countries situated on the Mediterranean; the majority of the others frequent the elevated countries of Western India.

In some, the first articulation of the knob of the antennæ is concave, and emboxes the others.

Pl. 4

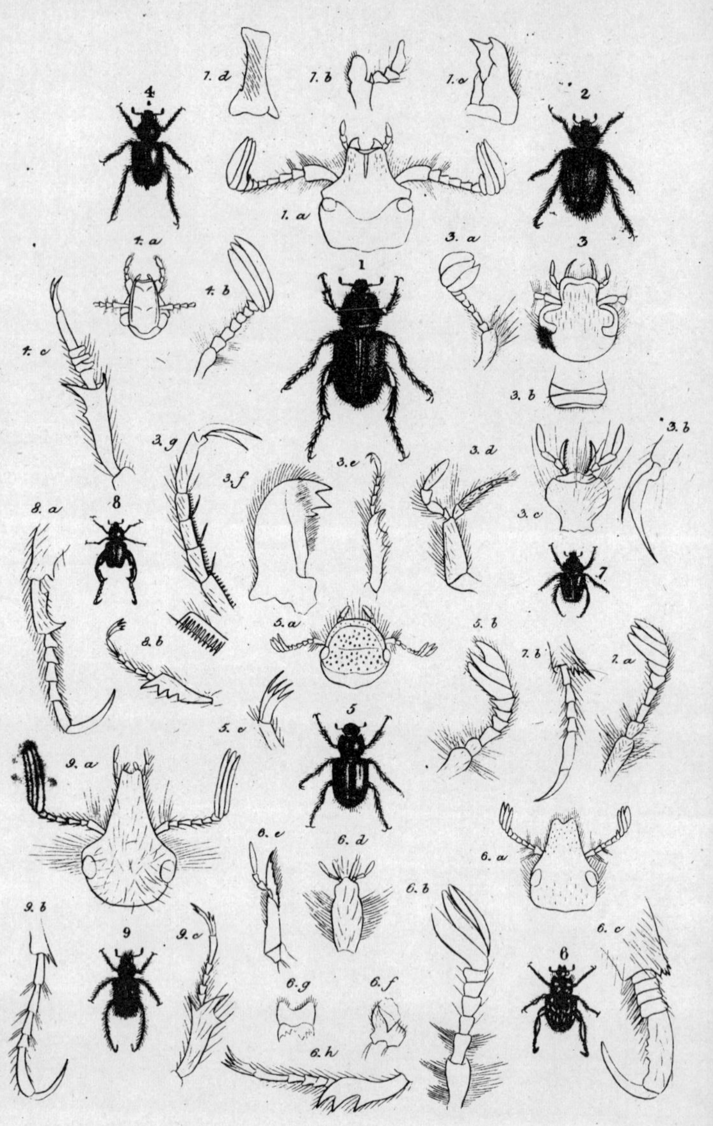

1. Glaphyrus *rufipennis*. *Gory.* 2. Amphicoma *bombytiformis*. *Pall.* 3. *Details of* Amph. *Lasserii*. 4. Anthipna *abdominalis*. *Esch.* 5. Chasmopterus *hirtulus*. *Illig.* 6. Pachycnemus *crassipes*. *Serv.* 7. Lepitrix *abbreviatus*. *Serv.* 8. Monochelus *gonager*. *Serv.* 9. Anisonyx *nasua*. *Wied.*

Glaphyrus, Latr.

Have the internal edge of the mandibles denticulated, and an acute angle at the other edge; the knob of the antennæ almost ovoid; the teguments firm, and the posterior thighs inflated. The maxillary palpi are sensibly larger than the labial, with the last articulation longer than the preceding. The internal lobe of the jaws is in the form of a tooth; the external or terminal is coriaceous. The corslet is oblong. The posterior feet are large.

Amphicoma, Latr.

Have the mandibles rounded and arched on the external side, without denticulations at the internal edge; the knob of the antennæ globular, the abdomen soft, and all the feet of the usual size.

The hood is very much edged. The anterior legs have three teeth at the external side. The first four articulations of their tarsi are strongly ciliate in the males.

In this sub-genus and the following, the jaws terminate in a membranous lobe, narrow, elongated, and in the form of a shred. Their palpi are scarcely longer than the labial, and the length of their last articulation little exceeds that of the preceding.

In others, such as

Anthipna, Escholtz.,

The knob of the antennæ is formed of free leaves, and is oval.

The hood is not edged or raised in front. The middle portion of the head forms with it a plate in a long square, edged laterally and posteriorly. The anterior limbs have two teeth at the external side. The first four articulations of the tarsi are dilated and tooth-formed in the males. These insects otherwise resemble Amphicoma.*

The others have the labrum and mandibles covered, or not projecting, and some, at least, of the hooks of their tarsi are bifid. The chin is elongated and hairy.

Sometimes all the tarsi have two hooks. The antennæ have never more than nine articulations. The hood is usually transverse. The palpi are but little elongated, with the last articulation ovaliform.

Sometimes the posterior feet differ but little from the others.

CHASMOPTERUS, Dej. *Melolontha*, Illig.,

Have all the hooks of the tarsi bifid; the terminal lobe of the jaws strait, elongated, with two teeth wide apart at the internal edge; the body almost ovaliform, with the corslet rounded, and the elytra of equal breadth throughout.

CHASME, Lepel. and Serv.,

Do not appear to differ from chasmopterus but in the crotchets of the two posterior tarsi, of which the thickest alone is bifid.

Sometimes the posterior feet have, at least in the males, the thighs very thick, denticulated, and the legs thick, and terminated by a very strong hook.

DICHELES, Lepel. and Serv. *Melolontha*, Fab. Oliv.

The body is short, but little hairy, with the elytra narrowed towards their extremity in an elongated triangle. The posterior feet are in part contractile. All the hooks of the tarsi are equal and bifid. The terminal lobe of the jaws is

* *Amphicoma abdominalis*, Latr. Gen. Crust. et Insect. II. p. 119; *M. Alpina*, Oliv. Col. I. 5. X. 112.

denticulated along the internal edge, as in hoplia, to which this sub-genus considerably approximates.

Sometimes the two posterior tarsi have but a single hook. Those of the others are unequal and bifid.

Some of them have, like the preceding, but nine articulations to the antennæ.

LEPITRIX, Lepel. and Serv. *Trichius*, *Melolontha*, Fab.

The body is short, with the corslet more narrow than the abdomen, almost squared, and a little narrowed posteriorly. The abdomen is broad, and the hinder feet large. The last articulation of the maxillary palpi is much longer than in the preceding sub-genera. The terminal lobe of the jaws is very small, in the form of a short triangle.

The others have ten articulations to the antennæ.

The body is short, very hairy, with the hood in the form of an elongated triangle, truncated, or very obtuse at the end; the palpi projecting, terminated by a long and cylindrical articulation, the maxillary lobe long, narrow, projecting at its extremity, and without teeth. The abdomen large, and the hinder feet long.

PACHYCHNEMUS, Lepel. and Serv. *Melolontha*, *Trichius*, Fab.

Have the elytra narrowed towards their extremity, the thighs and legs of the two hinder feet inflated, these last almost in a knob, with one of the two spurs of the end much stronger than the other.

ANISONYX, *Melolontha*, Fab.

Whose elytra form a long square, rounded posteriorly, in which the hinder legs are almost cylindrical, or in the form of an elongated cone, with the two spurs of the end of equal size.

The sixth and last section of scarabeidæ, that of Melitophili, is composed of insects whose body is depressed, most frequently oval, brilliant, without horns, with the corslet trapeziform, or almost orbicular; an axillary piece occupies, in the greatest number, the space comprised between the posterior angles and the exterior of the elytra. The anus is uncovered. The sternum is often prolonged in the manner of a point, or advanced horn. The hooks of the tarsi are equal and simple. The antennæ have ten articulations, the last three of which form a knob, always foliated. The labrum and mandibles are concealed, in the form of flatted laminæ, entirely, or almost entirely, membranaceous. The jaws are terminated by a silky lobe, in the form of a pencil, without corneous teeth. The chin is usually ovoid, truncated superiorly, or almost square, with the middle of the superior edge more or less concave or emarginate. The lingua is not projecting.

Some anatomical observations made upon many of these insects by M. Leon Dufour, lead us to the conclusion that they are, of all the scarabeidæ, those in which the alimentary tube is the shortest. The chylific ventricle has, commonly, its external tunic covered with very small superficial papillæ, in the form of projecting points. The inflation which terminates the narrow intestine is not cavernose like that of melolontha. The apparatus of generation in the males also differs from that of these last. The spermatic capsules are ten or twelve in number to each testicle. Their proper conduits do not all flow together to one point, for the formation of the deferential canal, but they communicate among themselves in divers manners. The number of vesiculæ seminales is one, or three pair. The ejaculatory conduit is turbinated, and swelled much before it penetrates into the copulatory apparatus. (*See* Annal. des Sciences Natur., tom. III. p. 235, and IV. p. 178.)

The larvæ live in rotten wood. The perfect insect is found on flowers, and often on the trunks of trees, from which a liquor distils which they suck.

This section is susceptible of being divided into three principal divisions, which correspond, the first, to the genus *trichius* of Fabricius; the second, to that of *Goliath* of M. de Lamarck; and the third, to that of *cetonia* of the first, but reduced and simplified by the retrenchment of the second genus, as well as that of *rutela* and other analogous groups.

The melitophili of the first two divisions have no sternal projection well marked. The lateral piece of the mesosternum, which we have designated by the epithet axillary (*epimeros* of Audouin), does not generally appear above, and occupies only a portion of the space comprised between the posterior angles of the corslet, and the exterior base of the elytra. The corslet does not widen from front to rear, as in the cetoniæ. The external side of the elytra is not abruptly narrowed a little below the humeral angles, as in these last mentioned insects. But a character which appears to us more rigorous, is that where the labial palpi are inserted in lateral fossets of the anterior face of the mentum, so that they are entirely discovered, and that the sides of this mentum out-edge them even at their origin, and protect them behind. In the first two divisions these palpi are inserted under the lateral edges of the mentum, or in the edges themselves, so that the first articulations do not appear, looking frontwards.

Some (*Trichidæ*) have the mentum either almost isometrical, or longer than broad, and leaving the jaws discovered. These are,

Trichius of Fabricius.

T. nobilis, *Scarabæus nobilis*, Lin. Oliv. Col. I. 6. iii. 10. About half an inch in length, of a golden green above, coppery, with hairs of a yellowish grey underneath. Found on umbelliferous flowers.

T. fasciatus, *S. fasciatus*, Lin. Oliv. *ibid.* ix. 84, a little smaller, black, with thick yellow hairs; cases of this last colour, with three black bands, transverse, and interrupted at the suture. Very common in spring, upon flowers.

T. Eremita, *S. Eremita*, Lin. Oliv. *ibid.* iii. 17. Large, of a brown black; edges of the head raised; three furrows upon the corslet. Found on the trunk of old trees, in the interior of which dwells the larva.

T. hemipterus (female), *S. hemipterus*, Lin. Oliv. *ibid.* IX. 83, xl. 103, and those of some other species of North America, are remarkable for the corneous auger, in the form of a dart, at the posterior extremity of the abdomen, which serves to introduce their eggs.

These species commonly remain upon the ground, where they walk very slowly. The last articulation of their maxillary palpi is proportionally shorter and thicker than that of other trichii. It has appeared to me that the first of the posterior tarsi considerably exceeded the following in length, while in the other trichii it is very little longer.

The second division (*Goliathidæ*) is distinguished from the preceding in the character of the chin, which is much more large, broad, and covers the jaws.

In some the chin is concave in its middle, having the figure of a widened heart, or of a transverse square. The interior extremity of the hood is neither denticulated nor corneous. The corslet is in the form of a heart, truncated at the two ends, and narrowed abruptly behind, or in the form of a transverse square, rounded laterally.

The first articulation of the antennæ is very large, triangular, or in an inverted cone. The palpi are short; the last articulation of the maxillary palpi is elongated. The external side of the first two legs presents two teeth.

Pl. 35

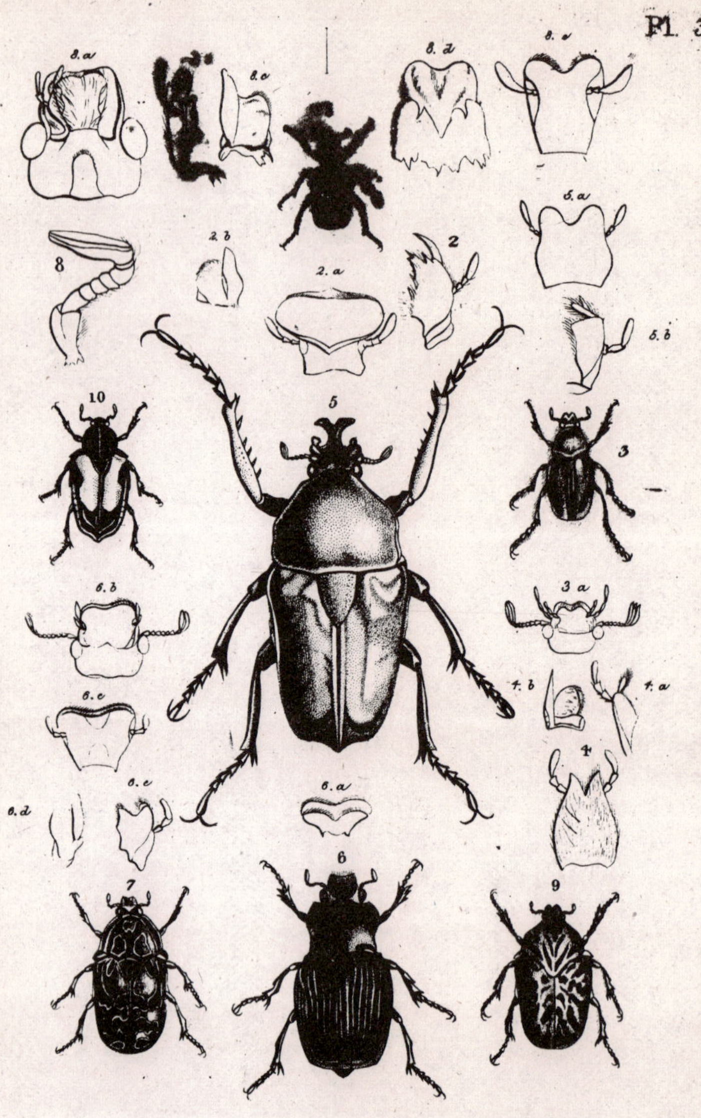

1. Cremastocheilus *hirtus*. *Gory & Perch. monogr.* 2. *Details of* Cremast. *elongatus. Oliv.* 3. Trichius *zebra. Oliv.* 4. *Details of* Trich. *fasciatus. F.* 5. Goliath *micans. G. & P. Oliv.* 6. Platygenia *zairica. Mac. Leay.* 7. Cetonia *Baxii G. & P.* 8. *Details of* Cet. *aurata. F.* 9. Gymnetis *nervosa. G. & P.* 10. Macronota *aegregia. G. & P.*

Platygenia, Macl.

Their body is very much flatted, with the corslet almost in the form of a heart, broadly truncated at the two ends; the jaws terminated by a bundle of hairs, and the internal lobe of which is triangular, emarginated at the end; the last articulation of the palpi is ovoid-cylindrical; the chin almost squared, emarginate at the middle of the superior edge, and a little upon the sides, and the posterior legs very hairy at the internal side.*

Cremastocheilus, Knoch.,

Whose corslet is almost in the form of a transverse square; whose jaws are terminated by a strong tooth, crooked or scythe-like, with silken hairs or small spines at the place of the internal lobe, which have the last articulation of the palpi very long and cylindrical, and the mentum in the form of a widened heart, or a reversed triangle, and rounded at the upper angles, without any sensible emargination.†

* Macl. Hor. Entom. I. part I. p. 151; *Trichius barbatus*, Schœnh. Synon. Insect. I. iii. App. 38.

† Latr. Gener. Crust. et Insect, p. 121. M. Dupont, naturalist to his Highness the Duke of Orleans, and whose collection of coleopterous insects is, after that of M. Dejean, the richest in Paris, has received from Lamana (French Guiana), an insect presenting all the essential characters of Cremastocheilus, but in which the axillary pieces are more apparent when the animal is observed from above. The anterior legs are arched, and have at the internal side a strong projection in the form of a tooth. All the tarsi are short, thick, cylindrical, and terminated by two very long hooks. The hood is raised at its interior extremity, in the manner of an almost square lamina. The posterior extremity of the head presents an elevation divided into two teeth, or tubercles. This insect is an inch long, with a red spot on the upper part of each elytron.

The *Cetonia elongata* of Olivier, appears to be a chemastocheilus.

GOLIATH, Lam., Kirb. *Cetonia*, Fab., Oliv.

A sub-genus which is composed, according to M. de Lamarck, of large and handsome species, some from Africa and the East Indies, others from equatorial America. MM. Lepeletier and de Serville (Encyc. Method., Article *Scarabeides*) have separated these last under the generic name of INCA. The axillary piece is not prominent. The two anterior feet have the thighs furnished with a tooth, and an emargination at their external base. The upper edge of the mentum is strongly emarginated in its middle; this piece in Goliath, properly so called, presents four lobes or teeth, two upper, and two others lateral. The labial palpi are inserted on its edges, in the emarginations of these last lobes. All the species with which we were acquainted were of large size. But M. Verreux, jun., nephew and fellow traveller of the late Delalande, and who has returned to the Cape of Good Hope, has just sent over a species which is not larger than *C. gagates*, which it otherwise resembles in its colours, and presents all the characters of Goliath. The *C. geotrupinus* of M. Schœnherr is perhaps also congenerous. The corslet of Goliath is less round and more narrow in front than that of Inca, and the legs have no emargination at the internal side.*

In the third division of the melitophili, a division corresponding to the family of cetonidæ of Mr. W. Macleay, the sternum is prolonged more or less into an obtuse point, be-

* *See* Encyc. Method., Article *Scarabeides;* Hist. Nat. des Animaux sans Vertebres, of M. de Lamarck; Observ. Entom., of M. Weber, and the twelfth volume of the Linnæan Trans., p. 407, in which Mr. Kirby describes two species. There is found in the Island of Java, an insect, which might be taken, at the first glance, for a Goliath, and which MM. Lepeletier and Serville have considered as such, but it has all the essential characters of cetonia; only the corslet is more rounded, and narrowed posteriorly. The male has a forked horn upon the head.

tween the second feet; the axillary piece always shews itself above, and occupies the void separating the posterior angles of the corslet from the base of the elytra. The corslet usually widens front to rear, and has the form of a triangle truncated anteriorly, or at its point. In some, however, it is almost orbicular, as *C. cruenta*, Fab.; *C. vencosa*, Schœnh, &c. The chin is never transverse, its upper edge is more or less emarginated at the middle. The terminal lobe of the jaws is silky, or in the form of a pencil. The body is almost ovoid and depressed.*

This division comprehends the genus

CETONIA of Fabricius,

The species excepted which belong to the preceding subgenus, and to that of Rutela. (Gener. Crust. et Insect.)

Some have the corslet prolonged posteriorly in the form of an angle, so that the scutellum disappears altogether. They form the genus GYMNETIS of Mr. W. Macleay, (Hor. Entomol. I. Part I. p. 152.) The new continent produces many species of them. The island of Java and other eastern countries of Asia present others, in which the corslet is equally elongated, but in which the scutellum, though very small, is still visible.† The chin is more deeply emarginated in the manner of an angle, and the last articulation of the labial palpi is proportionally longer. The hood is more or less bifid. Other species of the East Indies or New Holland, in which this

* M. Chevrolat, possessor of a very fine collection of Coleoptera, many of which belonged to that of the late Olivier, has shewn me a species found in the Island of Cuba, by M. Poe, having the appearance of Trichius, but with the axillary pieces and sternal prolongation of cetonia. Some species of this last genus (*C. cornuta*, Fab.) have the corslet provided with a small horn, and at the first glance resemble Scarabæi.

† *C. chinensis*, Fab. ejusd.; *C. regia;* the *C. plana*, *imperialis* of Schœnher.

piece is also bifid or armed with two horns in the males, whose body is proportionally more narrow and more elongated, with the abdomen growing remarkably narrow from front to rear, even almost triangular, and the knob of the antennæ is very much elongated, compose the genus *Macronota* of M. Wiedmann. But all these sections can only be well established when a particular study shall be made of the numerous species of the genus *Cetonia* of Fabricius.

Those of Europe are provided with a scutellum of the usual size. Such are

Cetonia aurata, *Scarabæus auratus*, Lin. Oliv. Col. I. b. 1. 1. nine lines in length, of a brilliant golden green above, of a coppery red underneath, with white spots on the elytra. Common on flowers, and often on those of the rose and elder-trees.

C. fastuosa, Fab. Panz. Faun. Insect. Germ. XLI. 16. larger than the preceding, of an uniform golden green, without spots, and with the tarsi blueish. South of France.

S. sticticus, Lin. Panz. *ibid*. I. 4. five lines in length, black, a little hairy, with white points; those of the belly arranged in two or three lines, according to sex. Very common on thistles.*

The second tribe of Lamellicornes, the LUCANIDÆ, thus named from the genus *Lucanus* of Linnæus, have the knob of the antennæ composed of leaves or of teeth disposed perpendicularly to the axis, in the manner of a comb. These organs have always ten articulations, the first of which is usually by far the longest. The mandibles are always corneous, most frequently projecting and larger, and even very different in the males. The jaws of the majority terminate in

* *See* the first division of the cetoniæ of Olivier; Latr. Gener. Crust. et Insect. I. iii. p. 126; Schœn. Synon. I. iii. p. 112; and the fourteenth volume of the Linnæan Trans., with respect to the genera *genuchus*, *schizorhina* and *gnathocera*, established at the expense of that of cetoniæ.

a narrow, elongated, and silky lobe; those of the others are entirely corneous and denticulated. The tongue, in the greater number, is formed of two small silken pencils, more or less projecting beyond a chin which is almost semicircular or squared. The anterior feet are most frequently elongated, with the legs denticulated all along their external side. The tarsi terminate in two equal simple crotchets, with a small appendage terminated by two silken hairs, between the hooks. The elytra cover all the upper part of the abdomen.

We shall divide them into two sections, which correspond to the genera *Lucanus* and *Passalus* of Olivier.

Antennæ strongly bent; smooth, or but little hairy; a labrum very small or confounded with the hood; jaws terminated in a membranaceous or coriaceous lobe, very silky, pencil-formed, without teeth, or presenting but one at the most; a tongue, either entirely concealed or incorporated with the chin, or divided into two lobes, narrow, elongated, silky, and more or less projecting beyond the chin, mark the first of these sections; the scutellum, moreover, is placed between the elytra.

This first section will form the genus

Lucanus.

We shall make a first division with those in which the knob of the antennæ is composed but of three or four articulations or leaves.

We shall commence it with insects almost entirely similar, with the exception of the antennæ, to the Oryctes, a sub-genus of the preceding tribe. The mandibles are concealed, without teeth, and similar in the two sexes. The chin is almost triangular, and entirely conceals the tongue, as well as the base of the jaws. The body is thick and convex above, almost cylindrical, and rounded posteriorly. The corslet is truncated and excavated in front. The head of the males is provided with a horn.

SINODENDRON, Fab.*

The knob of the antennæ is formed by the last three articulations.

Those whose body is thick, convex, ovoid, with the mandibles like a compressed pincers, and rising vertically in the males; the head much more narrow than the corslet, measured in its greatest breadth; and the legs, or at least the two anterior ones, broad, and in the form of an inverted triangle, form two sub-genera.

ÆSALUS, Fab.,

In which the mandibles, even in the males, are shorter than the head, and terminate above in the manner of a horn; in which the chin conceals the jaws; whose tongue is very small; whose body is short and gibbous, with the head almost entirely received in the emargination of the corslet, the legs compressed, triangular, and the sternum simple, and without projection. (*Æsalus scarabæoides*, Fab. Panz. Faun. Insect. Germ. XXVI. 15, 16.)

LAMPRIMA, Latr.,

In which the body is more elongated, with the mandibles much longer than the head, in the males, in the form of vertical angular laminæ, very much denticulated, and hairy internally; the jaws uncovered as far as their base; the

* *Scarabæus cylindricus*, Lin. Oliv. Col. I. 3, ix. 88. It is the only known species; the other sinodrenda of Fabricius, belong to other genera.

Pl. 51

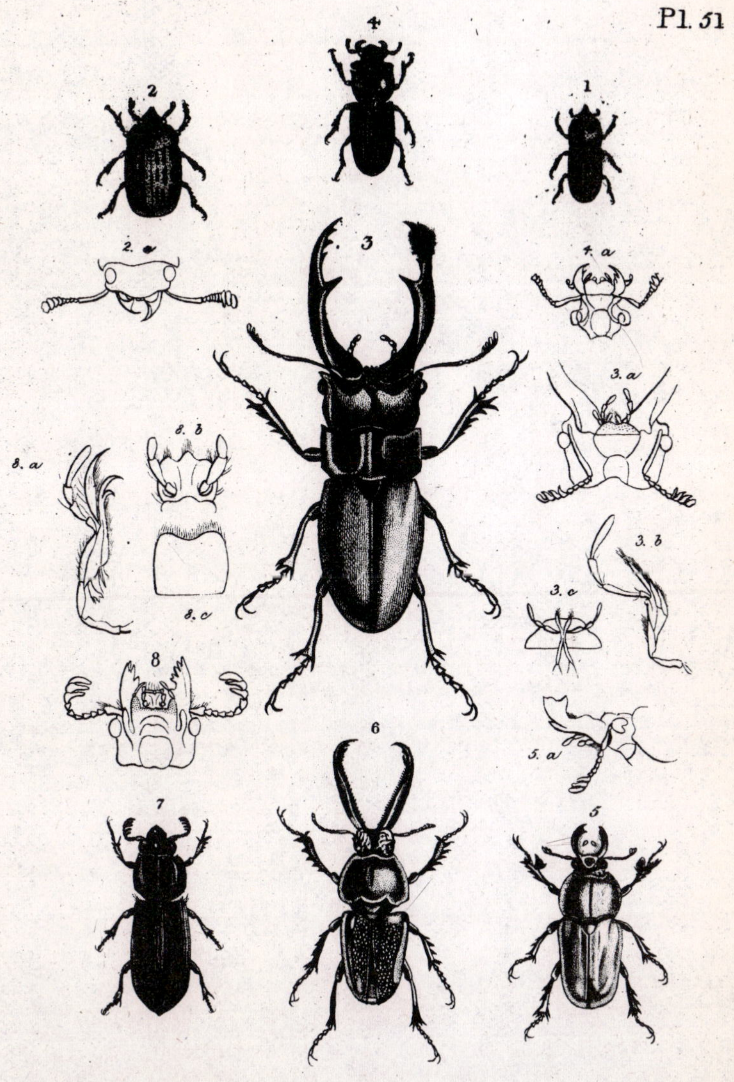

1. Sinodendron *cylindricum. L. Fab.* 2. Œsalus *scarabæides. Fab.* 3. Lucanus *cinnamomeus. Guer.* 4. Platycerus *auriculatus. Gory.* 5. Lamprima *ænea. Latr.* 6. Pholidotus *Humboldii. Sch.* 7. Passalus *pentaphyllus. Pals.-Bauv.* 8. *Details of* Pass. *interruptus. Fab.*

tongue very distinct; the labrum elongated; the two anterior legs widened, and presenting in the males a pallette-formed spur like a reversed triangle, and a sternal point.*

Two other sub-genera, established by Mr. W. Macleay, approach lamprima, in consequence of their mesosternum which is prolonged and advanced, but less, however, than in the preceding; of their head, much more narrow than the corslet, and their mandibles furnished with down at the internal side; but their body is flatted or but little raised, especially in the females. The labrum is concealed. The anterior legs are strait and without palette. The palpi and the lobes of the tongue are more elongated.

RYSSONOTUS,

In which the mandibles of the males form, as in Lamprima, pincers vertically compressed, angular, and denticulated.†

PHOLIDOTUS, Macl. *Chalcimon*, Dalm. *Lamprima*, Schœnh.,

In which the mandibles in the same sex are very long, narrow, arched, terminated in a hook curved inferiorly, and denticulated like a saw at the internal side.

The knob of the antennæ, formed by the last three articulations, is less pectinated than in the others, and almost perfoliate. The chin covers the jaws.‡

* Latr. Gener. Crust. et Insect. II. p. 132; *Lethrus æneus*, Fab. Schreib. Linn. Trans. VI. 1. See also as to this species and others, Macl. Hor. Entom. I. part I. p. 99.

† *Lucanus nebulosus*, Kirb. Trans. Linn. XII. xxi. 12; Macl. Hor. Entom. I. p. 1. p. 98.

‡ *Lamprima Humboldti*, Schœnh.; *Chalcimon Humboldti*, Dalm. Entom. I. p. 3; *pholidotus lepidosus*, Macl. Hor. Entom. I. part I. p. 97, the male, ejusd.; *Cassignetus geotrupoides*, the female.

In the following, the mesosternum makes no projection; the head is as large as, or even larger, (divers males) than the corslet. The mandibles are smooth, or at least without any thick down at the internal side. The body is always flatted.

In some, the eyes are not cut transversely by the edges of the head, the jaws are terminated by a very slender lobe in the form of a pencil, and without corneous teeth.

Lucanus (Proper), Lin.

The digestive canal is much less elongated than that of the scarabæidæ, but the œsophagus is much longer. The male organs of generation also differ much from those of the preceding, the testicles being formed by the circumvolutions of a spermatic vessel, and not by an agglomeration of capsules of this nature. The adipose tissue, almost nullified in the scarabæidæ, is here abundant and disposed in clusters, which converge to a medial line.

We presume that the larva of our great Lucanus, which lives in the interior of oaks, and passes some years there before it undergoes its final transformation, is the *cossus* of the Romans, or that animal having the form of a worm, which they regarded as a delicate viand.

L. Cervus, Lin. Oliv. Col. I. i. 1. Rœs. Insect. II. Scarab. I. iv. v. the male two inches long, larger than the female, black, with the elytra brown; the head broader than the body; mandibles very large, arched, with three very strong teeth, two of which at the end are divergent, and the other is at the internal side; they also have some small ones. The females, designated (in French) under the name of *biches*, have the head more narrow, and the mandibles much smaller. This insect flies in the evening during the summer solstice. Its size and mandibles vary. It is to one of these varieties that we must refer the *lucane chevre* of Olivier, or the *L. chevreuil* of Fabricius. The Lucanus, thus designated by Linnæus, is a species of North America, and very distinct from the preceding.

L. caraboides, Lin. Oliv. Col. *ibid.* II. 2. five lines in length, of a greenish brown, with the mandibles crescent-formed, and the length of which does not exceed, even in the males, that of the head.*

In others, the eyes are divided transversely, and integrally by the edges of the head. The jaws terminate in a shorter and less narrow lobe than in the preceding, and often present a corneous tooth at the internal edge.

PLATYCERUS, Latr.

The palpi, the maxillary lobes, and the tongue, are proportionally shorter than in the preceding sub-genera. The mentum forms a transverse square, while in the preceding it is often in a semicircle. It conceals on both sides the base of the jaws. The mandibles are generally short.†

The other Lucanides have the knob of the antennæ composed of the last seven articulations.

SYNDESUS, Macl. *Sinodendron*, Fab.

The corslet presents anteriorly a little horn, and in the same manner as that of most of the passali, a furrow in its

* To the Lucani I unite the *ceruchus* and *platycerus* of Mr. Macleay The proportions of the mandibles of the palpi, of the maxillary lobes, of the tongue, and the knob of the antennæ, cannot furnish constant and rigorous characters.

† The *Lucanus parallelipedus*, of Fab., a species, forming with another, the *C. dorcus* of Mr. Macleay. I again unite to the platyceri, the *nigidius*, *Œgus* and *Figulus*, of this learned entomologist.

middle. Its separation from the abdomen is also more marked than in the lucani. The two posterior feet are more thrown behind. The antennæ are less curved.*

The lucanidæ of our second section have antennæ simply arched, or but little bent elbow-like, and hairy; a labrum always uncovered, crustaceous, and transverse; mandibles strong and very much denticulated; but without any remarkable sexual disproportions; jaws entirely corneous, with two strong teeth at the least; a tongue equally corneous or very hard, situated in an upper emargination of the chin, and terminated by three points; the abdomen supported on a pedicle, presenting the scutellum above, and separated from the corslet by a very sensible interval. These insects compose the genus

PASSALUS, Fabr.,

Which Mr. Macleay confines to the species in which the knob of the antennæ has but three articulations, whose labrum forms a transverse square, and whose jaws have three strong teeth at the end, and two at the internal side, at the place of the internal lobe.

The species in which the knob has five articulations, in which the labrum is very short, and whose jaws have but two teeth, one terminal and the other internal, form his genus PAXILLUS.

Finally, he unites to the preceding, in his family of passalidæ, the *G. Chiron*, which we have placed in the tribe of coprophagi.†

These insects are foreign to Europe, and, as far as it appears, to Africa. It is in the eastern countries of Asia, and

* *Sinodendron cornutum*, Fab. Don. Ins. of New Holl. t. 1. 4; *Syndesus cornutus*, Macl. Hor. Entom. I. part I. p. 104.

† Hor. Entom. I. p. 105, et seq.

also more especially in America, that they are found. Mademoiselle de Merian says that the larva of the species, which she figures, feeds on the roots of the *patata.* The perfect insect is not rare in sugar plantations.*

* See Fabricius, Syst. Eleuth. II. p. 255; Web. Observ. Entom.; Palis de Beauv. Insect. d'Afr. et d'Amer.; Latr. Gen. Crust. et Insect. II. p. 136, and Schœnh. Synon. Insect. I. iii. p. 331, and Append. p. 143, 144.

SUPPLEMENT

ON THE

LAMELLICORNES.

The first division of this family begins with the tribe Scarabœides, Section Coprophagi, sub-genus Ateuchus.

We do not see on the hood of the insects of this sub-genus, the horns which are observed in this part in the other coprophagi. This is the character to which M. Weber, the founder of this genus, has made allusion. *Ateuchus* is derived from the Greek, and signifies "without arms," or "without defence." The ateuchi formerly composed a part of the *Scarabœus* of Linnæus, and the *Copris* of Fabricius.

These insects are found but little in Europe, beyond the fiftieth degree of north latitude. They appear proper to the warmer climates, and particularly to Africa. They live in ordure of all kinds, the excrements and dung of animals. Aristotle and Pliny have spoken of them, and designated them under the name of *Pilulariæ*, because they form, with the dung of animals, a tolerably thick ball, which they roll with their hinder feet. This ball, which encloses their eggs, is at first of a soft consistence, and irregular form, but by dint of being rolled it hardens, and becomes round. When it has acquired a sufficient degree of solidity, the insect pushes it with its hinder feet, as far as the hole which it has dug for its reception, with its anterior feet, the legs of which are strong, and armed with three or four denticulations, and pushes it in. It is in the commencement of Spring that those insects are observed to roll their pills. They serve both as food and lodging for the larva which issues from the egg. Sometimes many of them unite together, and roll in common. It happens pretty often, that during this labour, one of them loses its equilibrium, rolls down on one side, and the ball upon the other, and during the time that it takes to rise, the ball becomes the property of the first that can seize it. As soon as it has contrived to replace itself upon its feet, it proceeds in search of another pill, to supply the place of that which it has lost. If it can find none, it sets to work anew with indefatigable ardour in the formation of another. These insects are not very firm on their anterior feet and when on their back, have considerable difficulty in rising, but they fly tolerably well.

The *Ateuchus sacer*, has been so named because it formed an object of religious veneration among the ancient Egyptians. Its image is often to be met with amongst their hieroglyphics, and was symbolical of the world, the sun, and a courageous warrior. It was symbolical of the world, according to P. Valerianus, on account of the globular shape of its pills of dung, and from an odd notion that they were rolled from sunrise to sunset; of the sun, in consequence of the angular projections from its head similar to rays, and the thirty articulations of the six tarsi of its feet answering to the days of the month. It was considered the symbol of a warrior, from a most absurd idea that all these insects were of the male sex only. It was in consequence of this last symbol, that the Roman soldiers wore its image on their signets. Dr. Clark informs us, that as typical of the sun, the source of fertility, it is still eaten by the women to render them prolific. The attention of these insects, as well as that of other Scarabœidæ to their eggs, is so remarkable, that it attracted observation in the earliest ages, and is noticed by many writers, but with the addition of a number of absurd fables, such as we have already mentioned.

We cannot pass over the genus *Ateuchus*, of Latreille, without referring to a fine species, hereto described by our friend the Rev. W. Kirby, in the Zool. Journ., under the name of *Scarabœus femoralis*, with the following specific characters:—clypeus, with six teeth, prothorax dilated and sculptured. The anterior femora very large and toothed. This species, which belongs to the Rev. F. W. Hope, is a native of the Soudan. It is one inch and three quarters in length, and forms a distinct and new type in the genus scarabæus, which the last above-mentioned gentleman proposes to call *Pachylomera*.

We have figured a species of *Pachysoma*. The *P. Hippocrates, Kirby*. It is of a shining black colour, punctuated; the elytra with slightly punctured striæ, the anterior tibiæ beneath with series of tubercules. This species is from Africa.*

The insects comprehended under the name of ONITIS, inhabit the warm and sandy countries of the ancient continent, and live in the dung of animals. The females dig holes there, and sink into the earth to deposit their eggs, as well as the necessary provisions for the larvæ, which are to be disclosed from them.

The genus COPRIS (proper) was first separated from Scarabæus by Geoffroy, on the foundation that the latter insects have a distinct scutellum, while the others have none that is apparent. These insects have, moreover, the hood semicircular, and the abdomen shorter; so that the posterior feet are more approximate to the anus, and the parts of the mouth are different. They live on excrements, and make

* The species figured as *Æsculapius*, must be changed to *Hippocrates*; also the generic name of *Mnenatium*, to *Ateuchus*. (In plate 40.)

their sojourn in dunghills, and cow-dung. From this circumstance the French term them "*Bousiers*" (*Dung-beetles*). Foreign countries produce a great number of species, many of which are very remarkable from the fantastic forms of the corslet, and their prominences. Some are adorned with very rich and brilliant colours.

The LETHRI have very great relations with the Scaraboei of Fabricius, or Geotrupes of our author. The only species which is known, *Lethrus cephalotes,* lives in the arid plains of Tartary, Hungary, and Southern Russia. It is found in dry dunghills, in the dried excrements of animals, around the roots of living plants and underwood. The male and female, according to Scopoli, live together in a straight cylindrical hole, which they hollow in the ground. The larva, in all probability, lives in the earth, and feeds upon the roots of plants.

The insects of the genus *Geotrupes* were formerly classed with Scarabæus, from which, however, they differ most essentially in the characters of the mouth, and in their habits. In the supplement to his Systematic Entomology, Fabricius, in adapting the genus *geotrupes* established by Latreille, transfers this denomination of *geotrupes* to the insects left by this author in the genus scarabæus, and gives the name of scarabæi to those which form the genus *geotrupes.* Can any thing be more prejudicial to science than continual transpositions of this kind? In the case before us the synonimy of two genera is doubled, and habits appear to be referred to Scarabæus which belong only to geotrupes, the word signifying in Greek, *to dig the earth with the feet.*

The Geotrupes are insects of a middle size, which sometimes exhibit metallic colours of the highest brilliancy; nevertheless, the upper part of their body is almost always of a black, more or less mingled with brown or metallic green. They are found in summer, in spring, and in autumn, in the excrements of ruminating animals, but rarely in human excrements, or those of carnivorous animals. Some species prefer the habitation of the forests, and feed on mushrooms. Others, and they are the most numerous, are found abundantly in pasturages, and remain the entire day in heaps of cow-dung, from which they issue forth only in the evening to fly low, in a right line, and very heavily. The least shock throws them down. It is at this part of the day that these animals couple, and that the females seek to deposit their eggs in the dung, which appears to them proper to afford an abundant nutriment to the young larvæ which are to spring from them.

The larva of geotrupes very much resembles that of melolontha, but it is smaller. It is of a dirty white; its body is soft, folded on itself, and furnished with six scaly feet and a scaly head. This larva, after having lived some time on the matter with which it is surrounded, sinks into the earth, and sustains itself on roots. At the end of a year or two it is transformed into a nymph, and in a year after, this nymph changes into the perfect insect.

We have figured a new species, belonging to the singular genus *Bolboceras*, of Kirby, under the name of *Bolb. Bilbergii.* It is of a piceous colour, and is thickly but finely punctured, with the antennæ and legs redder. The thorax is broader behind than the elytra, and very much deflexed in front, having several oblique raised lines upon its disc; the elytra strongly striated. This insect is from Demerara, and is in the cabinet of Mr. Children.

The insects of the genus Trox differ from the Geotrupides in the number of articulations of the antennæ, and the habit of concealing their head between their anterior feet, and some other characters.

These insects are found on the earth, in the fields, in sandy and somewhat dry places. They are often observed to gnaw the tendinous parts which connect the bones of carcases, the flesh of which has been consumed and devoured long since. Those which inhabit Europe are to be found during the whole summer, but more especially in spring. As soon as they are touched, like *Hister*, *Byrrhus*, *Dermestes*, and *Anthrenus*, they glue their feet and antennæ against their body, cease their movements during some time, and appear, as it were, dead, until their fear is passed. They sometimes send forth a little sharp sound, which is occasioned by the friction of the base of the abdomen against the interior parietes of the corslet.

We are not acquainted with the larva of these insects, but it is probable that it lives in carcases, and generally in animal and vegetable substances in a state of putrefaction, or dried up.

Pallas discovered, in the arid deserts of Tartary, near the rivers Jaicus and Irtis, under carcases dried up by the heat of the sun, a species of Trox which, similar to the European species, gnawed and destroyed the tendinous portions of those carcases. In the description which this author gives of these insects, he names them *Scarabœi Silphioides*, doubtless on account of their manner of living similar to that of *Silpha*.

The *Scarabœus nasicornis* of Linnæus has been taken by M. Latreille as the type of his genus ORYCTES. Though this insect presents the general physiognomy and essential characters of the large exotic Scarabæi, it is not less true that it differs from them in its jaws, being rather coriaceous than corneous at their extremity, deprived of teeth, and simply furnished with hairs. The same is the case with those of the Scarabæi, *silenus*, *boas*, *tarandus*, &c. To such species is the genus *Oryctes* limited. The most common species in Europe (*Nasicornis*), which some authors have named *monk* (*moine*), lives in the half-rotten tan of the oak, and in the beds of gardens where this tan is employed. After coupling, which takes place in June or July, the female burrows under such places, for the purpose of depositing her eggs. They are oblong, about the size of a grain of hempseed, and of a clear yellow. The larva is disclosed at the end of five or six weeks. It is of a dirty yellow, mingled with grey, with the head of a lively red, sown with little points. In its form and attitude it otherwise resembles that of the other Scarabæidæ. It is believed not to arrive at its full growth, and pass into the nymph-state, until after the lapse of four or five years. It then constructs for itself an oval shell, elongated, and very smooth in the interior. The nymph remains there a longer or shorter time, constantly lying on the back. When it becomes the perfect insect, the animal, to give sufficient time for all its parts to acquire firmness, does not quit its retreat until the end of about a month.

Swammerdam, Roesel, &c. have given us some curious anatomical observations on this species, but rather too much detailed for us to follow here. Reaumur seems to be of opinion that the larvæ of this species would constitute very good eating, and recommends that they should be sought for "*dans les couches de fumier.*" In this recommendation we cannot exactly concur, though it is very probable that the grubs and larvæ of many coleoptera, as well as other insects, might be eaten with the greatest safety.

Pliny and other writers, have designated collectively various kinds of coleoptera, and even some orthoptera, such as the grylli, under the name of SCARABÆUS. This denomination is partly synonimous with that of *cantharos,* employed by the Greeks. The moderns, in preserving the former name, have assigned it to a single genus of insects only. The scarabæi were confounded by Linnæus with melolontha, trox, and some others. Geoffroy simply divided all these insects into two genera, preserving the denomination *scarabœus,* for the first, and that of *bousier* (*copris*) for the second. But these two genera, according to the characters assigned them by this author, would differ one from the other, only because the first had a scutellum, and the second none. This slight difference is hardly sufficient for the establishment of a genus, particularly as we know that the majority of the *bousiers* or copris have a very small scutellum. There is moreover the greatest analogy between some scarabæi with scutellum, and those which want it. All the insects ranged by Geoffroy in his genus *bousier* have, in truth, a mode of living peculiar to themselves. They have a form which differs from that of the other scarabæi, and which causes them to be recognized at the first aspect. They all live in the dung and excrements of animals. But the *Aphodii* live in the same manner, are found in the same places, have the mouth similarly formed, and do not differ from the bousiers very distinctly, but because they have a scutellum, and that all the feet are inserted at equal distances. The want then of this piece, or its existence, cannot be deemed a sufficient basis for the constitution of a *genus*.

Degeer has divided into three families the genus Scarabæus, such as it was established by Linnæus. The first family comprehends *Scarabæus*, *oryctes*, *synodendron*, *geotrupes*, *onthophagus*, *aphodius*, *copris* and *trox;* the second comprehends *melolontha* and *rutela;* and the third, *cetonia* and *trichius*. He has distinguished these families under the names of *Scarabæi of the earth*, *Scarabæi of the trees*, and *Scarabæi of the flowers*.

Voet has followed pretty nearly the divisions of Geoffroy, and those of Degeer. He has divided the genus Scarabæus of Linnæus, into *Scarabæus* and *Copris*. The first are subdivided into three families. The first comprehends the Scarabæi of the flowers, the second the Scarabæi of the trees, and the third, the Scarabæi of the earth, properly so called.

The second genus is the same as the *bousier*, or copris of Geoffroy. Schæffer has imitated the example of Geoffroy, in adopting two genera, *Scarabæus* and *Copris*.

Fabricius separated from Scarabæus, *melolontha*, *cetonia* and *trox*. But the genus Scarabæus, as founded by this celebrated naturalist, presented some very remarkable differences, which determined MM. Latreille and Olivier to form of it twelve genera, or large families, the first of which comprehended the scarabæi which possess mandibles, and no upper lip; the second contained the scarabæi which have mandibles and an upper lip; and in the third, those which have neither mandibles nor upper lip. In his Systematic Entomology he confines himself to detaching from his genus scarabæus, the species named *cylindricus*, to make of it that of *Synodendron*, but to which he refers erroneously some species of *Bostrichus*, an error which he subsequently left uncorrected. M. Latreille, in his summary of the generic characters of insects, established, with the scarabæi of the second division, a peculiar genus which he named *Geotrupes*, adopted that of *Copris*, and preserved the generic name of *Scarabæus* only for the species of the first division. These genera, and those which had been separated from the scarabæi of Linnæus before, composed his second family of coleoptera. Fabricius, in the Supplement to his Systematic Entomology, also admits the genus *Copris*, but he restrains it, establishing at its expense the genus *Onitis*. The *geotrupes* of M. Latreille, as we have already seen, were *Scarabæi*, in the system of the German naturalist, and the scarabæi of the former were geotrupes with the latter. Notwithstanding the inconvenience of this substitution of names, it was nevertheless received by the naturalists of Germany, so imposing is the authority of a great name. The same genus, *bousier* or *copris*, underwent some time after a new modification. M. Weber separated from it the *Ateuchus*, and Illiger

Aphodius. Such was in this respect the state of the science when M. Latreille published the third volume of his General History of Crustacea and Insects. The third division of the scarabæi mentioned above, there compose a peculiar family, that of the coprophagi, and which presents a new genus, that of *onthophagus.* Those of *Lethrus* and *geotrupes* form another family, that of *geotrupini,* a third, that of *scarabæides,* and to which he united the *Lucani* and *passali,* comprehend, beside the scarabæi of the first division as above, *hexodon* and *cetonia.* But this last genus is less extended there by the adoption of another, that of *Trichius,* and the establishment of another generic section, that of *rutela,* which is formed of the cetoniæ of the third division. M. Latreille also divides *Melolontha* into many small families, from considerations taken from the antennæ, the claws of the tarsi, and the form of the body. A new genus is also there presented, that of *glaphyrus,* which he detached from the preceding. Fabricius, in his system of *Eleutherata,* has not departed from the method which he adopted in his Supplement to the Systematic Entomology. Some subsequent changes were then made in this division by M. Latreille, and finally it was modified, as our readers have seen, in the "Regne Animal."

We shall not apologize for this slight notice of the history of the changes made in the classification of the *Scarabæi.* It is true that we do not very frequently enter into details of this kind, in this department of our labours; but in the instance before us we deem them of some importance in elucidating the researches, and assisting the studies of the zealous entomologist, though they may be passed over by such as merely take up Natural History as an amusement. In fact, without some examination into these matters, it is impossible to take up an author with any advantage, or to know precisely what subject he is treating on. The continual revolutions in nomenclature are greatly to be deprecated, but as far as they exist they must be studied. All the great authors which we have cited contain much that is of real value, but of which we shall be unable to avail ourselves, without some attention to the dry and painful subject of synonimy.

The scarabæi are met with on the ground, or flying from one place to another. They are usually found in rich and humid situations, in garden beds, or in fields, towards the root of old trees. The majority of them frequent dunghills, and fat and moist soils, for the purpose of depositing their eggs, but none of them live in the dung or excrements of animals.

It is in such places as we have last mentioned that the larva of these insects is to be found. It resembles a soft thick worm, usually curved into an arc. Its head is hard and scaly, and provided with two short and filiform antennæ. The body is composed of thirteen tolerably distinct rings, nine of which are provided with a stigma on each side. The nymph is buried in the earth, and shut up in a sort of shell constructed by the larva previously to its transformation. The skin which covers its body permits us to see all the parts which the perfect insect is to have. Their form is tolerably well designed under the skin which covers them, and which keeps them as it were swaddled up.

We shall not repeat here all the absurd puerilities which the greatest men of antiquity, such as Homer, Aristophanes, Theocritus, Isidorus, Aristotle, Lucian and Pliny have written on the *Scarabœi* or *canthari*, relative to their origin, habits, and sex. Should the reader be curious on this subject, he will find most of them detailed in Mouffet and Jonson. What we have already said under the head *Ateuchus*, may also, perhaps, apply to the insects we are now treating of, namely, that the ancient Egyptians believing that they were all males, and never coupled, sculptured those insects at the

Pl 14.

1. Scarabaeus *macropus.* *Sh. Nat. Mis.*

2. *Male.* 3. *Fem. of* Melolontha *chrysochlora.*

foot of the statues of their heroes, to express manly and martial virtue, exempt from all weakness, a virtue which, according to them, appertained to powerful minds—to men truly illustrious.

The perfect insect of the scarabæus, like most of the tribe, is not often in a situation immediately after its birth to make use of its organs. It has need to fortify itself, which often requires a tolerably long time. The internal organization has undergone great changes. The long intestine, all of one width, is no longer observed. The tracheæ are all vesicular.

The insects of the sub-genus Hexodon, have some relations with *melolontha*, *cetonia*, and especially *rutela*. They frequent trees and shrubs, and feed upon their leaves. The larva is not known, but it probably differs but little from that of melolontha, which we shall describe below.

In consequence of the singularity of the insect, and the rarity of the tract in which it was originally published, we have deemed the insect described by Mr. Francillon, under the name of the kanguroo beetle, or scarabæus macropus, to be very worthy of an introduction into our illustrations. This insect is supposed to have come from Potosi, in South America. It is of a bright green colour above, with the elytra smooth, and golden and copperish beneath, with the hind legs incrassated to a most extraordinary size.

We believe this unique insect is in the collection of Mr. Macleay. As an inhabitant of South America there can be but little doubt, since M. Humboldt has discovered a species very like it in form in Peru, which M. Latreille has described in a magnificent work published by himself and M. Humboldt, under the name of *melolontha chrysochlora*. This species is of a brilliant golden green, with the breast, the underside of the abdomen, and the femora and tibiæ of the posterior legs of a reddish copper colour. The head, thorax, segments of the abdomen rather smooth, with very fine punctures; but the elytra are strongly punctured, the punctures forming irregular ridges; the posterior legs are half as long as the body; the femora are rather arched; the interior angles of the posterior extremity are prolonged in the form of a strong spine. The last joint of the tarsi is very large and strong. This species is found in great numbers, at certain times of the year, somewhat after the fashion of our rose beetles (*Cetonia aurata*).

We have also figured a new species, under the name of *C. Kirbii.* It is oblong, of a brilliant shining golden green, the body darker underneath, pubescent, the elytra with numerous punctated striæ, the posterior legs long, with the femora and tibia large, the tarsi large and violet. It is from Brazil, and is in the cabinet of the British Museum.

This last species seems to form a link between the *Rutela* and *Chrysophora*, being related to the former in the form and general glittering appearance of the body, and to the latter in the form of the hind legs.

The Rev. W. Kirby has described an intermediate genus between *Chrysophora* and *Rutela*, and nearly allied to Pelidnota of Macleay, under the name of *Chrysina.* The characters are, the mandibles large, concave above, convex beneath, broad, externally rounded; the clypeus sub-elongate, with obtuse angles; labrum sub-trapezoid; the tip emarginate; the maxillary palpi four jointed; the first and third joints short, the second gradually larger; the terminal incrassated; the labial palpi three jointed, the second joint longer than the first, and the third longer than the second, and larger; the antennæ ten jointed, as in Arcoda and Pelidnota. Mr. Kirby mentions one species, under the name of *Peruviana*, which is in the collection of the Rev. F. W. Hope.

This beautiful species has one peculiarity which seems to

Pl. 46.

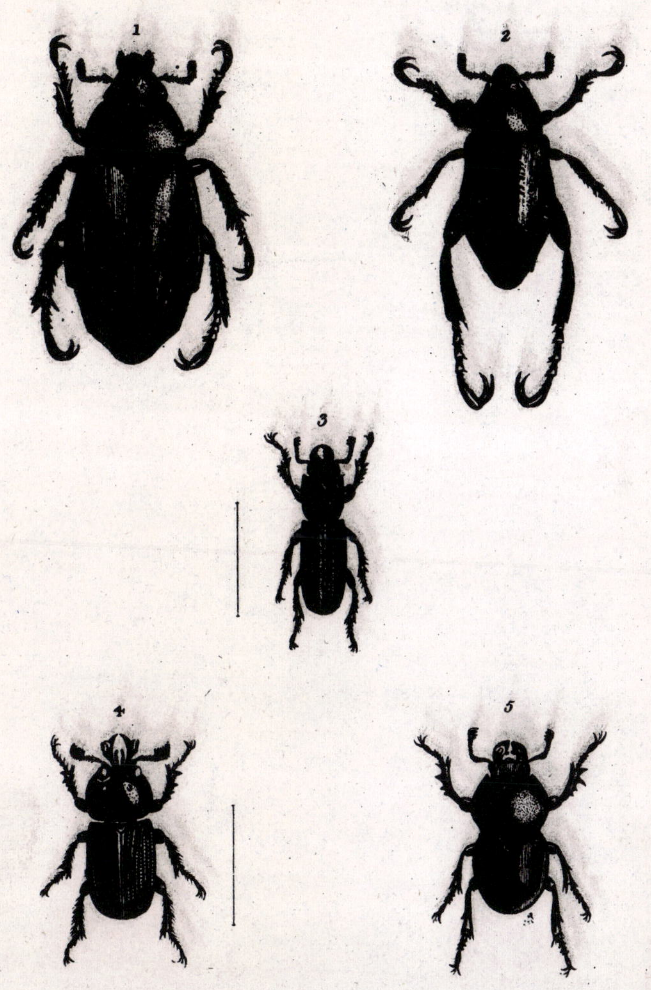

1. Chrysina *Mexicana.* *G.R. Gray.* 2. Chrysophora *Kirbii.* *G.R. Gray.* 3. Syndesus *cornutus.* *Macleay.* 4. Hexaphyllum *Brasiliensis.* *G.R. Gray.* 5. Colophon *Westwoodii.* *G.R. Gray.*

distinguish it from all its cognate tribes—its green colour, at least on its upper surface, is merely superficial, and may be removed by friction, when only black appears.

The species we figured is from Mexico, therefore we will call it *Mexicana.* It is of a rather dull golden green colour, with delicate punctated striæ, with minute punctures intermixed on the elytra, which have a lobe on the outer posterior angle. The antennæ ferrugineous, with the first joint golden green; the palpi ferrugineous; the tarsi violet; the penultimate segment of the abdomen with a transverse violet mark at the base above; the abdomen copperish green beneath; the legs are fulvous green, with the interior of the tibiæ copperish. The length is seventeen lines. There is also another species from the same place, which is about fourteen lines long, and is more strongly striated on the elytra, with some minute punctures between them, and very slightly copperish beneath the abdomen; the tarsi golden green. This species we name *Auripes.* Both these species are in the Collection of J. G. Children, Esq.

The Rutelæ were confounded by Fabricius and Olivier with the cetoniæ. They form, however, in the method of the last, a particular family. It has no triangular piece at the lateral base of the elytra.

These insects are proper to the New World, and especially to such of its countries as are situated between the tropics. We know nothing positive concerning their manner of life; but from the scaly consistence of their jaws, the numerous teeth with which they are provided, it is probable that their habits have great analogy with those of melolontha and scarabæus. Their brilliant or agreeable colours would seem to indicate that these insects do not shun the light like the latter, but that they are to be found in open day on flowers, or perhaps, rather, upon leaves.

Fabricius first separated the Melolonthæ from Scarabæus, with which Linnæus, Geoffroy, and some other naturalists had united them. They have subsequently given birth to the establishment of several genera or sub-genera; but notwithstanding all these reductions, the genus is still very extensive, and might become the object of a monograph so much the more interesting, inasmuch as many of these coleoptera, being a scourge to agriculture, demand but too urgently the attention of the naturalist.

Of all maleficent insects, there are very few indeed which are more mischievous and destructive than the melolonthæ. From their birth to their death these insects feed on vegetable substances, and occasion to them a most considerable degree of damage. In the larva state, they gnaw for two, three, or four consecutive years, the tender roots of annual plants, those of perennial plants, of shrubs, and even those of the hardest trees. In Europe, and in all cold and temperate climates, those larvæ cease their devastations during the winter, sink more deeply into the earth, in which they form a lodge, where they pass that season without taking any nutriment, and in a sort of lethargic state. When they have become perfect insects, the melolonthæ abandon the earth, and subsist no longer upon roots. But they then attack the leaves of trees and plants. Some years ago, the species which were found in the neighbourhood of Paris were so multiplied, that they despoiled, in a very little time, almost all the trees of the forest. The common Melolonthæ gnaw without distinction all kinds of roots, in their first state. When arrived to their final form, they attack and destroy the leaves of almost every kind of tree. A species, common in the southern parts of France, gnaws the buds and the tender leaves of the pines. The *Melolontha vitis*, thus named because it despoils the vine of its leaves, also attacks the willow, the poplar, and the majority of fruit trees.

This genus is very numerous in species; the majority of

them are greatly multiplied, and all are more or less hurtful. We shall proceed to sketch the history of that one with which it is most important to be acquainted, namely, the common species (*Melolontha vulgaris*) the *cock-chafer*, and we shall notice the means employed hitherto, if not to exterminate this devastating race, at least to lessen its population, and diminish the aggregate of mischief which it produces.

The melolonthæ pass the greatest portion of the day motionless, and in a state of lethargy, attached to the branches and leaves of trees. They rarely take their flight when the weather is hot and dry; but after the setting of the sun, being urged by the want of nutriment, and more especially the inclination to unite, they fly along humming from tree to tree, and the males pursue the females. The organs of generation are accompanied by a sort of pincers, the sides of which naturally approach each other, and cannot be opened without an effort. The flight of these insects is dull, heavy, and irregular. They strike at all the objects which they meet. They are often seen to fall at once, and rise again with sufficient quickness to resume their flight, unless the shock has been too rude, or unless they happen to have been thrown on their backs.

The duration of the life of the melolonthæ is very short in their final state. Each individual scarcely lives a week, and the species is rarely seen for the space of a month. A short time after they issue from the earth, the melolonthæ begin to couple, and their union endures for four and twenty hours. This appears to be the whole and sole object of the existence of these insects in their perfect state, as both die as soon as that object is attained. The mode of copulation is singular, but we shall not enlarge upon it here. The eggs are of a clear yellow, and in the form of an elongated spheroid. The union being ended, the male no more resumes his pristine vigour. He remains in a languishing state, takes no more nutriment, and perishes soon after. The female survives and lays her eggs.

As soon as the females are fecundated, they dig in the earth, with the assistance of their fore feet, a hole, half a foot in depth, in which they deposit their eggs, one by the side of the other. The oviposition being finished, they quit the hole, abandon the eggs, and return on the trees. They survive this operation but a little time, take scarcely any nutriment, and perish after having languished one or two days.

The larvæ which spring from these eggs at the end of about six weeks are soft, elongated, of a dirty white, approaching to yellow, and wrinkled; the posterior extremity of their body is curved underneath, and the excrements with which it is filled give it a violet or ashen colour. They have six short and scaly feet, a head thick and scaly, two antennæ composed of five pieces, and nine stigmata on each side. As yet they have no eyes, or at least those which they are to have are concealed under the envelops of which the larva is to disembarrass itself by small degrees. Their body is composed of thirteen rings, tolerably apparent. These larvæ, pretty generally known by the name of white worms, live three or four years in their first state, subsequently change into the nymph, and appear at the commencement of the third or fourth year, under the perfect form of the melolontha, or cock-chafer.

These larvæ are attached to the roots of plants and trees, which constitute their only nutriment. They do not eat, except during the fine weather. In autumn they bury themselves very deeply in the earth, and pass the winter in a state of lethargy, without taking any kind of nourishment, and without making the least movement. They issue from their retreat, and mount to the depth of about half a foot on the approach of the fine season, to feed anew on the roots of vegetables. They moult, or change their skin once every year, at the commencement of spring. Finally, when these larvæ have acquired their full growth, at the end of the summer of the third year, they cease to eat, bury themselves in the earth to the depth of a foot and a half, or two feet, construct a very even sort of lodge, very smooth within, and line it with their excrements, and with some silken threads. Their bodies become shortened and inflated, they quit their skin and change into a nymph, through the covering of which all the parts of the perfect insect are easily distinguished. The antennæ, the feet, the wings, the abdomen—in a word, all the parts are perfectly designed under the general envelope which covers them. In the month of February, the cock-chafer tears its envelope, pierces the shell, and issues forth under its final form. But the insect is then yellowish, and rather soft. It still remains for some time under ground, and gets rid of its superfluous humidity. It approaches by little and little to the surface of the earth, from which it does not issue forth entirely until it is attracted by the influence of a mild heat. The contact with the air completely fortifies it, and gives to its external parts their proper colour.

The ravages which the larva of the cock-chafers, and the cock-chafers themselves commit, would appear almost incredible, were they not attested by the most undeniable authorities. They sometimes produce mischief, nearly to as great an extent as those committed by the locust tribe. It appears from a paper printed in the Philosophical Transactions for 1697, from the pen of Mr. Molineux, that some particular districts in Ireland were infested by this insect in a very extraordinary manner. We shall here cite the account given by that gentleman.

"These insects were first noticed in this kingdom in 1688. They appeared on the south-east coast of Galway, brought thither by a south-west wind, one of the most common, I might almost say, trade-winds of this country. From hence they penetrated into the inland parts towards Heddford, about twelve miles north of the town of Galway. Here and there in the adjacent country, multitudes of them appeared among the trees and hedges in the day-time, hanging by the boughs in clusters, like bees when they swarm. In this posture they continued, with little or no motion, during the heat of the sun; but towards evening or sun-set, they would all disperse and fly about, with a strange humming noise, like the beating of distant drums, and in such vast numbers, that they darkened the air for the space of two or three miles square. Persons travelling on the roads, or abroad in the fields, found it very uneasy to make their way through them, they would so beat and knock themselves against their faces in their flight, and with such a force as to make the place smart, and leave a mark behind them. In a short time after their coming, they had so entirely eat up and destroyed all the leaves of the trees for some miles round, that the whole country, though in the middle of summer, was left as bare as in the depth of winter, and the noise they made in gnawing the leaves, made a sound much resembling the gnawing of timber. They also came into the gardens and destroyed the buds, blossoms, and leaves of all the fruit-trees, so that they were left perfectly naked: nay, many that were more delicate than the rest, lost their sap as well as leaves, and quite withered away, so that they never recovered again. Their multitudes spread so exceedingly, that they infested houses, and became extremely offensive and troublesome. Their numerous young, hatched from the eggs which they had lodged underground, near the surface of the earth, did still more harm in that close retirement than all the flying swarms of their parents had done abroad, for this destructive brood, lying underground, eat up the roots of corn and grass, and thus consumed the support both of man and beast. This plague was happily checked several ways. High winds and wet misling weather destroyed many millions of them in a day, and when this constitution of the air prevailed, they were so enfeebled, that they would let go their hold, and drop to the ground from the branches, and so little a fall as this was quite sufficient to disable, and sometimes perfectly to kill them. Nay, it was observable that even when they were most vigorous, a slight blow would for some time stun them, if not deprive them of life. During these unfavourable seasons of the weather, the swine and poultry of the country would watch under the trees for their falling, and feed and fatten on them, and even the poorer sort of the country people, the country then labouring under a scarcity of provisions, had a way of dressing them, and lived upon them as food. In a little time it was found that smoke was another thing very offensive to them, and by burning heath, fern, &c. the gardens were secured, or if the insects had already entered, they were thus driven out again. Towards the latter end of summer they retired of themselves, and so totally disappeared, that in a few days you could not see one left. A year or two ago, all along the south-west coast of the county of Galway, for some miles together, there were found dead on the shore such infinite multitudes of them, and in such vast heaps, that, by a moderate estimate, it was computed that there could not be less than forty or fifty horse-loads in all, which was a new colony, or a supernumerary swarm from the same place whence the first stock came in 1688, driven by the wind from their native land, which I conclude to be Normandy or Brittany in France, it being a country much infested by this insect, and from whence England has heretofore been pestered in a similar manner with swarms of this vermin; but these meeting with a contrary wind before they could land, were stopped, and tired with the voyage, were all driven into the sea, which, by the motion of its waves and tides, cast their floating bodies in heaps to the shore. It is observed, that they seldom keep above a year together in a place, and their usual stages or marches are computed to be about six miles in a year. Hitherto their progress has been westerly, following the course of that wind which blows most commonly in this country."

Mouffet, in his History of Insects, informs us, that in 1574, so great a number of cock-chafers were driven into the river Severn, that they altogether hindered the mills from working, and were with difficulty destroyed by the united efforts of the people, and the different kinds of hawks, ducks, and other birds, which devoured them with eagerness. In Normandy, he tells us, that they appear in general every third year. In England, the county of Norfolk seems to have suffered most from their ravages. In 1751, a great many crops were totally destroyed in that part of the country by these voracious insects.

The devastations of these animals have induced many agriculturists to seek out the means of destroying them at a small expense, or at least of diminishing their number, often so considerable, that the leaves of an entire forest are completely gnawed by these insects in less than fifteen days. We read in the Philosophical Transactions of the Dublin Society, that the inhabitants of some districts in Ireland had suffered so much from cock-chafers, that they determined to set fire to a forest of several leagues in extent, to cut off the communication with other districts which had not as yet been infested. Among the methods proposed for the destruction of the cock-chafers, the following are those which have most generally proved successful.

First Method.—Flambeaus are made of the thickness of

one's fist, composed of a well sulphured match, surrounded with pitch-rosin, and a slight external layer of yellow wax. When the cock-chafers begin to appear, that is to say, in the months of May and June, the hours are chosen when they remain in a state of repose on the leaves of trees and hedges, which is usually between nine in the morning and three in the afternoon, and the prepared flambeau being lighted, it is paraded underneath and around the trees and hedges, and held under in such a manner that the smoke, mixed with the odours of sulphur, resin, and yellow wax, suffocates the insects. It is sufficient to hold it in this manner, going and coming, for about seven minutes and a half, under the places where the cock-chafers have assembled. After this operation the hedges are shaken with sticks, and the fruit trees with crooks, or with the hand, taking care, however, not to suffer the flowers of these trees to fall, for thus the remedy would be worse than the disease. The cock-chafers, half stupified by the heat of the sun, and suffocated by the mingled odours of the flambeau, undergo a sort of lethargy which makes them fall more easily from the trees and hedges in which they were. When they have fallen they are collected together, placed in a heap upon a handful of straw, and set fire to, to prevent the possibility of their recovery or return; for this odour of the flambeau does not kill them, and it keeps them in a state of stupefaction scarcely for the space of an hour.

Second Method.—It has been proposed as a mode of preservation against the ravages of the *white worms*, or the larvæ of the cock-chafers, to cause the plough to be followed by children, to gather up in baskets such of these animals as the share might upturn. But, besides that all lands are not tilled at the same time, and that there still remain some in this state at the end of autumn and even during winter, the tracts planted with wood, or thickets, and those in which clover, trefoil, and the different kinds of grasses are sown, must naturally serve these insects for a retreat. Moreover, this feeble resource could not be adopted at any other time than in spring, and the commencement of autumn ; for about the end of this season these larvæ inter themselves deeply in the ground, for the purpose of sheltering themselves against the cold, and remain during the winter, at such a depth, that it would be quite impossible for the plough to get at them.

Third Method.—M. Gouffier having observed that the trees planted in espaliers, near which grew strawberry bushes, lettuces, &c., were less subject to these *white worms*, he judged that they would give the preference to these last mentioned plants, and which he found, in fact, to be almost all of them eat up. He then adopted the method of furnishing his espaliers with salad, and of planting thick tufts of strawberries, which he removed with the tuft on which they grew, to the foot of orchard trees. He was careful to visit them two or three times each day, and as soon as he observed a lettuce begin to wither, he dug at its foot with a little trowel, and always found one or more larvæ of cock-chafers gnawing at the root of it. As to the strawberries, he did not so soon perceive the sojourn of the larvæ in them ; but as their roots were numerous, they established for themselves there a sort of domicile, which made them forget the neighbouring trees.

This method, though very good for preserving the trees from these mischievous larvæ, can avail nothing against the ravages which they commit in the open country.

Fourth Method.—M. Gouffier, and many other agriculturists, have proposed to spread soot around the feet of the young trees, and frequently to stir up the earth there, and to cast in a field, peat, pit-coal, turf ashes, or even lime, to drive away or destroy these larvæ. According to the experiments tried by these agriculturists on a small scale, it appeared that these substances did in fact drive away the larvæ of the cock-chafers, when they were mixed in a tolerably large quantity with earth, in a pot or box. But these experiments did not succeed so well in a field, because the turf and the ashes had not been put there, either in sufficient quantity, or at a sufficient depth. The most proper time for spreading these substances, and mixing them with the earth, would be in spring, without doubt, at the moment in which the larvæ quit their retreat, and re-ascend to support themselves upon roots.

Fifth Method.—All the means hitherto mentioned have proved either useless or insufficient. The best would doubtless be, as Rosier informs us in his Course of Agriculture, to make, for several years in succession, a general chase of those insects, and to destroy them when in their final form. For this purpose women and children might be employed. The birds of the farm-yard, such as turkies, hens, &c., nocturnal birds, such as the different species of owls, &c., rats, weasels, ferrets, and all their congenerous quadrupeds, cause a great number of them to perish. Some *carabi* also destroy a great quantity of female cock-chafers, at the moment in which they are trying to descend into the earth for the purpose of depositing their eggs.

It is commonly believed that frost and the cold rains of spring, cause the cock-chafers to perish before they issue from the earth. It is also believed, according to the observations of Rœsel, that one may predict, from the number of cock-chafers in one year, their dearth or their abundance for the fourth year which will follow the prediction. But observation does not confirm these assertions, which are founded rather on conjecture than experience. It often happens on the one part, that after a rigorous winter, and a cold and humid spring, that the cock-chafers appear in great abundance in the month of May, and that they are sometimes less numerous after a mild winter and a dry and warm spring. On the other point, if their number for the fourth year was determined by their number in the year in which they were observed, one might, in following them for four years, determine their number for ever, a thing totally contrary to all observation. It often happens that these insects are very common for several years in succession, and that they are much less so during many other years, without there being any thing like a permanent order in the times of their appearances. A very abundant brood of eggs may be followed by few cock-chafers in the fourth year, if the causes connected with the growth and development of the larvæ should prove unfavourable. A less abundant oviposition may be followed by a considerable quantity of cock-chafers in the fourth year, if these causes, on the contrary, are very favourable, if the eggs disclose well and the larvæ do not perish. Each female cock-chafer, laying near a hundred eggs, it is easy to conclude that a small number of these insects may nevertheless leave a very numerous posterity.

Passing over the intervening sub-genera we come to the division *Hoplides*. The HOPLIÆ were thus named by Illiger, and until his time had been confounded with melolontha. From these last, however, they are very sufficiently distinguished.

The Hopliæ are small coleoptera, which appear to be peculiar to the western countries of the ancient continent. They are especially found in the hot and temperate parts. They remain tranquilly on the leaves of different vegetables, which they gnaw. Many species appear to prefer those which grow on the banks of streams, rivers, or in humid places.

The genus TRICHIUS, in the Entomology of Olivier, forms but a single division, the second of the genus *cetonia*, from which, in fact, it differs but little in the relation of the masticating organs. It may, however, be distinguished from it by the relative and comparative proportions of the palpi; also by the form of the corslet.

The metamorphoses and habits of the insects of these two genera, are otherwise almost identical. Some species of *Trichii*, those in which the females have the abdomen terminated by a corneous point, always prominent, hollowed into a gutter on its upper face, denticulated upon its edges, and serving as a wimble or augur, are a little different from the others in their manner of life. They are almost always to be found upon the ground. Some are found on the carious trunks of trees, others on roses and various flowers. The colours of *Trichius fasciatus* vary a little, according to the localities in which this insect is found, and according as the temperature is more or less cold. Thus black predominates more on the elytra in the individuals which are peculiar to alpine regions. This is more particularly to be remarked in the species which has been named *succinctus*.

Degeer had already separated, into a particular family of his genus scarabæus, the species with which Fabricius has formed those of Cetonia and *trichius*. He names them *scarabæi of the flowers*, and distinguishes them from the others because they have no teeth. This last writer associates to the cetoniæ, many insects which, though tolerably resembling them in other parts, have nevertheless very different masticating organs, and approximate in this point of view to melolontha, and the scarabæi of the text. They compose, in our author's system, the genus *rutela*—Olivier, who had very well appreciated this difference, since he forms with these species a peculiar division, still leaves them with the *cetoniæ*, as well as the *trichii*. M. Lamarck has separated from the cetoniæ of Fabricius, under the generic name of *Goliath*, some species in which the hood is deeply divided into two lobes, which imitate the forms of two horns. Of this last-mentioned genus we may remark, by the way, that there is nothing to be observed, except its size and beauty.

From the *trichii* it is more difficult to separate the cetoniæ. That they do differ, however, in some material points, is undeniable.

The cetoniæ are found during summer on umbellated flowers, on composite flowers, on willows, poplars, flowering shrubs, hedges, &c. They should not be confounded with the cock-chafers, those most mischievous of all destructive insects to the roots of every vegetable, and the leaves of every tree. The cetoniæ, in their larva state, scarcely do any mischief to plants, and in their final form they frequent flowers without doing them any injury. They content themselves with the melliferous fluid contained in the bottom of the corolla, and never attack either flowers or leaves.

Many naturalists had given the name of *platycerus*, which signifies *broad horns*, to some species of the genus LUCANUS, and this denomination was preserved to it by Geoffroy, who distinguishes it by very just characters from that of scarabæus with which it had been confounded. But the name *Lucanus*, which Scopoli gave to the same genus, in his Entomology of Carniola, printed in 1763, a year before the French naturalist had published his History of the Insects of the Neighbourhood of Paris, having been adopted by Linnæus, has most generally prevailed. The denomination of Geoffroy has, however, been applied to another generic section formed by M. Latreille, at the expense of the Lucani.

Pliny has employed the word *lucani*, in speaking of the *horn-beetles*. Fabricius tells us that he is unacquainted with the origin of the word. The etymology is not, however, very difficult. The ancients gave the name of *Lucas*, *Lucana*, to the *ox* and the *elephant*. It is said that Pyrrhus had thus named the elephant the first time that he saw it, because this word signified *ox* in his own language, and that he named it thus from the name of the largest animal which he had seen; Nigidius, according to Pliny, was the first who gave the name of Lucani to the horn-beetles. This name, as we see, corresponds to the vulgar name of *flying bull*, given in different languages to the *lucanus cervus*. Dalechamp thinks that the name *lucanus* was given to the horn-beetle only because this insect was very common amongst the Lucanians, a people of Italy. But it is probable, after what we have just said, that the Lucanians themselves were thus named, in consequence of the great numbers of oxen which they reared in their rich and abundant pasturages.

As our author has omitted a detailed description of these insects, it will be as well to give some notice of their external anatomy in this place.

The antennæ being bent, and the final articulations advancing on the internal side in the form of parallel teeth, or forming, when united, a knob more or less pectinate or serrated, cause the lucani to be very easily distinguished from the scarabæi of Linnæus. The length of the first articulation of these organs, and which, at its junction with the following articulation, forms a sort of elbow; the jaws terminated in the manner of a brush or pencil; the lingua concealed behind the mentum, and terminated by two pieces, each of which again imitates a small pencil, prevents us from confounding this genus with that of *passalus*. By some of these characters, as well as by its depressed form, it is also removed from those of *sinodendron* and *æsalus*. The jaws of the lucani are covered at their base by the lingua, which removes them from *Lamprima*, which, besides, has the antennæ terminated a little otherwise, the body more convex, and some other peculiar traits. But they are not distinguished rigorously from the platycerus of M. Latreille, but that the lateral and anterior edges of the head are prolonged a little on the surface of the cornea of the eyes, and seem to cut off a part of it.

The head of the Lucani is more or less bulky; that of the male is more so than that of the female. It is broader than long, angular, often irregular, with some elevations more or less projecting. The hood is tolerably large, and advanced into a point. The mandibles are very large, strong, corneous, arched, and denticulated internally. Those of the female are less long than those of the male.

The corslet is a little convex above, rounded at its sides, and more or less edged. The scutellum always exists, only it is but little visible in some species. The elytra are hard, of the length of the abdomen. They cover two membranaceous and folded wings, of which the insect frequently makes use for the purpose of flying. The feet are long; the anterior legs are denticulated laterally. All the tarsi are composed of five articulations, the last of which is armed with two hooks, accompanied by an intermediate appendage, terminated by two divergent silken hairs.

The larva is very thick; its body is curved into an arch, and composed of thirteen rings; its head is brown, scaly, armed with two strong jaws, of which it makes use to gnaw wood, which it reduces to a sort of tan. It has six scaly feet attached to the first three rings. Arrived at its final growth, it constructs, in the wood in which it has lived, a shell or cell, with the sawdust of the wood which it has gnawed. In this shell it becomes changed into a nymph, from which it does not issue until it has attained the form of the perfect insect. Rœsel believes that six years are required for the larva to attain to its perfect growth.

The lucani live but a short time in their final form. As soon as they have undergone their last metamorphosis they seek to couple, and lay their eggs. They then perish in a very short time after. They live, according to the observations of Degeer, on the melliferous juice which is spread over the leaves of the oak. It appears, that the females employ their mandibles in cutting the half-rotten wood, for the purpose of placing their eggs more deeply there. These insects cause very little injury to trees in their final state. But in that of the larva, the mischief which they produce is often very considerable. The larvæ not only gnaw the dead wood, but they also attack the living. They sojourn more frequently in the roots than in the trunks and branches, so that if the larvæ of the lucani do not cause the oaks quickly to perish, at all events they hasten their destruction. They advance the period of their dissolution by rotting the trunk, or a portion of the roots.

The lucani are seen to fly principally towards the evening around old trees. They form a numerous genus, the greater part of which is foreign to European climates.

The female of *lucanus cervus* is so much smaller than the male, and differs also so much in other characteristics, that some entomologists have doubted whether this insect was really of the same species. But M. Marechal, a celebrated painter of natural history, and a most accurate observer, has witnessed the coupling of these insects. Therefore no further doubt can remain upon this subject. The mandibles of these insects were formerly employed in medicine, under the name of the *horns* of *Scarabæi*. This remedy was administered as an absorbent, in case of pains or convulsions supposed to be produced by acidity in the *primæ viæ*. It was also, according to Pliny, suspended to the neck of children. The *lucanus parallelipipedus* is commonly found on the trunks of old willows, oaks, and on other trees.

To the lucanides we have added two new genera, 1st. *Colophon*, of Mr. Westwood's MSS. The characters are, the mandibles strongly curved, as in the genus *Lethrus*; the palpi short, filiform; the antennæ formed like the true lucanus; the thorax large, broader than the body, with three lobes anteriorly; the posterior square; the body very much rounded.

The species we have named *Westwoodi.* It is very smooth, of a shining black. The habitat is unknown. The length is twelve lines.

The next genus we add, is *Trictenotoma*, with the following characters. The antennæ as long as the body, the first joint club-shaped, the second very small, globose; the third, to the seventh, rather long and club-shaped, with the three last joints small, pectinated. The mandibles rather short, strong, with three small strong dentations; the maxillary palpi moderate, first joint very small, second the longest, third rather shorter, the last large, sub-securiform; the labial palpi shorter, and shaped like the last; the eyes reniform, the head rather square; the thorax broader than long, lobed anteriorly, the sides rounded; the scutellum rather triangular, with the anterior angles lobed, with the base of the elytra broader than the tip, which is rounded. The posterior tarsi with only three joints.

The species we have named *Childreni.* It is black, with the head, thorax, and elytra above, covered with a yellowish down, and beneath with a whitish down; the length is two inches and two lines, and the specimen is from the Tanesserim Coast in the East Indies. In the Collection of J. G. Children, Esq.

This insect was first made known by Mr. Westwood, in a note published in the Zoological Journal, which follows, and to which he has kindly added here some additional remarks:—

"Since the preceding observations upon a supposed connection between the Lucanidæ and Prionidæ were penned, I have casually examined a most interesting insect, contained in the Rafflesian Cabinet, which serves most satisfactorily to

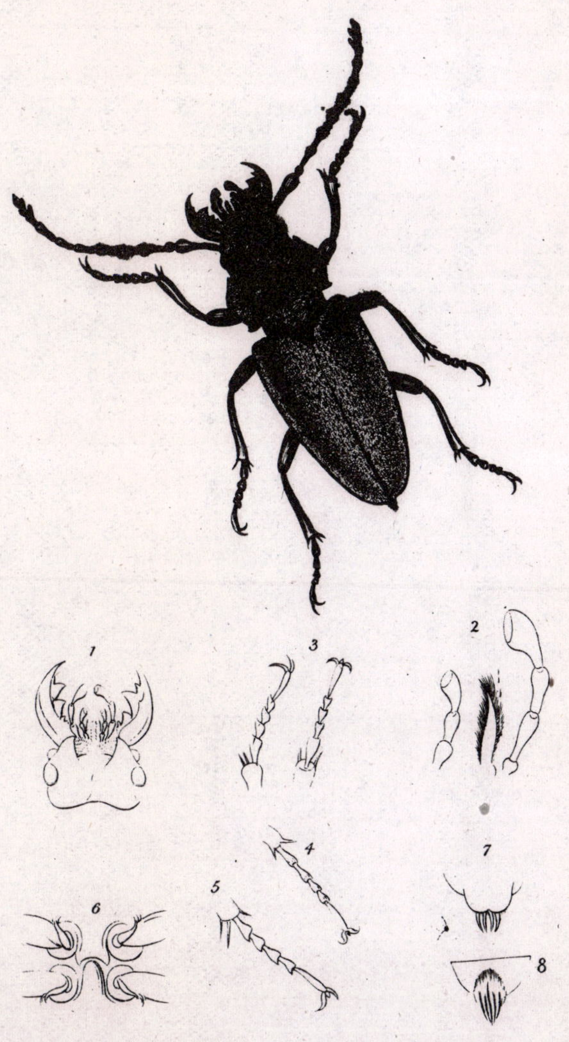

Coll. J. G. Children. Esqr

Trictenotoma *Childreni*. *G.R.Gray.*

connect the two families. Its general appearance is that of a Prionus, with short trigonate advanced mandibles, and moderately long antennæ; but the latter, on a closer examination, are decidedly those of one of the Lucanidæ. The basal joint is scarcely longer than the third, and a slight elbowing of the antennæ is observed at the second joint. The last three joints are not longer than the preceding joints, and very slightly produced on the inside, giving these organs the appearance of being almost setaceous. The joints of the tarsi are cylindric, but the insect is heteromerous ? ? ?

"Having had the good fortune, in conjunction with Mr. G. R. Gray, to discover a second specimen of this beautifully osculant insect in the Collection of J. G. Children, Esq., I consider myself happy in having been enabled to examine and figure that gentleman's specimen, although it is to be regretted, that the structure of the internal parts of the mouth could not be rigorously investigated without injury to the insect.

"In addition to and correction of the characters mentioned in the preceding note, it will be observed that from the casual examination which I was able to bestow upon the Rafflesian insect, some slight errors have crept into my description of the antennæ ; these I have been very careful to correct in the delineation of those organs in the figure of the insect, an examination of which will at once convince the student that it is legitimately lucanideous—notwithstanding it may be urged that the slight internal productions of the three terminal joints, is but an incipient attempt at the more perfect lateral developement observable in many of the Prionidæ. It must also be borne in mind that Mr. Children's specimen is a female, and we can but slightly conjecture the structure of the male antennæ or mandibles.

"In the formation of the Trophi there is not any very material difference existing between this insect and certain female

Lucanidæ, such as Rhyosonatus, Pholidotus, &c. We may likewise notice the transverse and slightly emarginated upper lip, the extreme woolliness of the transverse mentum (as in Pholidotus), and the non-sexertion of the laciniæ of the lower lip.

"It will be perceived that the eyes are reniform, as in the Prionidæ; instead of each being separated, and having the appearance of four eyes, as in the Lucanidæ.

"The clothing of down upon almost every part of the insect, the form of the scutellum, and the deeply emarginate terminal joint of the abdomen, or its under side, also deserve notice.

"Next to the antennæ, the tarsi may be considered as the most singular portions of the animal in question, not only from their being heteromerous, but also from the peculiar structure of the joints. These at first sight appear perfectly cylindric, as in the Lucanidæ, but on being more closely examined they are found to be slightly clothed with very fine short bristles, so as immediately to suggest the idea of an attempt to clothe them with the soft cushion with which the tarsi of the Prionidæ are furnished.

"Between the claws there is a very small coriaceous appendage, but without the additional minute pair of claws of the Lucanideous tarsi."

To the family of Passalidæ we add a new genus, under the name of *Hexaphyllum*—the antennæ with six laminæ; the mandibles nearly as long as the thorax, compressed, curved, with two external dentations near the middle. The maxillary palpi nearly as long as the mandibles, cylindrical; the head short, transverse, partly hid by the thorax; the eyes large, projecting; the thorax large, broader than the body, projecting anteriorly in the middle. The body broad, and shorter than the true Passali. The species we have named *Brasiliensis*. It is piceous black, with the antennæ reddish, the eyes white, the thorax channeled, the elytra with numerous longitudinal punctated striæ.

The PASSALI are removed from the lucani by many characters. Their antennæ are not elbow-formed. The labium is very different from that of the lucani. The mandibles, though salient and corneous, like those of the lucani, are thicker, always very much denticulated, and almost similar in the two sexes. The palpi are almost equal, with the last articulation cylindrical. The upper part of the head is unequal, and that of the male presents more decided eminences, or even a species of horn. The corslet is square, and separated from the abdomen by a very apparent pedicle. The elytra fall abruptly on the sides, to envelope those of the abdomen. The feet are nearly similar to those of the lucani, but shorter.

The passali are pretty large coleoptera, of a uniform and shining black, or sometimes of a marron brown. They inhabit the hot countries of both continents. Some are even found in New Holland. But they are more abundant in America, and particularly in Cayenne and Surinam.

Madame de Merian, who has given a figure of the *passalus interruptus*, tells us that she found its larva in the roots of the *battata*, a plant which grows in Surinam. She has also given the figure of this larva, which resembles a thick worm in its form. The body is very thick, the head small, the extremity of the body slender, and there are six scaly feet. As the passali have several relations with the lucani, we may believe that their larvæ live in the same manner, undergo the same metamorphoses, and are equally several years in arriving to their perfect state.

THE

SECOND GENERAL SECTION

OF

COLEOPTERA.

HETEROMERA.

THE second general section of the coleoptera HETEROMERA, has five articulations in the first four tarsi, and one less in the other two tarsi.

These insects feed all on vegetable substances. M. Leon Dufour (Ann. des Sciences Nat. vj. 181) has observed, that their male organs have a texture which approximates them to the scarabeides and the clavicornes; their testicle consists of spermatic capsules, or little bags.

We shall divide this section into four great families,* the first two of which have, in the apparatus for the excrementary secretion, discovered in many genera by the same M. Dufour, some analogy with the first pentamerous coleoptera; the chylific ventricle is often beset with papillæ. Many of these insects present, moreover, vestiges of another secretory apparatus, of which there are not many examples among the coleoptera, that which is distinguished by the denomination

* In a natural order, the fourth is allied to the first, through *Helops*, which Linnæus places in his genus, Tenebrio. It is evident that Tenebrio conducts to Phaleria, Diaperis, &c., or to our second family.

salivary. As in the pentamera, the hepatic vessels, with few exceptions, are six, and have two insertions distant from each other: at one end, says M. Dufour, they are inserted by six isolated nippers, round the convexity which terminates the chylific ventricle. The other ends open at the commencement of the cœcum, by trunks, whose number varies according to the family and genus.

Some, whose elytra are generally firm and hard, whose hooks of the tarsi are almost always simple, have the head ovoid or oval, capable of being withdrawn backward under the corslet, or narrowed sometimes behind, though not suddenly, and without a neck at its base. Many of these Heteromera are lucifugous. This division will include the three following families.

The first, that of

Melasoma,

Is composed of insects of a black or ash-colour, without mixture, whence the section is named; apterous, for the most part, and with the elytra oftentimes soldered together; the antennæ altogether, or in part granulated, nearly of the same thickness throughout, or a little enlarged at the extremity, inserted under the advanced edges of the head, with the third articulation in general elongated; the mandibles bifid or emarginated at the extremity, and having a careous tooth or hook on the inner side of the jaws. All the articulations of the tarsi are entire, and the eyes are oblong, and but little elevated, a character which, according to M. Marcel de Serres, indicates their nocturnal habits. Nearly all these animals live on the ground, either in sand or under stones, and often also in the low and dark parts of houses, as cellars, stables, &c.

According to M. Leon Dufour (Ann. des Sciences Nat. v. 276), the biliary vessels are inserted on the inner side of the cœcum, by a single tubular trunk, resulting from the confluence of two strong short branches, composed themselves of the union of three biliary vessels. The bile is yellow, sometimes brown or violet. The alimentary tube is long, and its length, in our first tribe (Pimelia), is three times that of the body; the œsophagus is long, and opens into a gizzard, smooth or glabrous on the outside, more developed in these last insects, where it forms an ovoid pouch lodged in the chest; it has on the inside fleshy folds, or longitudinal columns, terminating in some (*Erodia*, *Pimelia*,) at the side of the chylific ventricle, by a valve formed principally of four corneous oval apparently united pieces; the chylific ventricle is elongated, flexeous or folded, in general beset with small papillæ, similar to projecting points, and terminating in a pad, callous within, and in which the biliary vessels are inserted. The same naturalist has observed, in some sub-genera of the family (*Blaps*, *Asides*), a salivary apparatus, consisting of two vessels, or floating tubes, sometimes perfectly simple (*Asides*), sometimes irregularly branched (*Blaps*), and he has no doubt that these vessels exist in the other Pimeliaria. M. Marcel de Serres has studied with much care the texture of the tunic of the digestive canal. The adipose tissue is more abundant in these heteromera than in the following; hence they are able, even when pierced and fixed by a pin, to live nearly six months, without taking food, as I have seen in some species of akis.

We shall first divide this family, which forms, in the system of Linnæus, the genus *Tenebrio*, from the absence or presence of wings.

Among those deprived of these organs, and whose wing-cases are generally soldered, some have the palpi nearly filiform, or terminating in an articulation moderately dilated, and not forming a knob distinctly hatchet-formed or triangular: they will compose the first tribe *Pimeliariæ*, so named from the genus

PIMELIA, of Fab.,

Which is the most numerous of it.

Sometimes the mentum is more or less heart-shaped, with the upper edge either emarginated in the middle, and as if divided into two short and rounded lobes, or broadly emarginated or widened.

Here the last two articulations of the antennæ, or the tenth and eleventh, always distinct, sometimes unite to form an ovoid or pyriform body, or are evidently separated from each other. The upper edge of the mentum is round and emarginated in the middle, or as if divided into two festoons.

These have the anterior edge of the head nearly straight, or but little advanced in the middle, without a deep emargination, and fitted to receive the mentum, and the lateral edges simply and slightly dilated above the insertion of the antennæ; their head does not appear sensibly to be narrowed behind, nor enlarged and truncated in front. The corslet is not at all heart-shaped, deeply emarginated in front, and truncated behind.

We may separate from the last, those which have the anterior edge of the head straight, or nearly so, without angular or tooth-formed dilatation in the middle; the labrum is nearly square, of middle size, and entirely exposed; the corslet transverse, and the abdomen very voluminous and swollen.

Those with the body more or less ovoid or oval, with the corslet narrower, even at the base than the abdomen, generally convex, without sharp elongations to the posterior angles, or posterior projections to the presternum, compose the sub-genus, properly called,

PIMELIA, *Tenebrio*, Lin.

These heteromera belong to the countries which surround the Mediterranean, south-western Asia, and Africa. They are not, or at least have not yet been found, in the East Indies.

Some species, generally more elongated, have the mentum exposed, the antennæ slightly and insensibly thicker toward the end ; the last three articulations do by no means compose a knob, divided into two portions of equal thickness, the last of which formed by the tenth and the last articulation confounded together.

In some among them the abdomen is in proportion larger and more voluminous, the tarsi are relatively less elongated, the fore-legs are in the form of a reversed triangle, elongated, with the outer angle of their extremity prolonged, the spurs strong, and tarsi short.

M. Fischer (Entom. de la Russ.) has divided them into three genera, *Pimelia*, *Platyope* and *Diesia*, but whose characters, founded only on the more or less projection of the last articulation of the antennæ, and the indentures of the fore-legs do not appear to us sufficiently decided. The eleventh and last articulation of the antennæ is more distinct in the last genus. The fore-legs are very much indented on the outside in Platyope. Their corslet is squared transversely, with the base of the elytra straight, and the outer angles or shoulders a little advanced. With the Pimelia, properly so called, of this author, or those whose eleventh and last articulation of the antennæ are united, or nearly confounded, with the preceding, or the corslet nearly semilunar and convex, and whose abdomen is nearly ovoid or globular, is arranged a species, very common on the shores of the Mediterranean (*P. 2-punctator*, Fab. Oliv. Col. III. 59. 1. 1.). It is about eight lines long, of a shining black. Its corslet is shagreen, with two large deep points in the middle, united

Pl. 56

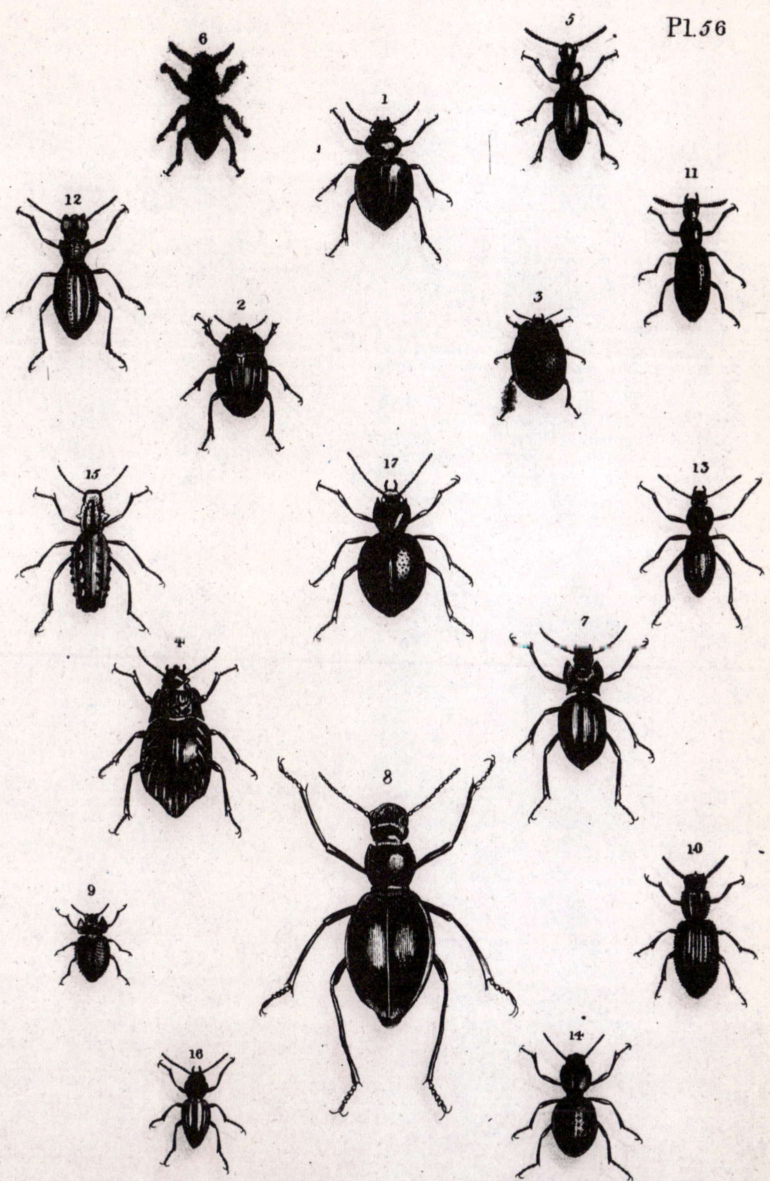

1. Pimelia *vestita*. *Gory*. 2. Erodius *gibbus*. *Fabr.* 3. Zophosis *testudinarius*. *Oliv.* 4. Nyctelia *Duczolii*. *Chevr.* 5. Hegeter *tagenioides*. *Gory*. 6. Tentyria *punctipennis*. *Lefebvre*. 7. Akis *Goryi*. *Guer.* 8. Elenophorus *americanus*. *Lacord.* 9. Eurychora *rugosula*. *Guer.* 10. Adelostoma *rugosa*. *Gory*. 11. Tagenia *orientalis*. *Gory*. 12. Psammeticus *costatus*. *Guer.* 13. Scaurus *rugosus*. *Latr.* 14. Scotobius *granosus*. *Lacord.* 15. Sepidium *vestitum*. *Gory*. 16. Trachynotus *vittatus*. *Latr.* 17. Moluris *luteipes*. *Guer.*

in some individuals into a transverse line. The elytra are in like manner shagreen, and have each, including therein the lateral keel, four elevated lines, not at all sensibly toothed, the two innermost of which are the shortest; the suture is elevated; *Ten. muricatus,* of Linnæus, is different. (Schœnh. Synon. Insect. I. Fab. iii. 9.)

Another very remarkable species, but peculiar to Upper Egypt, and which is found in the tombs there, is (*P. coronata,* Oliv. *ibid.* ii. 17) about fifteen lines long, blackish, beset with hair of a reddish brown, with a range of recurved bent spines on the lateral keel of each elytra.

M. Payraudeau has discovered, in the Island of Corsica, a new species (*Payraudii*) allied to the first, but with the abdomen more elongated, with the elytra more decidedly shagreened, and with the two internal elevated lines nearly effaced.

Other species (TRACHYDERMA, Latr.) have the abdomen in proportion narrower, and more elongated, oftentimes very compressed at the sides; the tarsi long, with the legs, not excepting the two anterior, slender, narrow, and terminating by two small spurs; these species are in general more from the south than the preceding.*

A last division of Pimelia (CRYPTOCHYLE, Lat.) is composed of species whose body is, relatively to the other, shorter, with the mentum covered by the presternum, and the antennæ terminating abruptly in a knob, divided into two portions, one formed by the ninth articulation, and the other by the two following confounded together. These species appear to be concentrated in the southern extremity of Africa.†

Under the generic denomination of Erodius, were at first

* *Pimelia longipes, hispida, morbilosa,* &c. of Fab.; *P. anomola,* of Fischer.

† *Pim. maculata, minuta,* of Fab. For the other species, see Oliv. Schœnherr, and Fischer.

united certain Pimelia, nearly allied to the last, but whose body is ovoid, short, arched or gibbous above, with the corslet short, as wide behind as the base of the elytra, terminated on each side by a sharp angle, and the presternum dilated behind into a point, supporting itself by the posterior end on the mesosternum.

These erodii form nevertheless three sub-genera.

Erodius, Proper, Latr.,

With the last two articulations of the antennæ united, and forming a little knob or bud; the fore-legs have a strong indentation near the middle of the outer side, and another at the end of the same side. The mentum is framed underneath, and covers the base of the jaws. Their body is in general swollen.*

Zophosis, Lat. *Erodius*, Fab. Oliv.,

In which the antennæ are nearly filiform, or thicker insensibly toward the end, with the second articulation very distinct from the preceding, a little larger, nearly ovoid; the fore-legs, as well as in the following, have no indentation near the middle of the outer side; the mentum is encased underneath, and covers the base of the jaws; the third articulation of the antennæ is scarcely longer than the second, and the ninth and tenth are nearly in the form of a top. (Lat. Gener. Crust. Insect. ii. 146.)

Nyctelia, Lat. *Zophosis*, Germ.,

Are nearly similar to Zophosis, but the third articulation of their antennæ is much longer than the preceding, and the following, as well as the ninth and tenth, are nearly globular.

* *Erodius bilineatus*, *gibbus*, *levigatus*, of Oliv. Col. III. No. 63. See Lat. Gener. Crust. et Insect. II. 145, et Cat. de lá Col. du Comte Dejean.

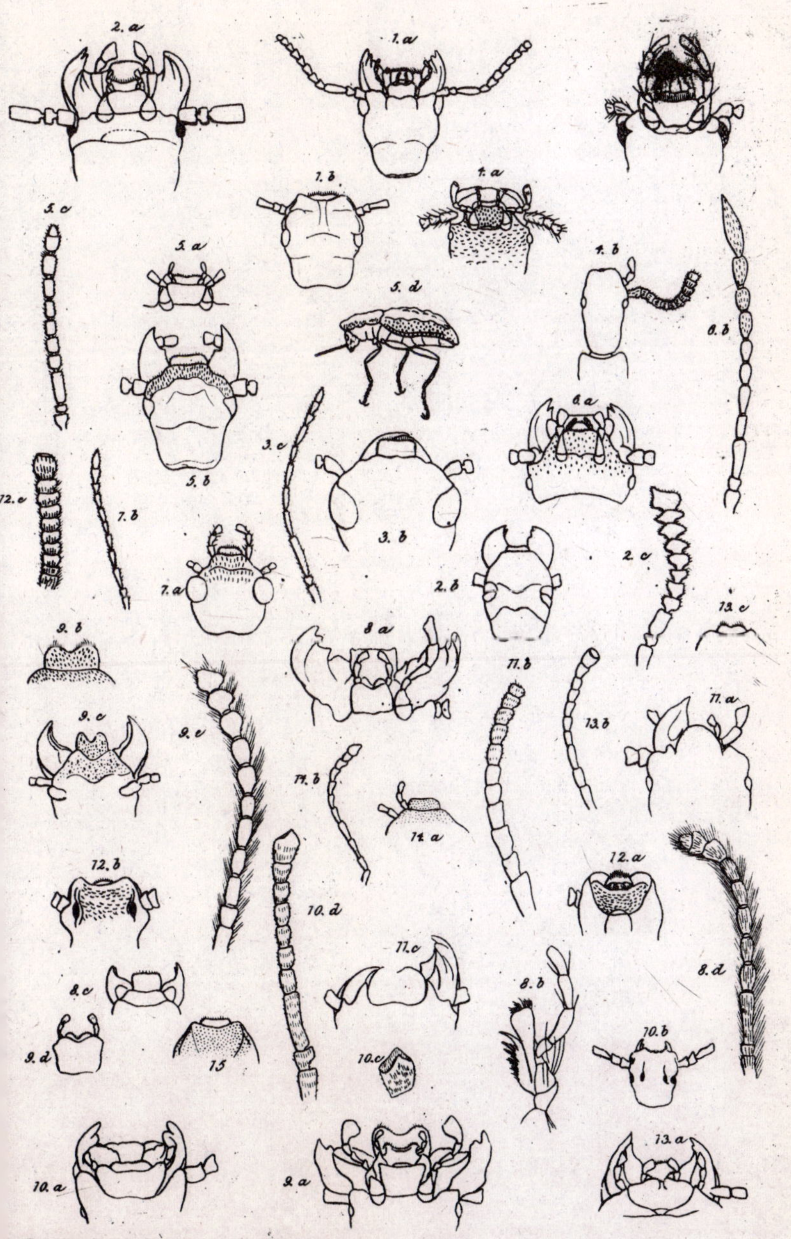

1. Psammeticus *costatus*. 2. Scotobius *granosus*. 3. Moluris *lutipes*. *Guer.* 4. Tagenia *orientalis*.
5. Sepidium *vestitum*. 6. Scaurus *rugosus*. 7. Trachynotus *vittatus*. 8. Pimelia *serica*. 9. Nyctelia *Luczotti*.
10. Eurychora *rugosula*. 11. Hegeter *tagenioides*. 12. Adelostoma *rugosa*. 13. Erodius *gibbus*.

The base of the jaws is uncovered. These insects are moreover proper to South America, while zophosis and erodius inhabit exclusively the western and southern parts of Asia, and the south of Europe and Africa. (*G. nodosa*, Germ. Insect. Spec. Nov.)

Other pimeliæ, terminating the subdivision of those whose labrum is not received into any deep emargination of the anterior edge of the head, and in which this last part of the body is neither truncated in front nor narrowed behind, are distinguishable from the foregoing by the following characters. The anterior edge of this part is advanced like an angle or tooth in the middle. The labrum, when the mandibles are closed, appear not at all, or but very slightly. The corslet is sometimes in the form of a trapezium, nearly as long as it is wide, sometimes nearly orbicular, or almost semicircular. The antennæ are filiform, and the eleventh and last articulation is always very distinct from the foregoing. The mentum is encased underneath, and covers the base of the jaws. The presternum is a little elongated into a point in many. These insects, as well as those proper to the two following subdivisions, are exclusively proper to the warm and western countries of the Old World.

HEGETER, Lat.

Have the corslet in form of a trapezium nearly as wide at the posterior edge as the base of the elytra, and applied along its whole length against it. The last articulation of the antennæ is a little smaller than the preceding.*

TENTYRIA, Lat. *Akis*, Fab.

Their corslet is nearly orbicular, sometimes narrower than

* Lat. Gener. Crust. et Insect. II. *Pimelia silphoides*, Oliv.; *Gnathesia glabra*, Fisch. Entom. de la Russ. II.

the abdomen, sometimes of its breadth, but rounded at the posterior angles, and leaving a void between them and the base of the elytra. The last articulation of the antennæ is at least as large as the preceding.*

Other pimeliæ differ from the foregoing in the form of the head and corslet. This former part is as if squared and more or less narrowed behind; the middle of its anterior edge has an emargination receiving the labrum. The dilatation of the lateral edges covering the base of the antennæ is larger and elongated to the anterior edge. These organs are always composed of eleven articulations very distinct, sub-cylindrical, except the last, with the third very long. The outside of the mandibles is strongly excavated in the middle, and the lower sides of the head, forming the lateral encasing of the jaws and chin, terminates in a point, or tooth-like. The corslet is in the form of a truncated heart, and much emarginated in front in most. These pimeliæ include a great part of the genus.

Akis, Fab.

These are now restrained to the species whose corslet is broader than the head, decidedly emarginated in front, short, widely truncated on the hind edge, with the lateral edges elevated. The first division of Akis of Fab. Fischer. Entom. de la Rus. I. xv. 7, 8, 9.

Another species, *A. collaris*, has the head measuring a little wider than the corslet, more elongated behind, and a little narrowed at the base in the manner of a neck; the corslet is much more narrow all its length than the abdomen,

* Lat. *ibid.* II. 154, *Akis glabra, punctata, abbreviata, angustata, orbiculata*, of Fab. I think that *Tagona*, of Fis. is also referable to this sub-genus.

small, convex-inclined, and not elevated at the sides. It forms the genus

ELENOPHORUS of Megerle and Dejean.

The antennæ are a little larger than in those of Akis; the eyes are more narrow and emarginated.

The last pimeliæ of the division of those whose mentum is emarginated, are distinguished from the foregoing by the manner in which the mentum terminates; instead of being round, and divided into two festoons, it is slightly emarginated or concave, with the lateral angles sharp, and in proportion shorter and more narrowed at the base, or more heart-shaped; it covers the jaws. The eleventh articulation of the antennæ is not apparent; the tenth, a little larger than the preceding, and in form of a top, truncated obliquely at the end, terminates them. As to the form of the head, of its anterior emargination, and often also as to the cut of the corslet, these insects are much assimilated to Akis, properly so called.

EURYCHORA, Thunb.

Have the body oval, with the edges sharp and ciliated; the corslet semicircular, and receiving the head into an anterior emargination; the abdomen nearly heart-formed, and the antennæ composed of linear articulations compressed or angular, and the third of which is the largest.*

ADELOSTOMA, Dup.

Have the body narrow and elongated, with the corslet nearly square, a little narrowed behind; the abdomen is a long square, rounded behind; the antennæ thick, almost perfoliated, with the articulations, except the last, nearly all len-

* Lat. Gener. Crust. et Insect. II. 150. Schœnh. I.

ticular and equal. The labrum, mandibles, and palpi are hidden.*

We shall terminate these pimeliæ by those whose mentum, squared, has neither emargination or void on the upper edge. Their body is always oblong, with the corslet sometimes nearly square, rounded, or dilated; sometimes narrow, elongated, sub-cylindrical, and the abdomen ovoid or ovaliform. The antennæ have always eleven distinct articulations. The anterior thighs are swollen and sometimes indented, or at least in one of the sexes. These insects evidently form the passage from this tribe to the following.

Sometimes the antennæ are entirely or nearly altogether granulated, or composed of short articulations, whether ovoid or globular, or in form of a top, or nearly hemispherical.

Among them some resemble the pimeliæ of the last subgenus, with reference to the dilatation and elongation of the lateral edges of the head. The labrum is very short and little advanced. The lateral edges of the corslet are straight, or simply arched and rounded, and without dilatation in form of an angle or a tooth. The eyes are but little elevated.

In these the corslet is narrow, whether cylindrical or heart-shaped, elongated and truncated at both ends. Such are

TAGENIA, Lat. *Stenosis*, Herbot. *Akis*, Fab.

The antennæ are almost perfoliated, with the third articulation scarcely longer than the following, and the eleventh or last very small, and united with the preceding. The head

* *Adelostoma sulcatum*, Duponchel Mem. de la Soc. Lin. de Paris, 1827; found about Cadiz, by the eldest son of this naturalist, and at Tangiers, by M. Goudat, the younger; but brought long since by Labillardiére, from Syria.

is elongated behind and carried on a sort of neck. The corslet is in the form of an elongated heart, truncated at both ends. The abdomen is ovaliform. (Lat. Gener. Crust. II. Herbst. Col. VIII.)

PSAMMETICHUS, Lat.,

With the antennæ composed of articulations top-shaped, the third of which is much longer than the following, and the eleventh or last as large as its preceding, and very distinct. The head and corslet are formed like a long square, of the same breadth throughout. The abdomen is nearly ovaliform and truncated at the base. This sub-genus is established on inedited species from Chili.

In these the corslet is at least as broad as the abdomen, and almost orbicular in form, or square, rounded laterally, and either isometrical or broader than it is long.

SCAURUS, Fab.

In these the last articulation of the antennæ is ovoidoconical and elongated; the corslet is almost isometrical, and the anterior thighs are swollen and often indented in the males. The legs are long and narrow. These insects belong to the hot western countries of the Old World.

SCOTOBIUS, Germ.

In these the last articulation of the antennæ is scarcely larger than the preceding, and in form of a reversed top; the corslet is obviously wider than it is long, much arched on the lateral edges; the thighs differ but little in thickness, and the fore-legs form an elongated triangle, and angular. These heteromera belong to South America.

Other pimeliæ, with the antennæ chaplet-formed, and the mentum entire, are remarkable for the lateral dilatations, angular or tooth-formed, of their corslet. The middle of the back has a furrowed keel, terminated before in a rounded and bilobed bunch. The labrum is entirely exposed, and of ordinary size. The eyes are more elevated than in the other pimeliæ. The antennæ, moreover, are hairy. The elytra are very unequal.

Sepidium, Fab.

They are found in the hot countries of Europe and Africa. (*Sep. tricuspidatum*, *variegatum*, *et cristatum* of Fab.)

The last pimeliæ, having, like the foregoing, the mentum without upper emargination, is removed from these in the form of the articulations of the antennæ. They are, for the most part, sub-cylindrical, or in form of a reversed and elongated cone; the last three or four only are rounded, whether ovoid, turbinated, or hemispherical. The labrum entirely uncovered, and the marginal projection of the head covering the origin of these organs, is but little elongated, as well as in Sepidium. These insects belong to the colony of the Cape of Good Hope. The eyes are nearly round or oval, entire, or but little emarginated and elevated; the corslet is depressed, sometimes dilated on either side like an angle, sometimes narrower, but furrowed and keeled above. The last articulation of the antennæ is obviously longer and thicker than the preceding. Such are

Trachynotus, Lat. *Sepidium*, Fab.*

In these the eyes are narrow and elongated, scarcely at all elevated. The corslet is convex, nearly orbicular, emarginated in front, truncated behind, without dilatations, lateral angles or dorsal keel. The last articulation of the antennæ is, at most, of the size of the preceding.

* *Sep. reticulatum, rugosum, vittatum*, of Fab.; *S acuminatum*, of Schœnhen. A species which the Count Dejean names *Curcoloides*, and figured by Degeer, forms a particular division.

Moluris, Lat. *Pimelia*, Fab. Oliv. *Psammodes*, Kirb.

The second tribe of Melasoma, Blapsides, receives its name from the genus Blaps of Fab. The maxillary palpi terminate in an articulation manifestly dilated like a triangle or hatchet. M. Dufour has observed, that in this genus, as well as in that of Asides, the stomach is less developed than in pimelia, and that valve to which it reaches behind is not at all formed of those four principal pieces, corneous and apparently united, which constitutes it in the preceding tribe, but by the approximation together of the fleshy columns of its interior. The chylific ventricle is in proportion longer, and the spermatic capsules are less numerous. The Blaps, according to the same naturalist, are provided with a double apparatus for the excrementary secretions, entirely different from that of the pentamera; it consists of two largeish vessels, oblong, and situated altogether under the viscera of digestion and generation, near each other, with the parietes thin and surrounded with vascular folds, adhering, and more or less turgid, and of which it is difficult to perceive the point of insertion, in consequence of the impossibility of unrolling them. The same must be said of the conduits destined to evacuate the secreted liquid; they are hidden by a sort of membranaceous diaphragm applied by means of a fleshy panicle on the last ventral segment. The secreted liquid is voided laterally, and not at the end of the last ring; it is ejected a distance of seven or eight inches, is brownish, of a very irritating acidity, and of a peculiar and penetrating odour.

This tribe will be formed of a single genus, that of

* *Pi. striata, unicolor, gibba,* of Fab. See Lat. Gener. Crust et Insect. II. 148.—*Psammodes longicornis,* Kirb. Lin. Trans. XII.

Blaps.

The Blapsides, whose body is in general oblong, with the abdomen embraced laterally by the elytra, which in general become narrow towards the end, and terminate in a point, or like a tail, whose tarsi are nearly alike in both sexes, and are without remarkable dilatation, will form a first division.

Some have the mentum small, or not occupying in width more than the third of that of the under part of the head, nearly squared or orbicular.

Here all the legs are slender, without crest or tooth, on the outer side. The corslet is never dilated in front, nor is it heart-shaped, broadly truncated.

Oxura, Kirb.

Have the body narrow and elongated, the corslet longer than it is wide, ovoid, truncated at both ends, and with the intermediate articulations of the antennæ long and cylindrical, (*Oxura setosa*, Kirb. Lin. Trans. XII.)

Acanthomera, Lat. *Pimelia*, Fab.

Have the corslet nearly orbicular, transversal; the abdomen nearly globular; the third articulation of the antennæ larger than the following, and cylindrical; these nearly of that form, and the last three at least granulated.*

Misolampus, Lat. *Pimelia*, Herbst.

The corslet is almost globular, and the abdomen almost ovoid; the antennæ have the third and fourth articulations equal, cylindrical; the eighth, and the two following a little

* *Pim. dentipes*, Fab, and some other species. The anterior thighs are swollen and indented; the body is very unequal and ashy; the spurs on the legs are very small.

thicker, and nearly top-shaped, and the eleventh and last larger and ovoid. (*Pim. gibbosa,* Herbst.)

BLAPS, Fab., properly so called.

Their corslet is nearly square, flat, or but little convex. The abdomen is ovaliform, truncated transversely at its base, and more or less elongated. The elytra of most of them are narrowed and elongated into a point, especially in the males. The third articulation of the antennæ is much longer than the following, and cylindrical; these, or the three last but one at least, are granulated; the last is ovoid and short.

With the species whose body and abdomen are in proportion less elongated and wider, whose elytra, in the female, terminate in a short point, and the corslet is nearly flat, and almost isometrical, are arranged

B. mortisaga, Oliv. Col. III. 60. *Tenebrio mortisaga*, Lin. About ten lines long, dull black, simply punctated beneath, with the corslet nearly square, and having on each side, on the posterior edge, vestiges of a small flatted border. The end of the cases forms a short and obtuse point. Found in dark and dirty places, near privies, and often even in houses.

B. lævigata, Fab., may form a peculiar sub-genus. Its body is much shorter than that of the other species, very convex or gibbous. From the fourth articulation the antennæ are granulated. The fore-legs terminate in a strong point or spine formed by a spur.

Fabricius states that the Turkish women inhabiting Egypt, where this insect is very common, eat the furrowed blaps, dressed with butter. It is said also to be efficacious in maladies of the ear, and the sting of the scorpion.*

There all the legs are angular, with longitudinal crests;

* *Blaps gages, sulcata,* of Fab. Dejean's Coleoptera.

the two fore-legs are broader, and strongly indented on the outside. The corslet is dilated before, like a heart, broadly truncated.

GONOPUS, Lat.

The third articulation of the antennæ is elongated and cylindrical, as well as the two or three following; those which come after are granulated, the last is ovoid or a little longer than the foregoing. The anterior edge of the head is concave, and the mentum in a transverse square. The inner side of the thighs is trenchant, with a furrow; the two anterior have a tooth, the four hind legs are straight, arched with some indentations; the tarsi are glabrous. (*Blaps tibialis*, Fab.)

The other insects of this tribe, and with feet alike in both sexes, differ from the preceding by their mentum, which occupies transversely the greater part of the under side of the head, and has the shape of a heart truncated at its base. The corslet is always transversal, emarginated or concave in front, and arched laterally, whether trapezoid, and larger behind, or much dilated laterally, and narrowed toward the posterior angles. The labrum is emarginated. Most of them are ash-coloured, and live in the ground, in sandy places.

Sometimes the corslet is enlarged in front, near the middle of its side, and narrowed behind. The base of the jaws is exposed.

HETEROSCELIS, Lat.,

Present, at the external side of the first four legs, two strong teeth, one at the middle, the other terminal. The posterior extremity of the presternum is prolonged in the manner of a flatted lamina, and received into an emargination of the mesosternum. The body is oval, rounded at the two ends, with the lateral edges of the corslet very much arched, and simply narrowed towards the posterior angles. The antennæ become slightly and gradually more bulky towards the end.*

MACHLA, Herbst.,

Have the antennæ terminated in a small knob, like a button, and formed by the last three articulations: they can lodge in the cavities which are formed under the sides of the corslet, which are very thick and rounded. (*Platynotus serratus*, Fab.)

SCOTIMUS, Kirb.,

With antennæ also terminated in a small knob, but the two last articulations of which are almost confounded, and otherwise unsusceptible of being lodged in particular cavities. The corslet is dilated in front.†

Sometimes the corslet is almost trapezoid, arched gradually in the whole length of its lateral edges, without any abrupt posterior contraction. The mentum covers the base of the jaws.

The two last articulations of the antennæ are united in a small knob.

ASIDA, Latr ‡

Now we come to blapsides, with oval and but little elongated body, in which the lateral fold of the elytra is

* *Pimelia dentipes*, Fab.; ejusd. *Platynotus reticulatus, pimelia obscura*, Oliv. Insects of the Cape of Good Hope.

† *Scotimus crenicollis*, Kirb. Lin. Trans. XII. xxi. 14, a sub-genus proper to the American continent.

‡ Latr. Gen. Crust. et Insect. II. p. 155. See the Cat. of the Collect. of Coleopt. of M. le Comte Dejean, p. 65. The *platynotus undatus* of Fab. is a very different species from the *A. grisea*. This author is, I believe, mistaken as to its habitat—*Platynotus lævigatus*, ejusd.

narrow, and extends but little underneath ; in which the corslet is always transverse, almost square or trapezoid, with the lateral edges arched, and more remarkable still from the several differences of the feet ; the two or four anterior tarsi are more dilated in the males. The under part of these organs are usually silky, or furnished with a brush.

These insects frequent sandy places. The two anterior legs are usually wider, dilated triangularly at the end, and proper for digging.

In some the anterior part of the head is always emarginated. The two anterior tarsi of the males are alone manifestly broader and more dilated than the following.

PEDINUS, Latr.

M. Megerle and M. le Comte Dejean, have sub-divided them into several other sub-genera, but without giving the characters.

Those in which the males have the first four articulations of the two anterior tarsi of the same width, with the radical one triangular, the three following transverse, and almost equal, all the limbs narrow and elongated, the corslet narrowed posteriorly, and terminated in acute angles, form the genus OPATRINUS of M. Dejean. These insects all belong to America.*

Those in which the same tarsi, and in the same individuals, have the first articulation, and especially the fourth, sensibly more narrow or smaller than the two intermediate, whose corslet is narrowed near the posterior angles, compose four other sub-genera; but whose characters are so slight and graduated that these sections may be united into a single one, that of DENDARUS, Meg. Dej.

* *Blaps clathrata*, Fab ; ejusd. *B. punctata*, may be also his *platynotus dilatatus*.

Some species have, as well as the Opatrini, the legs narrow, elongated, but little dilated at their extremity, and almost always identical in the two sexes; the corslet narrowed abruptly on each side, near the posterior angles, which form a sharp tooth. These are DENDARI, properly so called.*

In the following the four anterior legs, or at least the first two, are dilated triangularly at their extremity. The under part of the intermediate and the last two, and that even of the two posterior thighs, is silky in many males.

Sometimes the sides of the corslet are narrowed abruptly near the posterior angles, or are almost rounded, with a projecting tooth at this extremity. The body is oval. Such are HELIOPHILUS of M. le Comte Dejean. Sometimes the corslet terminates insensibly on each side, in a pointed angle. The body is proportionally shorter and broader.

Some species, with a large corslet, but little broader than long, strongly edged laterally, and whose body is but little gibbous above, compose the genus EURYNOTUS of Mr. Kirby.†

Others, whose body is sensibly more convex or more gibbous above, with the corslet transverse, and very slightly edged, are ISOCERUS, Meg. Dej.

In the males of the last pedini, the first three articulations of the two anterior tarsi, always very much dilated, diminish progressively in breadth, and the fourth is very small. The posterior thighs of the same individuals are concave and silky underneath, as are likewise those of the *Heliophili*. The body is oval, with the corslet but slightly edged out, widening from front to rear, or but slightly narrowed behind,

* See Dej. Cat. of Collect. of Coleop., p. 65. See the *platinatus excavatus*, and *crenatus*, of Fab.

† *Eurynotus muricatus*, Kirb. Linn. Trans. XII. xxii. 1. See *platinatus striatus*, Schœnh. Synon. Insect. I. i. tab. ii. 6.

always terminated posteriorly and insensibly by a pointed and prolonged angle. Such are the PEDINI, properly so called, of M. le Comte Dejean.

In others the anterior edge of the head is entire, or without emargination. The four anterior tarsi of the males are equally, or almost equally dilated. The form of the body, and that of the corslet, in particular, is still similar to that of the last pedini.

Those in which the anterior edge of the head still presents an emargination, form the sub-genus,

BLAPTINUS, Dej. (Dej. Cat. 66. *Blaps tibidens*, Schœn. Synon. Insect. I. 1, tab. ii. 8.)

Those in which it is entire or without emargination, are,

PLATYSCELIS, Latr.

We are now arrived to Melasoma, provided with wings. Their body is usually oval or oblong, depressed, or but little raised, with the corslet square or trapezoid, of the breadth of the abdomen at its posterior extremity. The palpi are thicker at their extremity, the last articulation of the maxillaries is in the form of a reversed triangle or hatchet. The mentum is but little extended in breadth, and leaves the base of the jaws uncovered.*

These melasoma will comprise the third and last tribe, that of the TENEBRIONITES, formed of a single genus, that of

* By reason of their jaws, armed at the internal side with a corneous tooth, the Epitragi ought systematically to belong to this tribe, and be removed from all the sub-genera, of which it is composed, by their chin much larger, and covering the base of the jaws; but in the natural order it appears to me, that these insects should be placed near helops.

TENEBRIO,

Such as Fabricius had at first formed it, and to which we shall re-attach that which he names *Opatrum*, and that of *Orthocera*. They will serve as types to so many particular divisions.

1st.—Those whose body is oval, with the corslet almost trapezoid, arched laterally, or in a truncated demi-oval anteriorly, wider at least at the posterior edge than the abdomen, but little or not at all edged out; the maxillary palpi terminated by a hatchet-formed articulation, or some form very analogous, and the antennæ thickening insensibly.

CRYPTICUS, Latr. *Blaps*, Fab.,

Have the body convex and smooth above, with the head uncovered, or but little sunk in the emargination of the corslet, without emargination at its anterior edge; the eyes external, or altogether outside of the anterior concavity of the corslet, and this last part of the body insensibly inclined in the sides, and but little emarginated in front. The antennæ are almost of its length, with most of the articulations in the form of a reversed heart or top, the penult alone being more rounded, or almost grained, but not transverse. The legs are always narrow and elongated, with the terminal spurs tolerably projecting.*

OPATRUM, Fab. Dej. *Phylam*, Meg.

Their body is generally less raised, and even often depressed. The head is received posteriorly, with the eyes in a deep emargination of the corslet, and its anterior edge pre-

* *Pedinus glaber*, Latr. Gen. Crust. et Insect. II. p. 164; *helops glaber*, Oliv. Col. III. 58. ii. 12; *Blaps glabra*, Fab., and some other unedited species of Spain, and the Cape of Good Hope.

sents a small one, in which the labrum is engaged. The corslet is depressed along its sides. The antennæ are shorter than it, for the greater part grained, with the last articulation lenticular and transverse. The elytra are rough or striated. The spurs of the legs are very small, and the anterior two are broader, and triangular in many species.

O. Sabulosum, *Silpha Sabulosa,* Lin. Oliv. Col. III. 56, i. 4.

2d.—Those whose body is narrow and long, almost of the same breadth, or broader posteriorly, with the corslet almost square, almost as long, at least as broad, and whose antennæ form a thick knob, or are dilated abruptly at their extremity.

Some have the antennæ thick, cylindrical or fusiform, perfoliated, hairy, appearing composed of only ten articulations, the eleventh or last being short, and but little distinct ; the second is as large as the following.

CORTICUS, Dej. *Sarrotrium*, Germ.,

Whose antennæ are cylindrical, and terminated by a larger articulation, forming a small knob. (*Sarrotrium celtis*. Germ. Ins. Spec. Nov. p. 146.)

ORTHOCERUS, Lat. *Sarrotrium*, Illig.,

In which these organs are broader in their middle, form a fusiform knob, very hairy, with most of the articulations transverse, and the last much more narrow than the preceding. (*Hispa mutica*, Lin. Panz. Faun. Ins. Germ. I. 8.)

The antennæ of the others are of the usual thickness. Simply grained, not sensibly perfoliated or hairy, and present distinctly eleven articulations.

CHIROSCELIS, Lam.

Have two strong teeth at the exterior side of the first two legs, and the antennæ terminating in a small knob, almost globular, transverse, and formed by the two last articulations.*

TOXICUM, Lat.,

With simple legs; the knob of the antennæ compressed and formed by the last three articulations, with triangular head, and corslet almost square and almost isometrical.†

BOROS, Herbst. *Hypophlœus*, Fab.

Having also simple legs, and the knob of the antennæ compressed, and formed by the last three articulations; but whose body is almost linear, with the head oval, narrowed posteriorly, the corslet ovaliform, truncated at each extremity, and the last articulation of the maxillary palpi in a truncated ovoid, a little inflated.‡

3d. Those whose body is equally narrow and elongated, with the corslet almost square, but whose antennæ are of the ordinary thickness, and do not terminate abruptly in a knob.

The two anterior feet have the thighs thick, and the legs narrow and curved or arched.

In some the penultimate articulation is perfectly similar both in form and size to the preceding, and this last, as well as all the others, is neither dilated nor channeled above.

CALCAR, Dej. *Trogosita*, Fab.

Have the corslet in a long square, the body linear, of the same breadth throughout, with the anterior edge of the head

* *Chiroscelis bifenestra*, Lam. Annal. du Mus. et Hist. Nat. No. 16. XXII. 2; *Tenebrio digitatus*, Fab.

† *Toxicum richesianum*, Lat. Gen. Crust. et Insect. II. p. 168, and I. ix. 9.

‡ *Boros corticalis*, Gyll. Insect. Suec. I. ii. p. 584; *Hypophlœus boros*, Fab.; *B. thoracicus*, Gyll. *ibid.* p. 586.

emarginated, and the three last articulations but one of the antennæ almost globular, and not sensibly transverse. (*Trogosita calcar*, Fab.)

Upis, Fab.

Have the corslet in a long square; the body narrow, but not linear; the anterior edge of the head strait, without emargination, and the articulations of the antennæ preceding the last, lenticular and transverse.*

Tenebrio (Proper), Lin. Fab.

Do not differ from upis but by having their corslet broader than long.

Is frequently found, especially towards evening, in the unfrequented places of our houses, in bake-houses, flour-mills, on old walls, &c.

The Meal-worm, (*Tenebrio molitor*, Lin. Oliv. Col. III. 57. i. 12.) seven lines long, of a brown, almost black above, marron-colour, and shining underneath. Corslet of the breadth of the elytra, square, with two posterior impressions; cases punctuated and striated.

Its larva is long, cylindrical, ochreous, scaly, and very smooth. It lives in bran and flour, is given to nightingales, and is transformed into a nymph in the substance, which serves it as food.

T. Grandis is found in Brazil, under the barks of old trees; shoots through the anus, and to the distance of more than a foot, a caustic fluid. Other species of the same country, but smaller, cover themselves entirely with this substance. I am indebted for these observations to M. de la Cordaire.†

* *Upis ceramboides*, Fab.; *N. saperdoides*, Basc.

† See, for the other species, the Cat. of the Collect. of M. le Comte Dejean, and Fabricius. But this genus, as it is actually composed, has

In others, the penultimate articulation of the tarsi is very small, in the form of a little knob, and received into a longitudinal gutter of the preceding, which is more dilated than the foregoing ones, and almost in the form of a heart.

The anterior edge of the head presents an emargination occupied by a portion of the labrum.

Heterotarsus, Latr.

A sub-genus founded on an insect of Senegal, having the character of tenebrio, but singular from its tarsi. At the first glance the four anterior tarsi appear to have but four articulations, and the other two, three.

need of purification, many species being referable to the Phaleriæ, or other sub-genera. Some may even form new ones.

SUPPLEMENT

ON THE

MELASOMA.

This family, which is the first of the second section of coleoptera (the Heteromera), is so named from the colour of the bodies of the insects which compose it being generally black, or uniformly dusky. They embrace a very large part of the genus *Tenebrio* of Linnæus, and in general avoid the light. Some abide in sand, others conceal themselves under stones, in the rubbish, or obscure and choked-up parts of houses. They do not quit their retreats but at night. All their movements are rather slow, and they gnaw different substances, and particularly vegetables, and such as are decomposed. Those of their larvæ which have been observed, are long, cylindrical, covered with an almost scaly, smooth, and shining skin, and furnished with six short feet. They are found in the places inhabited by the perfect insect.

The tribe of the Pimeliariæ is proper to the southern countries of Europe, to Africa, and the western part of Asia, that which is on this side of India.

They are afraid of light, and live, almost all of them, in saline and sandy situations. Many frequent the shores of the sea. Their body is generally heavy and thick, black or earthen colour. These insects often transpire a whitish humour, which forms on their bodies a dust of this colour.

The genus PIMELIA of Fabricius, formed of many apterous *tenebriones* of Linnæus, has suffered in succession divers modifications. Thunberg has made of the *Pimelia ciliata*, a peculiar genus, which he names *Eurychorus*. Herbst has adopted it, and has instituted two others, *Akis* and *Stenosis*, which he has similarly detached from the pimelia of Fabricius. M. Latreille, in his "*Precis des Charactéres génériques*," has united these first two sections into one, to which he has preserved the denomination of *Eurychorus*. Fabricius, in his *Eleutherata*, admits the genus *Eurychorus* of Thunberg, and that of *Akis*, but he refers to the latter the *Stenosis* of Herbst, and the *Pimelia* of the same author, whose corslet is orbicular. M. Latreille, afterwards, established along with them, the genus *Tentyria*. Some other pimeliæ of these two naturalists, tolerably analogous in the general form of the body to *Tentyria*, but which differ in the antennæ, have been united into another generic section called *moluris*

The habits of these insects is the same as those of the tribe in general. *Pimelia bipunctata*, which has been observed in a living state by M. Latreille, inhabits the sandy borders of the Mediterranean. It conceals itself in holes, which it excavates very promptly by means of its feet. We possess, otherwise, no fact relative to the habits and metamorphoses of this species or its congeners.

The genus Pimelia, though much more restrained than in its origin, is still composed of a great number of species, but the determination of which is very difficult. The authors who have described these insects, have neglected many characters in detail; such as the precise form of the articulations of the antennæ, those of the feet, their relative proportions, and the employment of which, seeing the multitude of species collected in the East, by Olivier, and M. M. Labillardière and Savigny, is absolutely necessary. Accordingly we find that many naturalists have referred to various species of Pimelia, the *tenebrio muricatus* of Linnæus, an insect of a different kind.

Fabricius and Olivier have confounded with the Erodii, insects whose body has the same form, but the first two legs of which are not palmated, and whose antennæ thickening insensibly, have their eleventh, or last articulation, very distinct. These species compose the genus *Zophosis.* The Erodii are found in the sands of the hottest countries of Europe, of Africa, and the western parts of Asia.

The insects of the sub-genus Tentyria were at first placed by Forskael and Pallas with Tenebrio. They afterwards passed into the genus Pimelia, and then into that of Akis. But they differ from these heteromera by some distinguishing characters.

The Tentyriæ inhabit the sandy countries of the south of Europe, Africa, and Asia, but a small number of their species is known. The *pimelia glabra* of Olivier, very common in the environs of Marseilles, on the shores of the sea, is of this genus, as well as those which he names *Scabriuscula* and *Striatula.* These are *Akis* with Fabricius.

The domain of the insects of the sub-genus Scaurus is but of small extent, and appears to be confined to the southern countries of Europe, to those of Africa which are situated between the ocean and the Mediterranean, to Egypt and Syria; so their country appears to be circumscribed by the basin of this internal sea. They live exclusively on the ground, in the sand, among rubbish, or under stones. They sometimes climb along walls. Their walk is heavy, like that of *blaps, akis*, and *tenebrio*, and other analogous coleoptera. But a small number of species are known, and some of them differ sexually in their anterior feet. Fabricius in his system of *Eleutherata* mentions four, but the third (*Sulcatus*) should be placed in the genus *Aristus*, in the family of the carnivorous coleoptera. The colours can hardly serve to distinguish the species, these insects being entirely black; sometimes, however, their tint is accidentally a little ashen, or earthy.

The genus Sepidium, established by Fabricius, has been adopted by all entomologists. These insects are found only in the hot climates of the Ancient Continent. Their habits are the same as those of the Pimeliæ. They are seen to walk upon the sand in dry and uncultivated places. Their larvæ are unknown. The *Sepidium cristatum* is found in Egypt, as is also another species, *triscuspidatum*, and on the coast of Barbary. It was found very common by M. Olivier, at the end of the winter, in the environs of Alexandria.

The genus Blaps has considerable relations with *pimelia*, *helops* and *tenebrio*. They are distinguished from the first by the last articulation of the antennulæ, which is almost filiform in the pimeliæ, and thicker than the others in *blaps*. They are distinguished from *helops*, inasmuch as the last articulation in the latter is broad, compressed, and of a crescent form. The antennæ, moreover, are composed of articulations almost conical. *Tenebrio* is distinguished from blaps by the third articulation of the antennæ not being so long as the third articulation of those of blaps; the antennulæ moreover are almost filiform. The majority of these insects are destitute of wings, and then the elytra hard, coriaceous, and convex, are united one to each other by a suture.

Though these insects, as we have just mentioned, are most of them wingless, nature has not given them the compensation of enabling them to run with any degree of celerity. The majority remain concealed during the day, under stones, or in holes. They issue forth at night to run about in various directions, and seek their food. They are sometimes found in cellars, in humid and uninhabited places. They emit a very fœtid odour, much stronger, though pretty nearly of the same description as that of the carabi, or the blattæ, which caused some ancient naturalists to class them among these last mentioned insects. The larvæ of the blaps are not known. It is probable that they are concealed in the earth, and that they differ but little from those of Tenebrio.

The *blaps sulcata* is found in Egypt, in gardens and fields. Fabricius reports that the Turkish women eat this insect, cooked with butter, for the purpose of making them fat; a thing which is neither true, nor probable. He also tells us that they use it in Egypt and the Levant, as a remedy for pains and maladies in the ear, and against the bite of scorpions.

We now come to the last tribe of the Melasoma, the TENEBRIONITES.

The insects of the genus TENEBRIO, have derived their name from their sombre and almost black colour, and the obscure, retired, sandy, or humid places which they frequent. All the family, united at first by Linnæus under the same genus, has been successively divided into a great number of others by Fabricius, and the authors who have written subsequently to his time. What most distinguishes the tenebriones, are the antennæ, slightly thicker towards the end, and the final articulations of which, the last of all especially, are globular; the third is elongated; the upper lip apparent, the last articulation of the palpi a little thicker than the preceding, cylindrico-conical, and compressed; the maxillaries advanced, and the chin almost square.

These insects have the body more or less elongated, smooth, and usually of an obscure colour. Their gait is lively, and they fly in general tolerably well, but rather in the evening and night than in the middle of the day. They are usually met with in houses, especially in garrets, kitchens, warm, or little frequented places. They conceal themselves in the clefts of wood furniture, or behind tapestry. Their larva resembles a scaly worm. It is about an inch long, and rather narrow. Its body is composed of twelve rings, and covered with a yellowish skin, tolerably hard and scaly. The head is oval, a little flatted, furnished with mandibles, antennæ, and antennulæ. The first three rings are provided with six scaly feet—the last is conical; we remark at its extremity two small scaly crooks, black and motionless. Between the juncture of this last ring with the one before, there issues forth, when the larva walks, a fleshy, whitish, and tolerably thick mass, furnished with two scaly nipples, a little elongated and mobile, which appear to be two small feet, of which, in fact, the larva makes use to advance, resting them on the plan of position. The anus is situated over this fleshy mass, between the two nipples which perform the office of feet.

These larvæ live in flour, bread, sugar, and even on dead and carious wood. Those of *Tenebrio molitor*, which are found in this last substance, serve to feed nightingales, and constitute almost the only bait by which these shy birds can be taken: a fact the more curious when it is considered that the nightingale, in a state of nature, can seldom or never see these larvæ. They are also used to feed cameleons which are exhibited. In English they are vulgarly termed meal-worms. The *Tenebrio molitor* is found throughout all Europe.

The insects of the genus Opatrum are almost all of an ashen or earthy gray above, a colour very analogous to that of the places which they inhabit, and which may to a certain degree protect them against their enemies, by deceiving their regards. They live in sandy arid soils, and in the fields. They must not be confounded with the other coleoptera of the same family, which have great relations to them, but which are apterous, such as *asida* and *pedinus*; nor with the *cryptici*, which are provided with wings, but in which the hood is not notched, and whose labrum is transversal.

The metamorphoses of these insects have not been observed.

The most common species, and which is to be met with from the earliest days of spring, is the *opatrum sabulosum*, Fab.; *silpha sablosa*, Linn.

The opatra are extended into all parts of the world; but in a greater number in the Ancient Continent. Olivier has described thirty-four species of them.

The only species known of the genus ORTHOCERUS, and which had been placed in various genera, is evidently connected from its habits with the opatra, and from its elongated form, as well as the parts of its mouth, with tenebrio.

This insect is winged. It is found in spring in dry and arid places, and in sand-pits. It walks slowly, and when it remains tranquil it is distinguished with difficulty; either because its colour is the same as that of the soil on which it remains, or on account of its littleness.

END OF VOL. XIV.